EXPLICIT BIOLOGY

A Revision Course

(3RD EDITION)

GCE, SSCE, JAMB (UTME),
POST - UTME & PRE - DEGREE

O.J. OLAOYE

Published by

Pedagogue House
Lagos.
Nigeria.
+2348069451567, +2348024761905

© O.J. OLAOYE
First Edition - January 2011
Second Edition - June 2012
Third Edition - January 2014

ISBN-978-978-919-842-9 (Nigeria)

ISBN-13: 978-1499325300
ISBN-10: 1499325304

Published in the United States of America (Third Edition)

Dedication

This book is dedicated to **Mr. Henry Inegbedion,** a teacher, a friend and a brother. And To the family of Mr. & Mrs S. O. Adepoju for their love and care from my sophomore till final year.

PREFACE TO THE THIRD EDITION

The text of this 3rd edition reflects the need for an illustrated approach to biology. This, you will notice in almost all the chapters.

In this 3rd edition, as in previous editions, the entire books has been thoroughly revised, with a view to eliminating errors, incorporating suggestions of readers, updating concepts and discarding materials that are no longer relevant. In this way, the book is kept up-to-date and accurate as possible.

The book continues to be 'the book' that everybody taking biology examination, particularly in JAMB and pre-degree, must have.

I thank all the students and others who took the time to write to me offering helpful criticisms and suggestions. Such comments are always welcome, and I solicit additional corrections and criticisms, which may be done via facebook, twitter, Whatsapp, 2go etc.

Lagos

Olaoye, OyeDele John.

January, 2014.

ACKNOWLEDGEMENT

In compiling this book, I consulted a number of textbooks listed on the reference page. I express my special appreciation to the authors of these books. I also appreciate the Joint Admission and Matriculation Board (JAMB) and the West African Examination Council (WAEC) for allowing the reproduction of their past questions.

For help at many stages of the preparation of the book I owe particular thanks to many people: my students of former years for their interest, stimulation and responsiveness; to my colleagues at Pedagogue House for their help and support; to Mr. Afolabi J.R., MD/CEO Dimensional Publishers, for his willingness to help at all time; to Mr. Sunny Odogwu, Proprietor of Corporate Comprehensive College, for his encouragement and support; to Mr. and Mrs. Oyebodunrin Olaoye, and the rest of my siblings for their encouragement and support; to my Mr. and Mrs. Ayobami Agbi, and the rest of my cousins for their motivation; to friends at Lagos State University College of Medicine (LASUCOM) who became enthusiastic about the book; and to my mother Mrs. F.B. Olaoye for her prayers. Speacial acknowledgements go to Professors O. A. Sofola, I. I. Olatunji-Bello and Engr. A. Adebola for taking out time to go through the manuscript and give their recommendations.

Finally, the stimulating environment provided by my family Elder and Deaconess Agbi has been extremely important in providing me with the help and motivation to write this book, and to them I am very grateful. **TO GOD BE THE GLORY.**

FOREWARD

The monograph "Explicit Biology: A Revision Course" for UTME, POST-UTME and other related examinations written by Mr. O.J. Olaoye is a complementary text to the various textbooks that have been written in the subject area. The book has five sections and twenty seven chapters on all the biology topics.

This book has become timely in view of the poor results that are being witnessed in the UTME examination especially in the sciences and mathematics, biology inclusive. These majorly are due to poor or inadequate access to good study materials.

Each chapter has annotated topics followed by copious specimen questions on some past UTME and POST UTME examination, for practice and emphasis. The topics in each chapter are adequately treated and with clarity. The illustrations are simple but effective.

The author has done a lot of work and put in efforts to come out with a very readable and useful text. It is highly recommended to all the students aspiring to undertake the UTME and POST-UTME examination, in order to improve highly their chances of success.

O.A. Sofola
B.Sc, M.Sc, M.B.B.S (Lagos), Ph.D (Leeds), FAS (Nig.)
Professor of physiology,
College of Medicine, University of Lagos.
Former Acting Vice Chancellor.
Olabisi Onabanjo University, Ago Iwoye, Ogun State.

CONTENTS

SECTION D: HEREDITY AND VARIATION

SECTION E: EVOLUTION

APPENDIXES

CHAPTER 1

LIVING ORGANISMS

Understand the following after reading this chapter:

> Characteristics of living organisms
> Prokaryotic and Eukaryotic cell
> Cell structure and functions of cell components
> Levels of organization.

CHARACTERISTICS OF LIVING ORGANISMS

1. Movement: an action by an organism or part of an organism causing a change of position, place,or aspect.

2. Assimilation of food (Nutrition): taking in of nutrients which are organic substances and mineral ions, containing raw materials or energy for growth and tissue repair, absorbing and assimilating them.

3. Respiration: the chemical reactions that break down nutrient molecules in living cells to release energy

4. Reproduction: the processes that make more of the same kind of organism

5. Irritability or Sensitivity: the ability to detect or sense changes in the environment (stimuli) and to make responses

6. Adaptation: the possession of factors (of structure or function) which enable organisms to live successfully and to survive in the respective environment

7. Growth: the ability to increase in size through acquisition of new protoplasm.

8. Excretion: removal from organisms of toxic materials, the waste products of metabolism (chemical reactions in cells including respiration) and substances in excess of requirements

Memory Tip: MARRIAGE

Viruses are category of extremely small microscopic parasites of plants, animals, and bacteria. Viruses are not cells but rather RNA or DNA molecules surrounded by a protein coating. Since viruses cannot reproduce without a host cell, they are not strictly speaking living organisms.Viruses are link between living and non-living things. The living properties of viruses are:
1. They reproduce by replicating in the living cell of a host
2. They possess characteristics which are transmitted from one generation to another. one generation to another.

Viruses and bacteria		
	Virus	**Bacteria**
Covered by	Protein coat	Cell wall
Cell membrane	No	Yes
Cytoplasm	No	Yes
Genetic material	DNA or RNA – only a few genes	DNA or RNA – enough for several hundred genes
Living or not?	Non-living unless in host	Living

CELL

There are two major types of cell. **Eukaryotic cells**, which have membrane-bound organelles, like a nucleus and mitochondrion; and **prokaryotic cells** which do not have these organelles. The best example of a prokaryotic cell is the bacteria cell.

1

	Prokaryotes	Eukaryotes
Typical organisms	bacteria	Protoctista, fungi, plants, animals
Typical size	~ 1-10 µm	~ 10-100 µm (sperm cells) apart from the tail, are smaller)
Type of nucleus	Nuclear body No nucleus	real nucleus with nuclear envelope
DNA	circular (ccc DNA)	linear molecules (chromosomes) with histone proteins
Ribosomes	70S	80S
Cytoplasmatic structure	very few structures	highly structured by membranes and a cytoskeleton
Cell movement	Flagellae/cilia made of flagellin	flagellae and cilia made of tubulin
Mitochondria	none	1 - 100 (though RBC's have none)
Chloroplasts	none	in algae and plants
Organization	usually single cells	single cells, colonies, higher multicellular organisms with specialized cells
Cell division	Binary fission (simple division)	Mitosis (normal cell replication) Meiosis (gamete production)

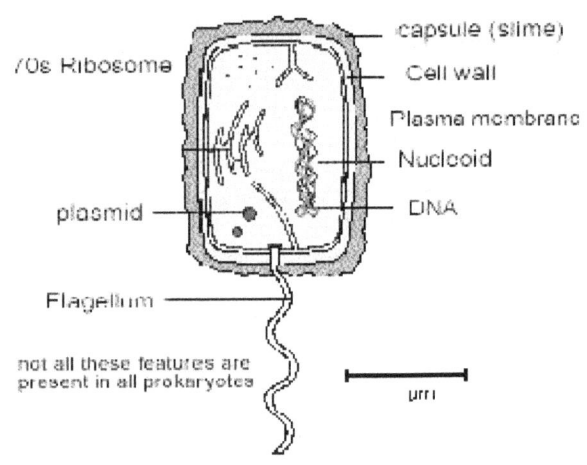

A prokaryotic cell

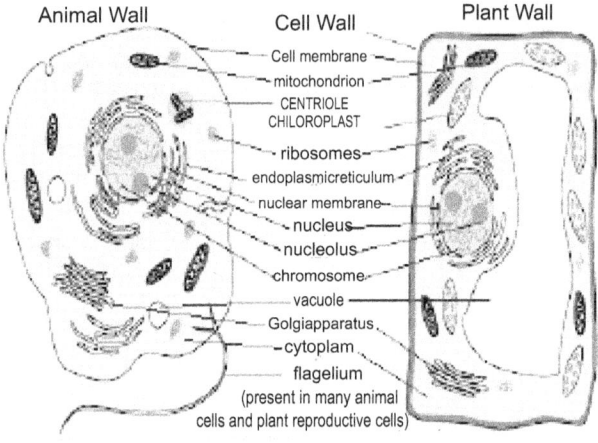

Animal and plant cell

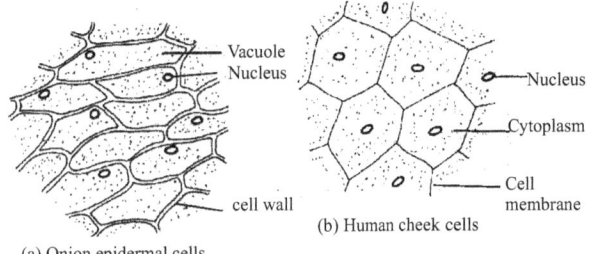

(a) Onion epidermal cells

(b) Human cheek cells

Epidermal cells

CELL COMPONENTS AND THEIR FUNCTIONS

Nucleus
The nucleus is the membrane-bound region which contains the DNA in a Eukaryotic cell. In the Eukaryotic cell DNA is found in chromosomes which are linear molecules joined together by histones. The nucleolus is a dark area in the nucleus which is where ribosomal RNA is being made.

Nuclear Envelope
The nuclear envelope is a double membrane which surrounds the DNA in a Eukaryotic cell. There are pores in the nuclear envelope which allows mRNA to leave the nucleus and enter the cytoplasm.

Rough Endoplasmic Reticulum

The Rough ER is is made up of a large number of connected compartments known as **cisternae**. There are ribosomes which are attached to the outside of the membrane. The role of the Rough ER is to make and then break up proteins which are then transported to other parts of the cell.

Smooth Endoplasmic Reticulum

The Smooth ER has a similar structure to the Rough ER but does not have the ribosomes. It is used to synthesize steroid hormones and cholesterol.

Golgi Apparatus

This organelle is involved in the modification of proteins. The proteins arrive here in vesicles from the rough ER and have carbohydrates or other molecules added to them to make them more complex and suited for their role. The modified proteins are then packaged into vesicles and transported out of the cell by the process of exocytosis.

Mitochodria

Mitochondria are sometimes known as the power house of the cell, it is here where aerobic respiration is used to release energy from organic molecules such as glucose and make ATP. The mitochondria has a double membrane, the inner of which is folded to provide a large surface area.

Lysosomes

These are vesicles which contain hydrolytic enzymes. They can be exported out of the cell by the process of phagocytosis or used in the cytoplasm to break down organelles or proteins which are no longer needed.

Ribosomes

There have the same role that they have in prokaryotic cells they read ribosomal RNA and make it into the proteins needed.

Microtubules

These are found in both prokaryotic and eukaryotic cells. They are found throughout the cytoplasm and are made out of a protein called **tubulin**. The role of microtubules is to assist in the transports of the cellular organelles and assist with cell division.

Chloroplasts

These are only found in plant cells where they contain chlorophyll which is used to transform light energy into carbohydrates. It is not unusual for chloroplasts to contain starch granules for energy.

Cell Wall

The roles of the cell wall in Plant cells are similar to the role of the cell wall in a bacterium. It prevents the cell rupturing due to osmotic pressures. In a plant cell the cell wall is made out of cellulose which is a very strong fibre but is fully permeable the role of controlling what leaves and enters the cell is down to the cell membrane which is the same as the one found in prokaryotic cells.

Cell Membrane

The cell membrane forms the outside of the animal cell, which acts as a barrier and controls the transfer of materials into and out of the cell. It is composed of a double layer of phospholipids and proteins.

Organelles with DNA

The Mitochondria and Chloroplasts have their own DNA

LEVELS OF ORGANIZATION

Living organisms are composed of cells organized at different levels.

1. **Cell**: the basic structural and functional unit of life e.g. *Amoeba*, rods and cones (in the retina).
2. **Tissue**: a group of similar cells organized into a structural and functional unit e.g. *Hydra,* bone.
3. **Organ**: a collection of tissues which are similar in structure and perform similar functions e.g. Onion bulb, eye.
4. **System**: a group of specialized and co-ordinated organs performing a distinct major function e.g. Nervous system.

CRITICAL THINKING

If you read that a new organism had been discovered, what would you know about the organism without examining it in terms of cells?

HIGHLIGHT

IT'S ALIVE!? . . . MAYBE.

An ongoing discussion in microbiology concerns the nature of viruses. Are they alive? Viruses are noncellular; that is, they lack cell membranes, cell walls, and most other cellular components. However, they do contain genetic material in the form of either DNA or RNA. (Few viruses have both DNA and RNA.) While many viruses use DNA molecules for their genes, others such as HIV and poliovirus use RNA for genes instead of DNA. No cells use RNA molecules for their genes. However, viruses have some characteristics of living cells. For instance, some demonstrate responsiveness to their environment, as when they inject their genetic material into susceptible host cells. Nevertheless, viruses lack most characteristics of life: they are unable to grow, reproduce, or metabolize outside of a host cell, although once they enter a cell they take control of the cell's metabolism and cause it to make more viruses. This takeover typically leads to the death of the cell and results in disease in an organism.

.1. CHARACTERISTICS

1. UME 79/1
All living organisms
A. Photosynthesize   B. Respire
C. Move   D. Feed
                          E. Transpire

2. UME 94/2
The smallest living organisms which share the characteristics of both living and non living matter are
A. Bacteria   C. Virus
B. Fungi   D. Protozoa

3. PCE 2001/15 TYPE: 4
Which of this is a living thing?
A. air   C. spirogyra
B. water   D. star

4. PCE 2002/13 TYPE: 5
Which of this is a living thing?
A. Water   C. Bread
B. Sun   D. None of the above.

5. UME/2009/2
A characteristic exhibited by all living organisms is
A. sexual reproduction
B. aerobic respiration
C. the ability to move from one place to another
D. the ability to remove unwanted substances

6. UME/2008/2
A characteristic that can possibly be shared by both living and non-living organisms is
A. locomotion   B. irritability
C. increase in biomass   D. increase in size

1.2. CELL STRUCTURE AND FUNCTION OF CELL COMPONENTS

7. UME/2008/3
The cell of an onion bulb can be differentiated from a cheek cell by the presence of
A. plasmalemma   B. chloroplast
C. cell wall   D. nucleus

8. UME/2006/30
The role of the Golgi complex in a eukaryotic cell is to
A. conduct ions in and out of the cell
B. transport genetic materials out of the cell
C. provide attachment for ribosome granules
D. transport organic materials in and out of the cell

9. UME/2009/3
In a cell, the genes are carried by
A. nuclear membrane
C. lysosomes
B. chromatin threads
D. mitochondria

10. UME/2010/2/TYPE: D
Which of the following characterizes a mature plant cell?
A. the nucleus is pushed to the centre of the cell
B. the cell wall is made up of cellulose
C. the nucleus is small and irregular in shape
D. the cytoplasm fills up the entire cell space

11. UME/2010/3/TYPE: D
Which of the following is NOT a function of the nucleus of a cell?
A. it translates genetic information for the manufacture of proteins
B. it stores and carries hereditary information
C. it is a reservoir of energy for the cell
D. it controls the life processes of the cell

12. UME 1982/2
Which of these combinations is common to plant and animal cells?
A. Centriole, middle lamella, Golgi bodies, mitochondrion
B. Cytoplasm, sap vacuole, starch grains, leucoplasts
C. Plasma membranes, chromosomes, mitochondria, lysosomes
D. Nucleus, nuleolus, cellulose cell wall, endoplasmic reticulum
E. Cytoplasm, centriole, cellulose cell wall, nucleolus

13. UME 1987/1
The function of endoplasmic reticulum is
A. Protein synthesis
B. Intracellular transport of materials
C. Digestion and destruction of foreign bodies
D. Production of energy from glucose

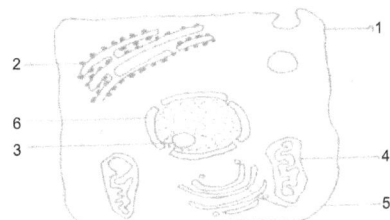

14 UME 89/1
The structure labeled 5 is the
A. Nucleolus   B. Golgi body
C. Mitochondrion   D. Vacuole

15. UME 89/2
Which of the labeled parts of the cell contains ribonucleic acid?
A. 1   B. 2   C. 3   D. 5

16. UME 89/3
Which structure is known as the power house of the cell?
A. 6   B. 5   C. 4   D. 5.

17. UME 93/1
On what structures are the units of inheritance situated?
A. Golgi bodies   B. Ribosomes
C. Chromosomes   D. Endoplasmic reticulum

18. UME 94/1
The membrane surrounding the vacuole in a plant cell is called the
A. Plasmalemma
B. Tonoplast
C. Nuclear Membrane
D. Endoplasmic reticulum

19. UME 2002/4 TYPE: Y
The structures found only in plant cells are
A. Cell membrane and cytoplasm
B. Chromatin and nucleus
C. Cell wall and Chloroplast
D. Cell membrane and lysosome

20. PCE 91/1
The structure in a cell concerned with the synthesis of proteins is the
A. Endoplasmic reticulum
B. Ribosome
C. Nucleus
D. Plasma membrane

21. PCE 91/2
Mitochondrion is regarded as the "power house" of a cell because it
A. Contains numerous enzymes involved in respiration
B. Controls the synthesis of fatty acids
C. Contains numerous enzymes involved in photosynthesis
D. Contains numerous enzymes involved in metabolic activities

22. PCE 97/1
The cytoplasm in unicellular organisms is made up of
A. Ectoplasm and Endoplasm
B. Ectoplasm and Nucleus
C. Endoplasm and Nucleus
D. Protoplasm and Nucleus

23. PCE 99/4 TYPE: E
The organelle associated with detoxification of harmful substances is
A. Lysosome B. Mitochondrion
C. Golgi body D. Endoplasmic reticulum.

24. The cell membrane consists of
A. Carbohydrate and Lipids
B. Vitamins and Proteins
C. Lipids and Proteins
D. Water and Sugar
E. Starch and Cellulose

1.3. LEVELS OF ORGANIZATION

25. PCE 96/1
Which of the following is the correct hierarchy of biological organization
A. Cell→ Organ→ System→ Tissue→ Multicellular organism
B. Cell→Tissue→ System→ Organ→Multicellular organism
C. Cell→ Tissue→ Organ→ System→ Multicellular Organism
D. Cell→ Organ→ Tissue→ System→ Multicellular Organism

26. UME 85/12
A group of similar cells performing the same function is
A. an organ   C. a tissue
B. a system   D. an organelle   E. an enzyme.

27. UME 2002/1 TYPE: Y
An Amoeba and an Unlaid chicken egg are
A. Animal tissues   B. Organelles
C. Single cells   D. Organisms.

28. UME/2004/29 TYPE: Y
The rods in the retina of the eye are examples of
A. Organs   C. Systems
B. Cells   D. Tissues

29. UME/2009/1
What is the level of organization of an onion bulb?
A. tissue   B. organ
C. systemic   D. organism

30. PCE 91/3
Hydra is said to be at the tissue level of organization because
A. it is unicellular
B. it has many cells performing different functions
C. it has many cells performing the same function
D. it is cells are differentiated into tissues

31. PCE 2006/31/TYPE:A
Cellular level of organization is found in the phylum
A. nematode
B. coelenterate
C. protozoa
D. pletyhelminthes

MISCELLANEOUS QUESTIONS (1)
32. Which of the following does a virus have in common with animal cells?
A. Nucleus   B. DNA
C. Glycogen   D. Cytoplasm (UME 95/1)

33. Virus are considered to be living organisms because they
A. possess transmittable characters
B. move from one place to another
C. respond to stimulation
D. inject food materials (UME 90/1)

34. Which of these organelles are likely to be present in cells that are actively respiring and photosynthesizing?
A. Nucleus and centrioles
B. Mitochondria and chloroplasts
C. Lysosomes and ribosomes
D. Golgi apparatus and endoplasmic reticulum
   (UME 95/2)

35. The nucleus is considered the central organelle of a cell because it
A. contains the genetic material
B. contains the nuclear sap
C. is bounded by the nuclear membrane
D. is located at the centre of the cell (UME 98/1)

36. All living cells require water because it
A. is a medium that neutralizes acids in cells
B. is the main source of energy for the cells
C. preventing the development of diseases in cells
D. is a medium for all reaction (UME 2004/19)

37. Which of the following is likely to have a higher concentration of mitcohondria?
A. sperm cell
B. white blood cell
C. egg cell
D. red blood cell (UME/2005/2)

38. The tiny organelles on the walls of a rough endoplasmic reticulum are the
A. lysosomes
B. microvilli
C. microtubules
D. ribosomes (PCE/2005/44)

39. The organelle common to both plant and animals cells is the
A. centriole
B. plasmalemma
C. cell wall
D. chloroplast (UME 2005/1)

40. The component of a bacterial cell that is not organized into a nucleus as in other cells is the
A. deoxyribonucleic acid
B. ribonucleic acid
C. internal membrane
D. cell wall   (PCE/99/8/TYPE: E)

41. Non-living things differ from living things in that they
A. are composed of one or more cells
B. have a relatively low energy content
C. obtain and use energy
D. have a high degree of organization
   (PCE/06/29/TYPE:A)

CHAPTER 2

EVOLUTION AMONG ORGANISMS

Understand the following after reading this chapter:

> Monera(prokaryote) e.g. bacteria and blue-green algae
> Protista (protozoan and protophyta) e.g. Amoeba and Euglena
> Fungi e.g. mushrooms and bread moulds
> Plantae (plants)
> Animalia (animals)

EVOLUTION AMONG LIVING ORGANISMS

Classification and diversity of living organisms
The system of classifying living organisms on a hierarchy is known as taxonomy. Below is the taxonomical hierarchy; Kingdom, Phylum, Class, Order, Family, Genus, Species.

Memory Tip: *(King Play Card On Fine Green Sand)*

Binomial system: a system of naming species in which the scientific name of an organism is made up of two parts showing the genus (starting with a capitol letter) and species (starting with a lower case letter), written in italics when printed (therefore underlined when handwritten) e.g. *Homo sapiens*

The five kingdoms:
Animal: Multi-cellular ingestive heterotrophs (eat living organisms)
Plant: Multi-cellular photosynthetic autotrophic (make their own food) organism with a cellulose cell wall.
Fungi: Single celled or multi cellular heterotrophic organism with a cell wall not made of cellulose, saprotrophs (feed off dead organisms) or parasites
Monera: Single celled organism with no true nucleus
Protista: Single celled organism with a nucleus

CLASSIFICATION OF PLANTS
Plants are classified into division based on their structural differences, presence or absence of chlorophyll, presence or absence of flower, presence or absence of vascular bundle and life cycle.

1. The Schizophytes: These are unicellular and microscopic plants. They lack definite nucleus and possess cell wall without cellulose. Reproduction is asexual e.g. bacteria.

2. The Thallophytes: These have a simple body structure. They lack roots, stems and leaves. The two divisions are: the
a. **Algae** e.g. *Spirogyra, Volvox and Chlamydomonas*
b. **Fungi** e.g. *Rhizopus, Mushroom* and *Mucor.*

3. The Bryophytes: These are multicellular non-vascular and non-flowering plants. They possess root-like structures called rhizoids. They are found in damp places, on the barks of trees, floor of grass-lands and forests e.g. mosses, hornworts and liver-worts.

4. The Pteridophytes: These are non-flowering vascular plant having structures resembling roots, stems and leaves. They reproduces by means of special gametes and spores e.g. ferns and selaginella.

5. The Spermatophytes: These are the seed producing vascular plants. They have true roots, stems and leaves. They are subdivided into two;

a. **Gymnosperms:** They bear naked seeds inside scale – like structures. They bear naked seeds inside seed boxes called cones e.g. conifers, cycads, whistling pine etc.
b. **Angiosperms:** These are flowering seed producing vascular plants. Their fruits are covered with fruit wall.

They are subdivided into two namely;
i. **Monocotyledons:** These have parallel leaf venation fibrous root system, scattered vascular bundles in the stem and only one seed coat e.g. maize, rice, grasses, bamboo etc.
ii. **Dicotyledons:** These have net or reticulate venation ringed or organized vascular bundles in the stem, taproot system and two seed coats e.g. bean, mango, hibiscus etc.

Memory Tip: Samuel The Biologist Planted Sugarcane

CLASSIFICATION OF ANIMALS
Major phyla
Phylum : Protozoa
This phylum includes a great diversity of small, microscopic organ- isms. These are single celled eukaryotes. Their locomotion happens using pseudopodia, cilia or flagella.
The nutrition is either autotrophic or heterotrophic. They reproduce either asexually or by sexual methods. Eg: *Amoeba, Paramoecium, Plasmodium.*

Phylum : Porifera

These are multicellular, aquatic organisms. They have a cellular grade of construction without the occurrence of tissues. The sponges belonging to this phylum are characterised by the presence of a canal system in their body. The body wall contains spicules. They can reproduce both by asexual and sexual methods. Eg : Sponges.

Phylum : Coelenterata or Cnidaria

All coelenterates are aquatic animals. They are mostly marine. The body is **radially symmetrical**. The body wall is of two layers of cells. The outer layer is called the **ectoderm.** The inner layer, **entoderm** is seperated from the ectoderm by a non-cellular **mesogloea.** The mesogloea is a jelly-like sub- stance. Due to the presence of two layers in the body wall, these are said to be **diploblastic animals.**

Many coelenterates exhibit polymorphism. In this phylum, organisms exist in two different body forms namely, a polyp, and a medusa.The ectoderm contains stinging cells called **nematocysts (cnidoblasts).** These cells when triggered can explosively penetrate prey and inject poison. The layers in the body wall contain several cells and tissues such as muscle cells epithelial tissues, gland-cells and sensory cells.

They reproduce both asexually and sexually. They are divided into three classes, namely **Hydrozoa, Scyphozoa** and **Anthozoa.**

In **Hydrozoa,** the animal has a dominant polyp body form and a reduced medusa stage. (e.g) Hydra, Obelia.

In **Scyphozoa** the medusa form is permanent. This group includes jelly fishes such as Aurelia. They swim in the surface waters. They have a bell shaped medusa stage.

The **Anthozoans** mostly remain as polyps. Their body cavity is divided by large radial partitions called mesenteries. (eg) sea-anemone and corals.

All animals of subsequent phyla show the following general characters.

1. All of them have three layers in the body wall. They are named as outer ectoderm, middle mesoderm, and inner endoderm. Thus they are called as Triploblastic animals.
2. The body is bilaterally symmetrical.

Phylum: Platyhelminthes

This phylum includes flatworms. These are acoelomates, without a body cavity called coelom. The alimentary canal is either absent or very simple. Excretion and osmoregulation occur through flame cells.These worms are mostly hermophrodites, having both male and female reproductive organs in a single individual. Most of the members are parasites. It is divided into three classes, namely Turbellaria, Trematoda and Cestoda.

Class Turbellaria :-These are free living aquatic flatworms. The Planaria of this class shows characteristic regeneration.

Class Trematoda :- These are flukes living as parasites inside a host (en- doparasites). A protective cuticle covers the outer surface of the body. Flukes have suckers for attachment to the host tissues. The examples are Fasciola (liver fluke), *Schistosoma* (blood fluke).

Class Cestoda :- It includes all tape worms. These are internal parasites with a complex life history. The life cycle involves two hosts.

Their body characters are adaptations for parasitic life. Mouth and alimentary canal are absent. Food is absorbed through general body surface. The head is called the scolex. It has a ring of hooks and suckers for attach- ment to the host tissue. The body consists of several segments called Proglottids. (eg) sheep and cattle tape worms.

Phylum : Nematoda

These are the popular round worms. The body is narrow and pointed at both the ends. There are no body segments. The body is covered by a thin cuticle. The body cavity is considered as a pseudocoelom. The alimentary canal is a straight tube. They reproduce sexually and the sexes are seperate. There are several free living soil nematodes. Others are parasites. (eg) Ascarislumbricoides.

In subsequent Phyla the animals show following general characters

1. There is a coelom within the mesoderm. Hence these are called as coelomates.
2. The body consists of a series of compartments. This phenomenon is called as metameric segmentation. They have a circulatory system pro- viding internal transport.

Phylum: Annelida

These are worm like animals. The body segments are rings externally. Internally the segments are seperated by septa. Externally the body is protected by a cuticle. Excretion and osmoregulation are acheived by ciliated tubules called nephridia. There is a central nervous system. The brain is formed of ganglia in the head region. The nerve cord is ventral in position. For the first time head formation or cephalization happens. These are bi- sexual and hermophroditic. The larva is called the trochophore.

This phylum includes three Classes, namely Polychaeta,Oligochaeta and Hirudinia.Thepolychaetes are marine worms. They have a distinct head. There are pairs of lateral projections called parapodia. The examples are Nereis (ragworms), Arenicola (lugworm).

Earthworms are included in the Class Oligochaeta. The Class: Hirudinia includes leeches. These are blood suckers and ectoparasites. They have well developed suckers for attachement at anterior and posterior ends.

Phylum : Arthropoda

These are the most successful group of animals. They outnumber all other animals in population strength. The body is segmented. It is covered by a hard exoskeleton made of chitin. During growth the exoskeleton is shed (moulting of ecdysis). The legs or paired appendages are jointed. The head region has a pair of prominent compound eyes. Each compound eye is made up of several photoreceptor sub units called Ommatidia.

They have an open circulatory system without vessels. The body cavity is filled with a fluid called haemolymph. Such body cavity is known as haemocoel. These are unisexual, exhibiting sexual dimorphism. The young forms produced are invariably called the larvae. The larvae undergo metamorphosis and develop into adults.

This Phylum comprises five Classes:

Class Onychophora: - It includes small worm like Peripatus. Peripatus shows Annelidan and Arthropoda characters. Hence this may be considered as a connecting link between the two groups.

Class Crustacea :- The examples for this class are prawns, crabs and lobsters. The dorsal body surface is covered by a sheild like carapace.

Class Myriapoda :- It includes centepedes and millipedes. These organ- isms have a distinct head and simple eyes. The centepedes have a pair of poison claws. The body consists of numerous segments, bearing pairs of legs.

Class Insecta :- It comprises the common insects. The body is divided into head thorax and abdomen.In several insects, the adults have two pairs of wings on the thorax. Respiration happens through the tracheal system.

Class Arachnida :- It includes scorpions, spiders, ticks and mites. The body is divided into cephalothorax and abdomen. There are four pairs of legs attached to the cephalothorax.

Phylum Mollusca It is a very successful and diverse group of animals. Considered to be the second largest group of animals with regard to species number. These are soft bodied animals without segmentation. The body is divided into head, muscular foot and visceral mass. The body is covered by a mantle and a shell.

Respiration happens through gills (ctinidia) in the mantle cavity. The most common larva is a trochophore larva.

There are seven classes of which three are more prominent.

Class Pelecypoda or Bivalvia :- These are aquatic molluscs having bivalves. They burrow in mud and sand. The body is laterally compressed. (eg) mussels, clams, oysters.

Class Gastropoda :- These are either aquatic or terrestrial molluscs. They posses a spiral shell.

The foot is large and flat. They have well developed head with tentacles and eyes. (eg) snails, slugs, and limpets.

Class Cephalopoda :- These are mostly marine. They are adapted for swimming. The foot is modified into eight to ten long tentacles in the head region. The shell is either internal or absent. (eg) Octopus, Loligo, Sepia.

Phylum Echinodermata :- These are marine organisms. While the adults are radially symmetrical the larvae remain bilaterally symmetrical. The mouth is on the lower surface. They have a water vascular system with tube feet. eg. star fishes, brittle stars, sea urchins and sea-cucumbers.

Phylum Chordata

This phylum derives its name from one of the common characteristics of this group namely the notochord (Gr. noton, back + L. chorda, cord). The animals belonging to all other phyla of the Animal Kingdom are often termed 'the non -chordates' or 'the invertebrates' since they have neither notochord nor backbone in their body.

The backboned animals (vertebrates), together with a few closely re- lated animals which do not possess a backbone, are included in this phylum. Most of the living chordates are familiar vertebrate animals. The chordates are of primary interest because human beings are members of this group.

Classification.

The Phylum Chordata is classified into four sub phyla:

Sub phylum 1. Hemichordata,
Sub phylum 2. Cephalochordata
Sub phylum 3. Urochordata
Sub phylum 4. Vertebrata.

First three sub phyla are collectively known as Protochordates. Since the members of these sub phyla do not have a cranium or skull they are also referred to as Acrania.

Protochordata (Acrania)

The protochoradates are considered as the fore runners of vertebrata The classification of the protochordates is based on the nature of the noto- chord.

Sub phylum :Hemichordata.

These are exclusively marine organisms. They are solitary or colonial forms. They mostly remain as tubiculous forms. The body is soft, vermiform, unsegmented, bilaterally symmetrical and triploblastic. The body is divisible into three distinct regions namely proboscis, collar and trunk. The body wall is composed of single layer of epidermal cells. The dermis is absent. They have no

endoskeleton. A projection from pharynx, projecting inside the proboscis may be considered as notochord. They have a spacious coelom lined by coelomic epithelium. The alimentary canal is a straight tube running between mouth and anus. They are ciliary feeders. Sexes are separate.
Examples :Balanoglossus, Saccoglossus.

Sub phylum :Cephalochordata.
Cephalochordates are small fish like marine chordates. The persistent notochord extends forward beyond the brain. Hence these are called cephalochordates. The epidermis is single layered. Paired fins are absent. Muscles, nephridia and gonads are segmentally arranged. The pharynx is large with numerous gills. It is a filter feeder.
Example : Amphioxus.

Sub phylum : Vertebrata (Craniata)
This group is characterized by the presence of brain case or cranium and a vertebral column which forms the chief skeletal axis of the body.

The sub phylum vertebrata may be classified into two groups (i) Pisces and (ii). Tetrapoda.

Class : Pisces
Fishes are poikilothermic, aquatic vertebrates with jaws. The body is streamlined. It is differentiated into head, trunk and tail. Between head and trunk, the neck is absent. Locomotion is effected by paired and median fins.
Examples: Shark, Catla.

Tetrapoda
The vertebrates with two pairs of limbs adapted for locomotion on land are known as tetrapods. The limbs are of pentadactyl type. The tetra- pods are identified by a cornified outer layer of skin and nasal passages communicating with mouth cavity and lungs. The super class Tetrapoda is divided into four classes namely. Amphibia, Reptilia, Aves and Mammalia.

Class :Amphibia
The living representatives of this class include frogs, toads, newts, salamanders and limbless caecilians.

Amniota
The tetrapods like reptiles, birds and mammals are referred to as amniotes. The amniotes have certain membranes associated with embryos inside the egg. It is an adaptation in terrestrial forms during development. These membranes are the amnion, chorion and allantois.

Class :Reptilia
Reptiles are represented by lizards, snakes, turtles, tortoises, alligators, crocodiles and the tuatara lizard, Sphenodonpunctatum.

Class : Aves
Birds are one of the most interestingand widely known group of animals. Birds as a group exhibit a characteristic uniformity in structure.
Aves are warm blooded vertebrates with an exoskeleton of feathers forming a non-conducting covering to keep the body warm. The feet are cov- ered with scales. The forelimbs are modified as wings and provided with feathers for flight. The hindlimbs are attached far forwards to balance the weight of the body. Examples : Pigeon, parrot, crow, sparrow, peacock, ostrich, penguin.

Class : Mammalia
The term "mammalia" was given by Linnaeus (1758) to that group of animals which are nourished by milk from the breasts of the mother. They are a successful group, for they adapt themselves readily to new situations and to new food habits. The class Mammalia is subdivided into three subclasses namely:

1. Sub class :Monotremata or Prototheria
These are primitive egg laying mammals Example : Spiny ant-eater, duck billed platypus.

2. Sub class :Marsupialia or Metatheria
These are popularly called as marsupials or pouched mammals. The young ones are born in an immature stage and migrate into the pouch on the mother's body. Further development is completed in the pouch or marsu- pium.
Example : Kangaroo

3. Sub class :Placentalia or Eutheria
In this group eggs develop within the uterus. The developing embryo receives nutrition through maternal blood circulation via the placenta.
Example : Elephant, tiger, lion, man, monkey, dog, cat , rat, bat.

Order Primates :
It is an order coming under the subclass Eutheria. This order is of interest because it includes man, besides lemurs, tarsiers, monkeys and apes. They inhabit chiefly the warmer parts of the world. This group stands first in the animal kingdom in brain development. However, most of them are unspecialized and tree dwelling (arboreal).

Revision Questions

2.1. MONERA (Procaryote) e.g. Bacteria and Blue – green algae

1. UME 98/2
The prokaryotic cell type is characterized by a
A. Complex cytoplasm in which different regions are poorly defined
B. Localization of different regions of the cell into tissues
C. Collection of organelles and macromolecular complexes
D. Simple cytoplasm with well defined regions

2. UME 2003/31
The pioneer Organisms in ecological succession are usually the
A. Lichens   B. Algae
C. Ferns   D. Mosses

3. UME 2003/4
The similarity among organisms belonging to the same group will be least within each
A. order   B. family
C. species   D. kingdom

4. UME 2010/8/TYPE: D
Which of the following groups of cells is devoid of true nuclei?
A. Algae   B. Monera   C. Fungi   D. Viruses

5. UME 2007/39
The chromosomes of members of the kingdom monera are within the
A. nucleoplasm   B. cytoplasm
C. nucleus   D. nucleolus

6. UME 2009/7
A blue-green alga is not a protophyte because
A. it is aquatic
B. its cells are prokaryotic
C. it cannot move
D. it is not a green plant

7. The bacteria are known for their:
A. being photosynthetic   B. being holophytic
C. being unicellular   D. being non-nucleated
E. None of the above

8. Bacteria are quite common except:
A. in the atmosphere   C. in the soil
B. in salt water   D. in the alimentary tract of animals
E. None of the above

9. The cell of bacterium differs from that of a typical plant by:
A. the possession of semipermeable membrane
B. the absence of the cytoplasm
C. the absence of cellulose
D. the presence of a nucleus
E. None of the above

10. Which of the following statements are true?
A. Bacterial add to the soil certain nitrogenous materials
B. Vinegar is produced by bacterial activities
C. The taste of butter is due to bacteria
D. Bacterial activities are involved in preparing and preservation of silage (food for cattle made from maize stems, leaves and cobs)
E. All of the above

11. Useful as bacteria may be, yet they cause a number of diseases. Which of these are not caused by bacteria?
A. Tuberculosis   B. Pneumonia
C. Lockjaw   D. Malaria
E. None of the above

12. The bacteria that produce food poisoning are
A. The floating bacteria   B. Putrefying bacteria
C. Denitrifying bacteria   D. Fruit bacteria
E. None of the above

13. The poisonous substance produced by bacteria is known as:
A. Auxins   B. Antibodies
C. Antotixins   D. Toxins E. Tetanus

14. After one has successfully resisted an attack of a particular disease such as chicken pox, a large quantity of antibodies remain in the body to prevent further attack by the same disease. This power to resist attacks is known as:
A. Innoculation   B. Vaccination
C. Resistance   D. Immunity

2.2. PROTISTA (Protozoan and protophyta) E.g. Amoeba and Euglena

15 UME 87/2
Which of the following features of Euglena is found only in animals
A. Paramylum granules   B. Flagellum
C. Pellicle   D. Pyrenoid

16. PCE 91/5
The structure in Amoeba which performs a similar function as the mammalian kidney is the
A. food vacuole   B. cytoplasm
C. plasmalemma   D. contractile vacuole

11

17. PCE 94/3
An organisms that possesses both plant and animal characteristics is
A. Spirogyra   B. Amoeba
C. Paramecium   D. Euglena

18. UME 88/7
Spirogyra, Euglena and Chlamydomonas share many characteristics EXCEPT
A. Nutrition   B. Reproduction
C. Mobility   D. Irritability

2.3. FUNGI E.g. Mushrooms and Bread moulds
19. UME 2003/1
The umbrella-shaped fruiting body of a fully developed mushroom is the
A. mycelium   B. basidium
C. pileus   D. stipe

20. UME 87/3
An organism found on a bare rock surface has features of algae and fungi. The organism is
A. an epiphyte   B. a lichen
C. a bryophyte   D. a fern.

21. UME 95/3
One common characteristic of fungi, algae, mosses and ferns is that they
A. show alternation of generations
B. reproduce sexually by conjugation
C. produce spores that are dispersed
D. posses chlorophyll in their tissues.

22. UME 81/39
One common feature of the fungi, algae, mosses and ferns is that they
A. Are photosynthetic
B. Show alternation of generation
C. Reproduce by means of conjugation
D. Can survive dry conditions
E. Have no seeds.

23. UME 91/4
Which of the following are non-green plants?
A. Euglena   B. Fungi
C. Spirogyra   D. Angiosperms.

2.4. PLANTAE (PLANTS)

24. UME 89/6
A multinucleated body without internal cell boundaries is a characteristic of
A. bryophytes   B. fungi
C. algae   D. gymnosperms.

25. UME 94/5
Which is the correct order in an evolutionary sequence for the following plants groups?
A. Bacteria→ Ferns →Algae → Mosses → Seed plants
B. Bacteria → Ferns→Mosses →Algae→Seed plants
C. Bacteria→ Algae→ Mosses → Ferns → Seed plants
D. Bacteira→ Mosses→ Algae → Ferns → Seed plants

26. UME/2000/5
Which of the following groups is the most advanced?
A. Pteridophytes   B. Bryophytes
C. Thallophytes   D. Gynosperms.
27. PCE 2002/10 TYPE: 5
The first group of plants to live exclusively on land are the
A. pteridophytes
B. angiosperms
C. bryophytes
D. gymnosperms

2.4.1. BRYOPHYTA (Mosses and Liverworts)

28. UME 84/3
Mosses, liverworts and ferns can be grouped together because they
A. are all aquatic plants
B. all grow in deserts
C. are seedless plants
D. have undifferentiated plant bodies
E. all produce colourless flowers.

29. UME 85/8
Bryophytes are different from flowering plants because they
A. live in moist habitats
B. are small plants
C. reproduce sexually and asexually
D. have small leaves
E. have no vascular tissues.

30. UME 2004/27 TYPE: 4
The absence of special food and water-conducting systems restricts the body size in
A. Bryophytes and the Pteridophytes
B. Thallophytes and the Pteridophytes
C. Liverworts, Mosses and Ferns
D. Algae, Liverworts and Mosses.

31. UME 2006/33
The dominant phase in the life cycle of a bryophyte is the
A. gametangium   B. sporophyte
C. gametophyte   D. prothallus

2.4.2. PTERIDOPHYTA (Ferns)

32. UME 90/5
The algae, bryoplytes and pteridophytes are similar in that they
A. are sea weeds
B. have no vascular tissues
C. require moisture for fertilization
D. are macroscopic plants.

33. UME 91/6
Which fof the following are differentiated into true roots, stems and leaves?
A. Algae   B. Schizophyta
C. Pteridophyta   D. Bryophyta

34. UME 2009/4
Alternation of asexual and sexual modes of reproduction is found in
A. blue-green algae   B. *Euglena*
C. fern   D. maize

35. UME 2006/3
The evidence that supports the advancement of ferns over mosses is derived from
A. comparative anatomy
B. molecular records
C. biochemical similarities
D. physiological records

36. UME/2010/4/TYPE: D
The dominant phase in the life cycle of a fern is the
A. prothallus   B. sporophyte
C. antheridium   D. gametophyte

2.4.3. GYMNOSPERMAE (Conifers)

37. UME 85/3
Which of the following is seed bearing?
A. Mosses   B. Whistling pine
C. Algal filaments   D. Liverwort E. Fern fronds.

38. UME 86/2
Which of the following pairs are fully adapted to terrestrial life?
A. Ferns and algae
B. Ferns and mosses
C. Bryophytes and flowering plants
D. Flowering plants and conifers

39. UME 86/4
Which of the following has cones?
A. Angiosperm   B. Gymnosperm
C. Pteridophyte   D. Bryophyte

40. UME 95/2
Production of naked seeds is a distinctive feature of the group of plant called
A. grasses   B. conifers
C. legumes   D. palms

41. PCE/03/23/TYPE: N
The production of naked seeds is a distinctive feature of
A. palms   B. grasses
C. conifers   D. legumes

2.4.4. ANGIOSPERMAE (Flowering Plants)

42. UME 87/7
Double fertilization is a unique feature of
A. angiosperms   B. bryophytes
C. pteridophytes   D. algae

43. UME 91/3
Angiosperms and gymnosperms belong to the plant group known as
A. Schizophyta   B. Bryophyta
C. Pteriodphyta   D. Spermatophyta

2.5 ANIMALIA (Animals)

2.5.1   INVERTEBRATES
44. UME 88/6
Which of the following sets of organism represents the correct trend from simple to complex structural organization?
1. Mollusca   2. Platyhelminthes
3. Nematoda   4. Protozoan
A. 4→1→2→3   B. 4→3→2→1
C. 4→2→1→3   D. 4→2→3→1

45. UME 2008/8
The most abundant group of organisms in the animal kingdom is
A. mammalian   B. aves
C. annelid   D. insect

2.5.1.1. COELENTRATES (E.g. Hydra)
46. UME 86/3
Which of these animals is radially symmetrical?
A. Squid   B. Hydra
C. Snail   D. Cockcroach

47. UME 89/6
A good example of a diploblastic organism is
A. Amoeba   B. Hydra
C. Earthworm   D. Roundworm.

48. UME 89/5
A characteristic of the phylum coelenterate is that
A. Most of them are marine
B. they possess a gut with a single opening
C. they posses numerous pores in their body
D. They are bilaterally symmetrical

49. UME 90/3
Hydra removes undigested food by
A. passing it through the anus
B. passing it through the mouth
C. means of contractile vacuole
D. egesting it through the body surface

50. UME 90/4
Which of the following groups of invertebrates reproduces by budding?
A. Arthropoda   B. Annelida
C. Mollusca   D. Coelenterata.

51. UME 91/5
Sting cells are normally found in
A. Flatworms   B. Hydra
 C. Snails   D. Paramecium

52. UME 97/5
The cnidoblast cells found in Hydra are used for
A. Reproduction   B. Offence and defence
C. Locomotion and nutrition   D. Food collection

53. UME 2000/2 TYPE: M
Coelum is absent in the class of animals termed
A. Mollusca   B. Reptilia
C. Arthropoda   D. Coelenterata

54. UME/2006/38
One primitive feature of the coelenterates is the possession of
A. a dorsal mouth   B. radial symmetry
C. bilateral symmetry   D. a false root

55. Hydra is a coelenterate. It is because it has:
A. Tentacles   B. Mesogloea
C. Cnidoblast   D. Hypostome
E. Enteron

56. That structure through which the mouth of the hydra opens is:
A. The hypostome   B. the oral cone
C. the bulge   D. the tentacles
E. the bud

57. The testes of the hydra are usually located
A. At the foot of the hydra   B. At the basal disc or foot
C. One the tentacles   D. Below the buds
E. Below the tentacles

58. The ovary is usually located:
A. A little distance above the basal disc or foot
B. on the hypostome
C. Above the buds
D. Just below the tentacles
E. Above the tentacles

59. The cnidoblasts, structures of offence and defence of the hydra are found in large quantitites on:
A. the ectodermal cells
B. The endodermal cells
C. The hypostome
D. The tentacles
E. The foot

60. The intestitia cells of hydra are responsible for
A. Digestion B. Defence
C. Replacement of lost cells
D. Elongation of the hydra
E. None of these

61. The hydra receives impulses and stimuli through it:
A. Nerve net   B. Sense cells
C. Nematocysts   D. Flagellae
E. None of the above

62. Hydra lives in fresh water. Since the osmotic pressure of its body fluid is greater than that of the surrounding water, much water is taken into the body by osmosis. The excess water is removed by
A. The contractitle vacuoles   B. Cnidoblasts
C. Enteron   D. Flagellea
E. None of the above

63. The small green plants that live inside hydra are called
A. *Zoochlorellae*   B. *Chlamydomonas*
C. *Spirogyra*   D. *Euglena*
E. None of the above

64. One important method by which the hydra moves from place to place is:
A. Running   B. Crawling
C. Looping   D. Rolling
E. Sliding

65. The gametes of the hydra are formed from:
A. the nematocyst cells
B. the gland cells
C. the intestitial cells
D. The roving cells of the mesogloea
E. None of the above

66. Hydra guards against self-fertilization by being
A. Protogynous   B. Hermaphroditic
C. Bisexual   D. Monoecious
E. Protandrous

2.5.1.2. PLATYHELMINTHES

67. UME 82/40
Flatworms and roundworms are said to be invertebrates because
A. they are small animals
B. they can live inside the vertebrates
C. some of them are unicellular
D. they have no backbones
E. they are parasitic

68. UME 2003/5 TYPE: A
Hermaphroditic reproduction can be found among the
A. Annelids and molluscs
B. Pisces and amphibians
C. Coelenterates and platyhelminthes
D. Arthropods and nematodes

2.5.1.3. NEMATODES (Roundworms)

69. Annelids differ from nematodes in that they
A. exhibit bilateral symmetry
B. are triploblastic
C. are metamerically segmented
D. possess complete digestive system

2.5.1.4. ANNELIDS (Earthworms)
70. UME 89/8
Which of the following phyla have members with both internal and external segmentation?
A. Plathyhelminthes   B. Nematoda
C. Annelida   D. Mollusca

71. UME 91/10
In the earthworm, the cocoon is secreted by the
A. chaeta   B. prostomium
C. peristomium   D. clitellum.

72. UME 94/4
The soil swallowed by the earthworm to form the worm cast in ground up in the
A. Clitellum   B. Prostomium
C. Mouth   D. Gizzard

73. UME 99/5 TYPE: D
The organism that has a hydrostatic skeleton is
A. Tilapia   B. Hydra
C. Mosquito larva D. Earthworm

2.5.1.5. ARTHROPODS.

74. UME 85/4
Each of the following is an arthropod EXCEPT the
A. Crab   B. Spider
C. Snail   D. Millipede
E. Cockroach

75. UME 93/7
The most successful group of animals in terms of diversity of species is
A. Mollusca   B. Arthropoda
C. Mammalia   D. Platyhelminthes

2.5.1.6. CRUSTACEAN

76. UME 86/7
Which of the following lacks chaetae, tentacles and antennae?
A. Snail   B. Crab
C. Millipede   D. Earthworm

77. UME 87/8
The crayfish is an arthropod because
A. its body consists of a cephalothorax and an abdomen
B. it has a pair each of antennae and antennules
C. every segment of its body carries a pair of appendages
D. its body is covered with an exoskeleton made of chitin.

78. UME 91/11
The function of maxillipeds in Crayfish is to aid
A. walking   B. swimming
C. feeding   D. respiration

79. UME 78/23
Which of these is NOT true of the insect? The possession of
A. two pairs of antennae   B. jointed appendages
C. exoskeleton   D. three pairs of legs
E. segmented bodies

80. UME 82/26
Which of these statements is NOT true of insects?
A. They are arthropods
B. Their body is divided into three distinct regions of head, thorax and abdomen
C. Their thorax comprises three segments. Only two of which bear a pair of appendages each
D. Respiration is by means to trachea.
E. They undergo metamorphosis.

81. UME 98/7
The ability of the cockroach to live in cracks and crevices is enhanced by the possession of
A. Wings and segmented bodies
B. Compound eyes
C. Claws on the legs
D. Dorso – ventrally flattened body.

2.5.1.7. ARACHNIDS

82. UME 88/13
Lungbooks are used for respiration in
A. Spiders   B. Insects
C. Millipedes   D. Snails.

83. UME 89/11
The peripalp in spiders is used for
A. grasping   B. walking
C. feeling   D. web spinning.

84. UME 99/7/TYPE: D
The group of arthropods that has no antennae is the
A. Crustacea
B. Chilopoda
C. Arachnida
D. Diplopoda

85. UME 83/23
A centipede differs from a millipede by its
A. Colour
B. Numerous abdominal segments
C. Paired legs on each abdominal segment
D. Poison claws
E. Cylindrical body.

86 UME 88/14
Insects and millipedes have many features in common EXCEPT
A. Exoskeleton
B. Jointed appendages
C. Compound eyes
D. Segmented body.

2.5.1.9. MOLLUSCS

87. UME 87/13
Which of the following is NOT a characteristic of snails?
A. Bilateral symmetry
B. Chitinous exoskeleton
C. Muscular foot
D. Soft unsegemented body in mantle

88. UME 88/5
 Parasitic forms are NOT found among
A. Platyhelminthes   B. Nematodes
C. Molluscs   D. Annelids.

89. UME 89/12
The body of a snail is divided into head,
A. thorax and abdomen
B. visceral mass and abdomen
C. thorax and foot
D. visceral mass and foot

90. UME 91/12
The respiratory organ in the land snail is the
A. radula   B. Mantle
C. tentacle   D. foot

2.5.2. VERTEBRATES

91. PCE/95/3
The correct sequence that represent verterbrate evolutionary trend is
A. fish→amphibians→birds→reptiles →mammals
B. fish→reptiles→Amphibians→ birds →mammals
C. fish→birds →reptiles → amphibians →mammals
D. fish→amphibians →reptiles → birds →

92. UME 2001/2/TYPE: 1
Amphibians are normally found
A. on dry land and in water
B. in water and on moist land
C. on moist land   D. in water

93. UME 2001/3/TYPE: 1
Viviparity occurs mainly in the
A. mammals   B. reptiles   C. aves   D. amphibians

94. UME/2009/5
The first terrestrial vertebrates evolved from
A. pisces   B. reptilian
C. amphibian   D. mammalian

95. PCE/03/16/TYPE: N
Vertebrates having only one functional ovary are the
A. mammals   B. toads   C. reptiles   D. birds

96. PCE/04/26/TYPE: 9
Which of the following vertebrates by the highest number of eggs?
A. Birds   B. Mammals
C. Reptiles   D. Fishes

MISCELLEANEOUS QUESTIONS (2)

97. Euglena may be classified as a plant because It
A. has chloroplasts   B. has a gullet
C. lives in pond   D. possess a flagellum
E. has a pellicle   (UME/78/7)

98. Which of these animals is radially symmetrical?
A. squid   B. Hydra   C. Snail   D. cockroach
(UME/86/3)

99. Which of the following organisms does not exits as a single free living cell?
A. paramecium   B. volvox   C. amoeba
D. chlamydomonas (UME/2001/9/Type 7)

100. The   hyphal wall of fungi is rigid owing to the presence of
A. cell wall   B. lignin
C. cellulose   D. chitin
(UME/2005/3/Type D)

101. Angiosperms and gymnosperms belong to the class
A. schizophyta   B. spermatophyta
C. pteridophyta   D. bryophyta
(UME/2005/4/Type:D)

102. The leech and the earthworm belong to the
A. Molluscs   B. crustaceans
C. arachnids   D. annelids (UME/2005/6/type D)

103. The correct order of the evolutionary sequence is
A. bacteria→ mosses→ algae→ seed plants
B. bacteria→ algae→ mosses→ ferns→ seed plants
C. bacteria→ ferns→ algae→mosses→seed plants
D. bacteria → ferns→ mosses→ algae→ seed plants (PCE 03/18/Typ: N)

104. In evolutionary trends, the pteridophytes are higher than the bryophytes because of their possession
A. Spore – bearing bodies
B. Xylem and phloem vessels
C. Photosynthetic pigments
D. Parenchymatous cells (PCE/2004/2/Type:9)

105. Which of the following phyla has the basic characteristic of the body being divided into head, visceral hump and muscular foot?
A. protozoa
B. chordate
C. mollusca
D. platyhelminthes (PCE/93/6).

106. Which of the following group of animals is completely oviparous?
A. Amphibia   B. Aves
C. Mammalia   D. Reptilia (PCE/93/8)

107. A water medium is necessary for fertilization in
A. fungi   B. conifers
C. ferns   D. angiosperms

108. Which of the following exists in the form of a colony?
A. spirogyra   B. chamydomonas
C. volvox   D. euglena (PCE/91/4)

109. Crab, grasshopper and millipede belong to the same phylum but different classes because they have
A. exoskeleton with varied number of appendages
B. endoskeleton with varied number of appendages
C. exoskeleton with jointed appendages
D. endoskeleton with jointed appendages
(PCE/91/6)

110. Which of the following features is common to all insects?
A. possession of stalked compound eye
B. complete metamorphosis in the life cycle
C. possession of jointed appendages
D. possession of two pairs of wings (PCE 91/7)

111. Which of the following indicates the correct order of increasing structural complexity?
A. Amoeba →Planaria→ Hydra→ Lumbricus
B. Amoeba→ Planaria→ Lumbricus → Hydra
C. Amoeba → Hydra → Planaria→ Lumbricus
D. Amoeba→ Lumbricus → Planaria → Hydra
(PCE/98/4)

112. Diploblastic body organization is found in
A. Amoeba   B. Obelia
C. Lumbricus   D. Periplaneta (PCE/98/5)

113. The earthworm is more advanced than the roundworm owing to its possession of
A. pseudocoelum   B. an alimentary canal
C. true segmentation   D. a hydrostatic skeleton
(PCE/04/3/TYPE: 9)

114. A feature of annelids that distinguishes then from nematode is their possession of
A. a three-layered body   B. a segmented body
C. an enlongated body   D. a cylindrical body
(PCE/06/27/TYPE: A)

115. Which of the following flatworms is free living?
A. Tapeworm   B. Planaria
C. Liver fluke   D. Ascaris (PCE/07/4)

116. Seed-producing vascular plants belong to the
A. phylum thallophyta   B. phylum spermatophyta
C. phylum pteridophyta   D. phylum bryophyta
(PCE/07/06)

117. In terms of the number of individuals, which of the following taxa is most inclusive
A. order   B. family
C. class   D. species (UME/2008/1)

118. unique feature of the gymnosperms is the production of
A. naked seeds   B. fruits without seeds
C. multicellular spores D. seeds within the fruit
(PCE/06/34/TYPE: A)

CHAPTER 3

STRUCTURE AND LIFE HISTORIES OF SOME ORGANISMS

Understand the following after reading this chapter:

> Spirogyra or any green filamentous algae
> Rhizopus or Mucor
> Brachymenium and Marchantia or any other Moss/Liverworts.
> Dryopteris or Neprolepsis or any other fern.
> Flowering Plants
> Amoeba and Paramecium
> Tapeworm (Taenia)

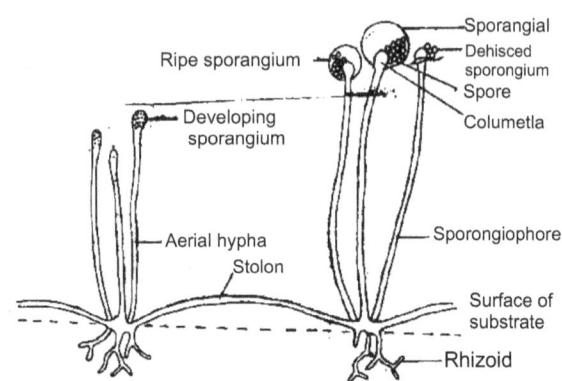

A rhizopus

SPIROGYRA OR ANY OTHER GREEN

FILAMENTOUS ALGAE

Spirogyra is an algae. It floats in masses in ponds, lakes and slow-moving stream. It's a green plant which is composed of large similar cells joined end to end to form a filament. Each cell is cylindrical and contains one or more green ribbon-shaped chloroplasts formed into a characteristic helix. The chloroplast contains numerous colourless dense bodies called pyrenoids around which starch granules are found. Nutrition is autotrophic by photosynthesis. Reproduction is by fragmentation (asexual) and conjugation (sexual).

HIZOPUS OR MUCOR

Rhizopus is a fungus. It is a saprophyte on moist substrates rich in carbohydrates e.g bread, soil and other decaying matter.

The body is a **mycelium** which is made up of white fine threads called **hyphae**. The hypha is coencocytic (i.e has no cross walls) and encloses a mass of cytoplasm with many nuclei and oil droplets. The mycelium is differentiated into branched **rhizoids** that penetrates the substrate, horizontal hyphae known as **stolons** that spread over the surface of the substrate, and erect sporangiophores (sporangial stalks) that develop in tufts above the rhizoid and bear the asexual reproductive structure, the sporangia. The mould spreads rapidly by producing stolons which penetrate the substrate at intervals to produce new rhizoids. Respiration is aerobic and is by diffusion of gases. Asexual reproduction is by spore formation and sexual reproduction is by conjugation.

MOSS/LIVERWORTS

They are found in damp shaded areas. The stem is erect and surrounded by thin leaves. At the base of the stem are hair-like out-growths, rhizoids. These anchor the plant and absorb water and mineral salts.

Reproduction (sexual): The sex organs are found at the apex of the stem. These are: **antheridia**, which produce large numbers of ciliated male gametes called **antherozoids**; **archegonia**, each of which contains a single female gamete, ovum.

When matured the antherozoids are released and they swim through the film of moisture on the leaves. They are attracted to the ovum by chemicals it secretes. The female gamete is eventually fertilized by a single male gamete. The resulting zygote develops within the female organ on the parent plant into a spore bearing structure called **sporophyte**. The sporophytes are produced in the wet season; each consists of a long stalk bearing an enlarged structure called the capsule. The capsule contains a spore-bearing tissue covers by a ring of tooth-like structures (peristome). The capsule dehisces at the tip when dry. Dried spores are dispersed by the peristome and are carried by air currents. The spores, which land on suitable moist habitat, develops into a branched filamentous structures called protonema. Protonema develops into new upright leafy moss plant.

Reproduction (asexual): could be by regeneration of leaf or other parts.

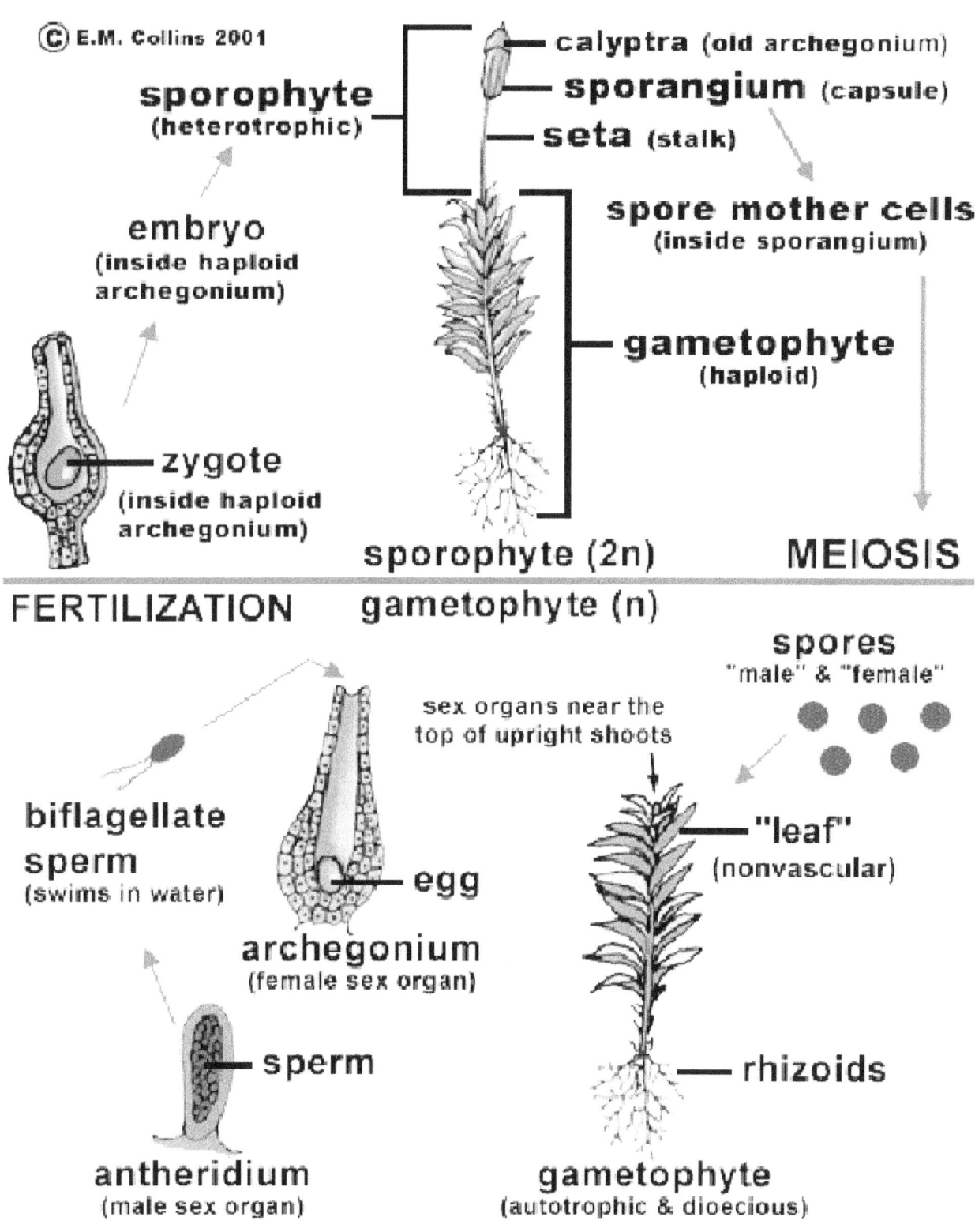

© E.M. Collins 2001

sporophyte
(heterotrophic)

calyptra (old archegonium)

sporangium (capsule)

seta (stalk)

spore mother cells
(inside sporangium)

embryo
(inside haploid
archegonium)

gametophyte
(haploid)

zygote
(inside haploid
archegonium)

sporophyte (2n)

MEIOSIS

FERTILIZATION

gametophyte (n)

spores
"male" & "female"

sex organs near the
top of upright shoots

"leaf"
(nonvascular)

biflagellate
sperm
(swims in water)

egg

archegonium
(female sex organ)

sperm

antheridium
(male sex organ)

rhizoids

gametophyte
(autotrophic & dioecious)

Life cycle of a moss

19

FERN

They are found in wet shady areas usually near streams or as epiphytes on palms or other trees.

The leaves called fronds are large and feather-like showing pinnate arrangement. They arise from buds on underground rhizome. The young leaves are tightly rolled showing circinate venation. The rhizome bear true root.

Reproduction: Groups of spore-bearing sporangia are borne on the underside of mature leaves. Each group is called sorus and is covered by an umbrella like indusium.

At maturity the sporangium dehisces and releases many spores which are dispersed by air currents. If a spore lands on moist soil, it germinates and develops into a small, flat, heart-shaped structure called **prothallus**.

The prothallus is green and bears many rhizoids. **Archegonia** develops towards the notch of the heart-shaped prothallus while the **antheridia** develops towards the conical part. The male gametes swim by the aid of flagella to the female ova. A fertilized ovum develops into a new fern plant.

Alternation of Generation

Both mosses and ferns show alternation of generation. This means that the asexual spore-bearing phase (sporophyte) alternates with the sexual gamete-bearing phase (gametophyte) in the life cycle. In the moss, the sporophyte grows on the gametophyte. In fern, the prothallus is the gametophyte while the fern plant is the sporophyte; both exist independently.

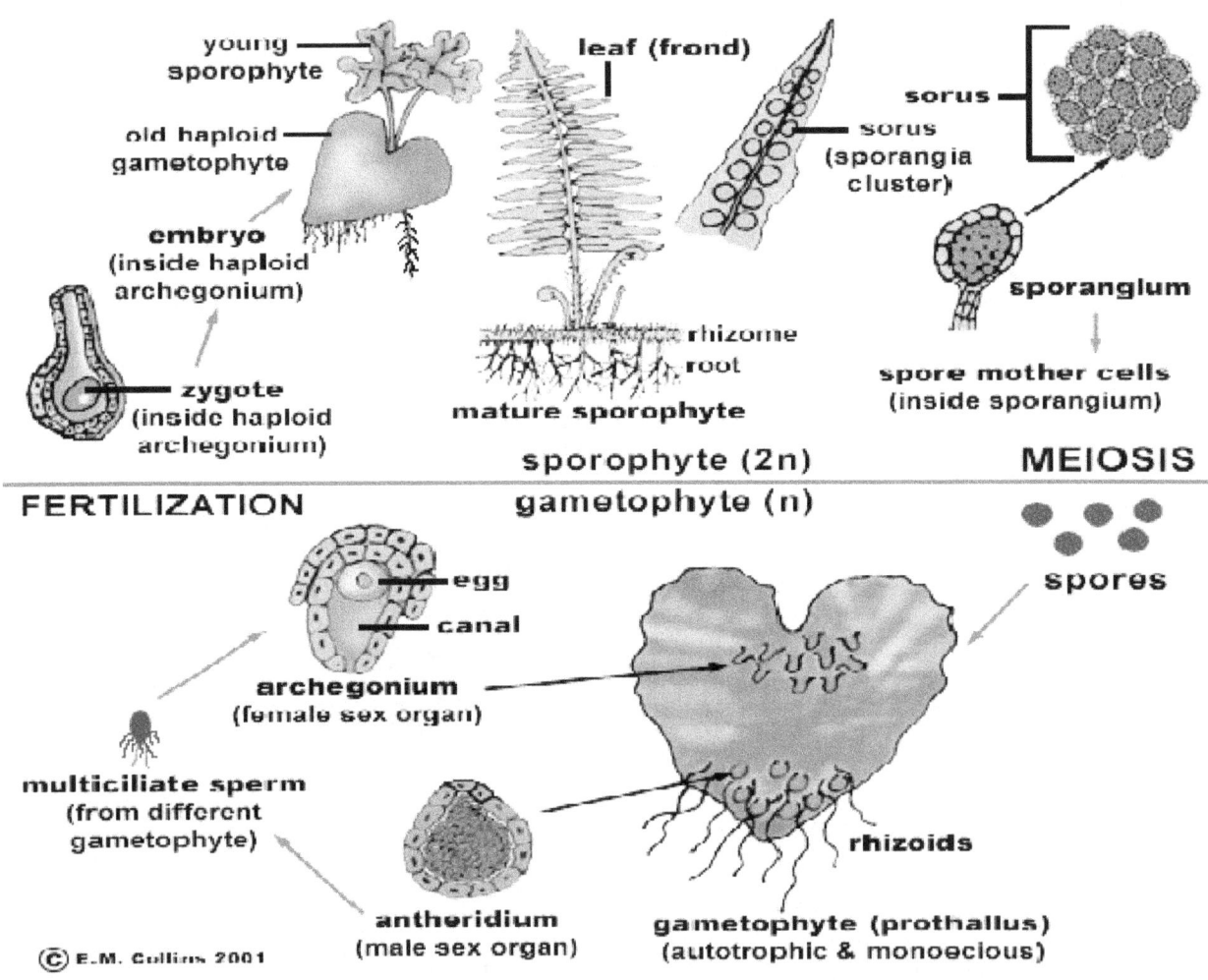

Life cycle of a fern

	AMOEBA	PARAMECIUM
	Variable shape	Constant shape ensured by pellicle
Nucleus:	One	Two, each with a separate function
Locomotion:	By pseudopodia, giving slow speed	By cilia, co-ordinated by sensory organelles giving rapid speed
Food-getting:	Pseudopodia surrounding food	Cilia in oral groove creating feeding current
Ingestion:	At any point	Only through the mouth-pore
Egestion:	At any point	Only through anal -pore
Digestion:	Any part of cytoplasm	Only along a definite path
Excretion:	One contractile vacuole, variable location	Two contractile vacuole, each with six collecting canals, constant location
Reproduction:	By fission (asexual only)	By fission and conjugation (asexual and sexual)
Organ of defence:	None	Trichocysts

TAPEWORM (Taenia)

The tapeworm (Taeniasolium) is a flat, ribbon or tape-like organism. It has a minute knob-like scolex or head. The scolex has four muscular suckers on the sides. The tip of the scolex is elevated to become rostellum. The rostellum is surrounded by hooks. A short "neck" connects the scolex with the rest of the body. The body consists of a series of proglottides about 1000 in number. Each proglottid is separated from the next by a transverse septum. The proglottides are identidal in structure, but those at the posterior end are larger and most matured.

A ripe proglottide with fertilized eggs breaks off from the rest of the tapeworm and passes out with faeces to soil. A man is the primary host. The eggs remain viable for weeks in the soil. If a pig which is the secondary host eats the eggs, the eggs hatch into larvae in the small intestine. The larvae bore through the wall of the small intestine and enter the general blood and lymphatic circulation and carried to the muscles where they develop into cyst-like "bladderworms" or cysticerci.

A man becomes infected when he eats raw or inadequately cooked meat infected with bladderworm. The bladder worm passes into the small intestine and grows into an adult in 5 to 12 weeks and attaches itself into the wall of the small intestine by means of its hooks.

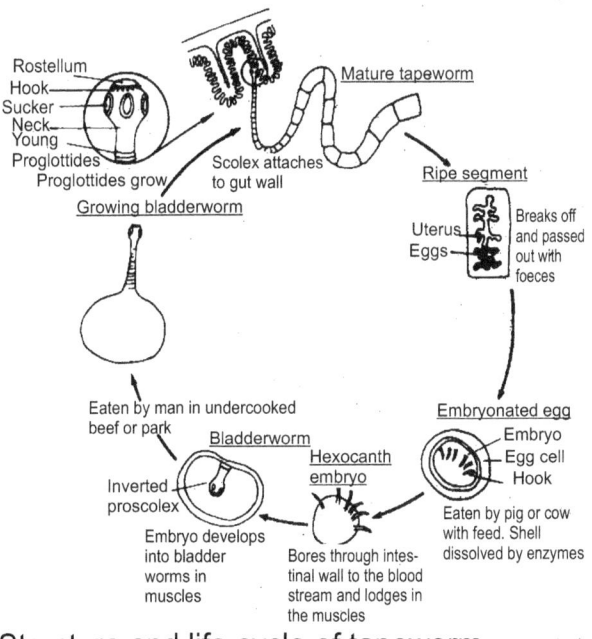

Structure and life cycle of tapeworm

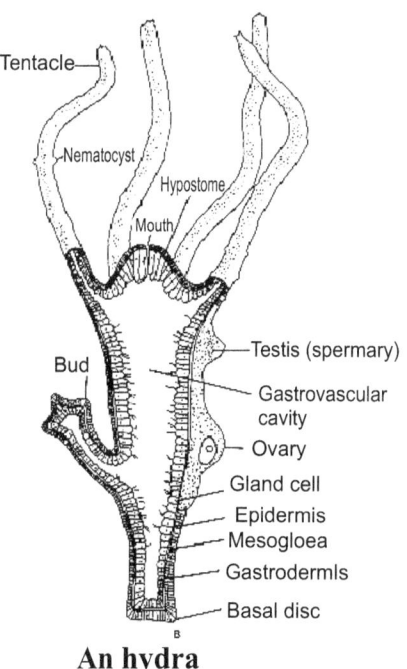

An hydra

21

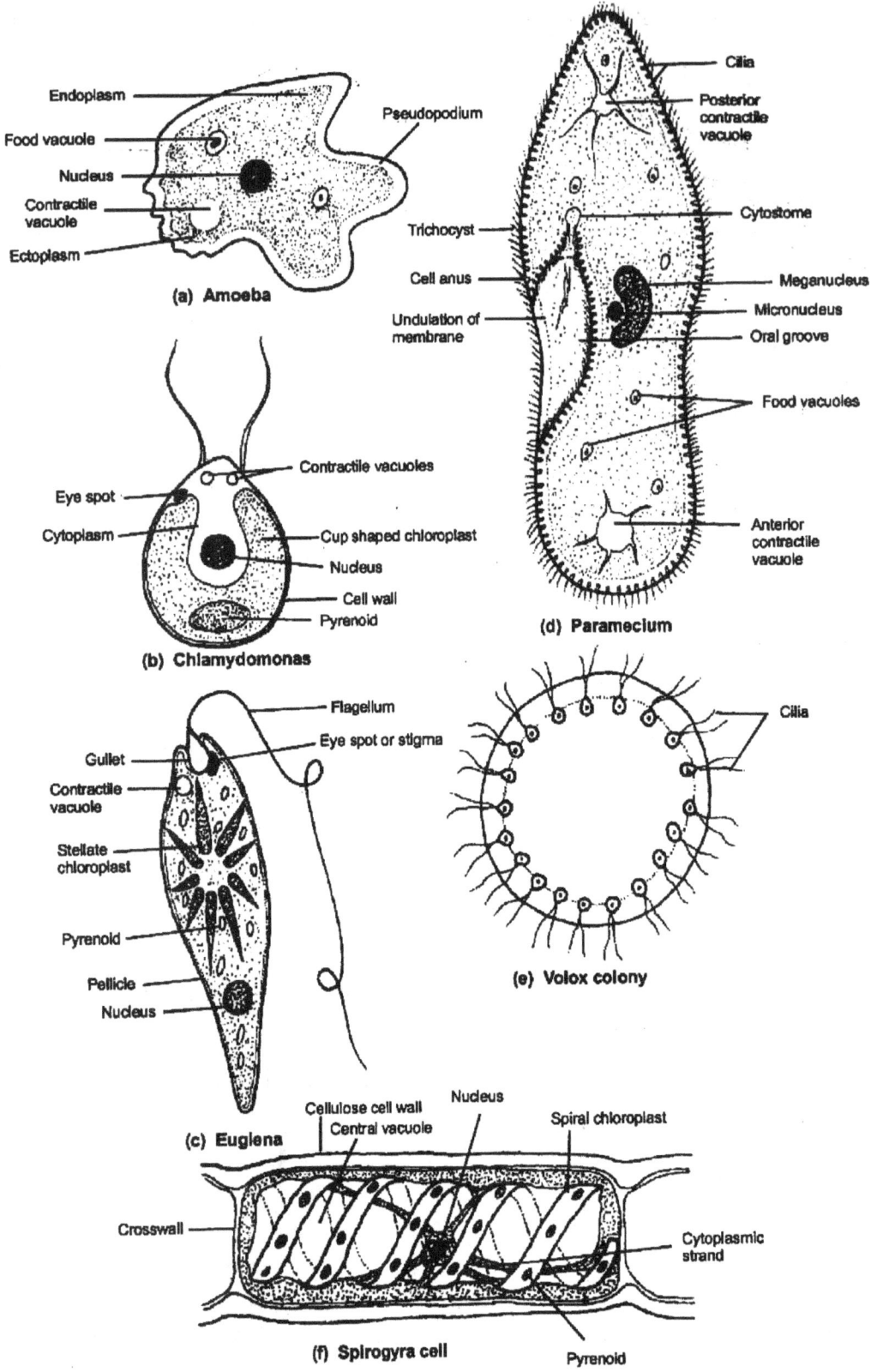

Endoplasm
Food vacuole
Nucleus
Contractile vacuole
Ectoplasm
Pseudopodium

(a) Amoeba

Eye spot
Cytoplasm
Contractile vacuoles
Cup shaped chloroplast
Nucleus
Cell wall
Pyrenoid

(b) Chlamydomonas

Cilia
Posterior contractile vacuole
Cytostome
Trichocyst
Cell anus
Undulation of membrane
Meganucleus
Micronucleus
Oral groove
Food vacuoles
Anterior contractile vacuole

(d) Paramecium

Flagellum
Gullet
Eye spot or stigma
Contractile vacuole
Stellate chloroplast
Pyrenoid
Pellicle
Nucleus

(c) Euglena

Cilia

(e) Volox colony

Cellulose cell wall
Central vacuole
Nucleus
Spiral chloroplast
Crosswall
Cytoplasmic strand
Pyrenoid

(f) Spirogyra cell

Unicellular organisms

3.1. SPIROGYRA OR ANY GREEN FILAMENTOUS ALGAE

1. UME/99/8/Type D

In which of the following organisms does each cell combine the functions of nutrition, reproduction and growth?
A. Rhizopus
B. Dryopteris
C. Brachymenium
D. Spirogyra.

2. UME/86/6

Which of the following is NOT true of spirogyra?
A. Reproduces by conjugation
B. Reproduces by fragmentation
C. Consists of branched filaments
D. Consists of unbranched filaments

3. UME/85/1

In spirogyra, the pyrenoid
A. excretes wastes products
B. is suspended to cytoplasmic strands
C. is mainly used for respiration
D. makes the plant skinny to touch
E. usually contains starch

4. UME/80/2

One of the following statements is NOT true of spirogyra
A. It is a simple multicellular plant.
B. During conjugation, it gametes, which are structurally and physiologically similar fuse to form a zygote
C. It possess spiral chloroplast which enables the plant to photosynthesize
D. It cells are protected by a layer of mucilage
E. There are pyrenoids along the chloroplast.

5. Some types of Spirogyra can be attached to rocks by means of:
A. Roots
B. Rhizoids
C. Tentacles
D. Sticky hairs
E. Mucillage

6. The slimy and slippery nature of the Spirogyra is due to a covering of a substance known as
A. Jelly
B. Gelatin
C. Mucus
D. Mucilage
E. Moss

7. The chloroplast of Spirogyra differ from other types in being:
A. Spirally coiled with indented edges
B. Brown with dark sports
C. Ladder-like with many nucleic
D. Green and clustered around the nucleus
E. None of the above

8. Nutrition in Spirogyra is said to be
A. Holozoic
B. Saprophytic
C. Parasitic
D. Symbiotic
E. Holophytic

9. In the asexual reproduction of spirogyra the growth of the cross wall to divide the mother cell into two daughter cells begins from
A. the wall of cell and grows inward
B. from the centre of cell
C. from the side wall
D. from within the cell
E. None of the above

10. In the sexual reproduction of spirogyra, as the conjugation tube is formed the cell contents contracts from the cell wall to form:
A. the zoospores
B. the zygospores
C. the gametophytes
D. the gametes
E. the sperms

11. Spirogyra is regarded as a multicellular plant because
A. Its cells are linked together by cytoplasmic strands
B. Its cells are joined to form organs
C. The cylindrical cells are linked end to end
D. It contains large vacuole
E. None of the above

12. The nucleus of spirogyra is suspended by:
A. the vacuoles
B. the chloroplasts
C. the cytoplasm
D. pyrenoids
E. none of these

3.2. RHIZOPUS OR MUCOR

13. The part of Mucor within the substance that it grows is:
A. The roots
B. The mycelium
C. The rhizoids
D. The holdfast
E. The nodules

14. The fine thread-like filaments that make up what could be loosely termed the root system' of Mucor are:
A. the hyphae
B. the root hairs
C. the rhizoids
D. the mycelia
E. the hyphen

15. The cell-wall of Mucor is made of
A. Cellulose
B. Hemi-cellulose
C. Fungal-cellulose
D. Plant cellulose
E. Celluloid

16. In sexual reproduction of Mucor, fusion takes place between
A. the two swollen heads of the hyphae
B. the two gametangia
C. the plus and negative strains
D. two nuclei, one from each gametangium
E. the gametes

17. The difference between the hypha and sporangiophore is that:
A. The hypha has nuclei while the sporangiophore has not
B. Structurally, there is no difference
C. The hypha is more durable than the sporangiophore
D. The sporangiophore has funal cell wall, while the hypha has not
E. None of the above

18. UME 85/7
Which of the following is NOT true of Mucor? It
A. contains chlorophyll
B. grows saprophytically
C. bears spores in sporangium
D. consist of hypae
E. reproduces by conjugation

19. UME 87/9
The hyphae of Rhizopus is said to be coenocytic because it
A. does not contain chlorophyll
B. has no cross walls
C. is vascularised
D. stores oil globules.

20. UME/93/9
The spores of mucor are dispersed by
A. water   B. wind
C. insects   D. explosive mechanism

21. PCE/ 2000/7/Type A
The hypha of Rhizopus is
A. septate and uninucleate
B. non-septate and uninucleate
C. non-septate and multinucleate
D. septate and multinucleate

22. UME/80/1
In mucor or Rhizopus carbohydrate is absorbed in the form of
A. starch   B. sucrose   C. glucose
D. glycogen   E. arabinose

23. UME/83/43
In Rhizopus, carbohydrate is stored in the form of
A. glucose   B. paramylon
C. glycogen   D. starch
E. oil

3.3.BRACHYMENIUM AND MARCHANTIA OR ANY OTHER MOSS/LIVERWORTS
24. UME 85/10
In lower plants like Mosses, the structure which performs the functions of roots of higher plants is called
A. root hairs   B. rootlets
C. hyphae   D. rhizoids
E. thalli

25. UME 90/7
In bryophytes, sex organs are produced in the
A. gametophyte   B. rhizoid
C. protonema   D. sporophyte

26. UME 87/11
In the reproduction of mosses, water is essential because
A. they live in moist habitats
B. they cannot reproduce without water
C. the male gametes must swim to fertilize the ovum
D. they produce spores

27. PCE 95/8
The asexual stage in the life cycle of a moss plant is known as
A. sporophyte generation
B. gametophyte generation
C. bryophyte generation
D. pteridophyte generation

28. PCE 2000/5
The male and female reproductive organs of mosses and liverworts are respectively called
A. antheridia and archegonia
B. antheridia and oogonia
C. paragymous and amphigymous
D. spermatocyte and oocyte

3.4. DRYOPTERIS OR NEPROLEPSIS OR ANY OTHER FERN
29. UME 92/3
Which of the following correctly summarizes the life cycle of fern plant?
A. Spores → prothallus → thallus → sporangium
B. Male and female gamentangia → zygospore sporangium → spores
C. Spores → thallus → spermatozo + ovum sporangium
D. Prothallus → spermatozoid + egg cell → leafy plant → sporangium → spore

30. UME 97/7
Which of the following perform similar functions?
A. Ascospores and ascocarp
B. Antherozoids and rhizoids
C. Sorus and indusium
D. Strobili and inflorescence

31. UME 97/8
In ferns, the sporophyte
A. develops from a haploid zygote
B. reproduced sexually to produce spores
C. is haploid and dependent on the gametophyte
D. is diploid and independent of the gametophyte

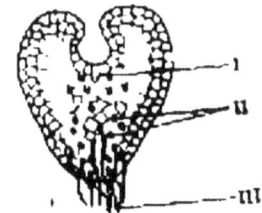

32. UME/2001/10/Type T
The structures labeled II and III respectively are
A. female organs and rhizoid
B. male organs and rhizoid
C. sporophyte and sori
D. annulus and stalk of sporangium

33. UME/2001/11/Type T
In ferns, the structure above is equivalent to the
A. zygote of moss
B. gametophyte generation of a moss
C. sporophyte generation of a moss
D. sporangium of a moss

34. UME/84/6
Which of the following is INCORRECT
The prothallus of a fern.
A. is a flattened heart – shaped structure
B. is green because its cells contains chloroplasts
C. is the dominant plant
D. bears the sexual organs
E. is attached to the ground by numerous rhizoids.

35. UME/87/4
In a plant exhibiting alternation of generations, the diploid multicellular stage is known as
A. gametophyte   B. spermatophyte
C. holophyte   D. sporophyte

36. PCE/93/4
Which of the following describes the function of an antherozoid?
A. It facilitates the growth of the antheridium
B. It regulates the formation of gametophyte
C. It forms the hair – like structures called paraphyses
D. It fertilizes the ovum in the archegonium to produce zygote

37. PCE/97/6
The spores that germinates from the sporangium of a fern grows into
A. a thallus   B. a prothallus
C. as indusium   D. an antheridium.

38. PCE 98/3
Which of the following has vascular tissue?
A. *Funaria*   B. *Dryopteris*
C. *Spirogyra*   D. *Rhizopus*

39. The sporangium of the fern is found in a chamber of the sorus covered by a structure known as
A. the epidermis   B. the indusium
C. the placenta   D. the ramenta
E. none of the above

40. The annulus of the sporangium of the fern assists in the dispersal of spores by:
A. Losing water from its cells
B. Taking up water from the air
C. Losing the entrapped air of its cells
D. Collapsing in the face of strong winds
E. Being attacked by insects

41. The sporangium is
A. Biconcave in shape   B. Plano-convex in shape
C. Biconvex in shape   D. Circular in shape
E. None of the above

42. The stomium has
A. Highly sclerotized cells walls
B. thin-celled walls
C. Overlapping cell walls
D. Tough rectangular cell walls
E. Trap-like walls

43. The spore of the fern usually germinates into:
A. the fern plant   B. the prothallus
C. the sporangium   D. the sporophyte
E. None of the above

44. The life history of the fern illustrates the phenomenon known as:
A. Sporophyte generation
B. Gametophyte generation
C. Antheridial generation
D. Archegonium – antheridial generation
E. Alternation of generation

45. The gametophyte generation is
A. the antheridia
B. the spermatozoa
C. the prothallus
D. the ovum or egg
E. the zygote

3.5. FLOWERING PLANTS
46.   UME 94/3
Green plants are distinguished from other living organisms by their ability to
A. make use of water
B. make use of oxygen
C. respond to sunlight
D. manufacture their own food

47.	UME/92/6
The flowering period of plants in a habitat is determined by
A. duration of sunlight
B. intensity and duration of rainfall
C. relative humidity of the atmosphere
D.temperature of the habitat.

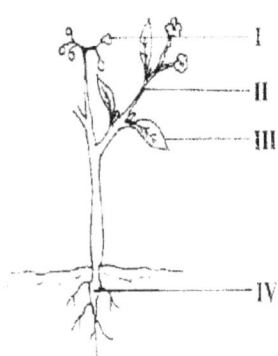

48. UME 2005/8/Type D
The angle between the parts labeled II and III is the
A. bud	B. node
C. internode	D. axil

49. UME/2005/7/Type D
The part of the plant where photosynthesis is least likely to take place is
A. I	B. II
C. III	D. IV

50. UME/2000/3/Type M
Which of the following features are all associated with monocots?
A. Fibrous root system, branched network of veins and one seed leaf
B. Fibrous root system, two seeds leaves and floral parts in threes
C. One seed leaf, petals in threes or groups of threes and parallel venation of leaves.
D. One seed leaf, net-veined leaves and petals in threes or multiples of three

51. The root hairs are found in:
A. The region of cell division of roots
B. Near the root caps
C. The region of cell elongation
D. The region of maturation
E. The region of growth.

52. Some roots are modified for storage purposes. What is the name given to such roots?
A. Rhizomes	B. Tuberous roots
C. Corms	D. Swollen roots
E. None of the above.

53. Which of these plants have storage roots?
A. Yam	B. Onion	C. Cocoyam
D. Cassava	E. Canna Lily

54. Which of these plants produces breathing roots?
A. The oil bean tree	B. Silk cotton tree
C. Para-rubber	D. White mangrove
E. None of the above.

55. Epiphytes produce aerial roots which can absorb water from the atmosphere. This feat is possible because the roots posses a spongy-like structure called:
A. Xylem	B. Spongen	C. Velamen
D. Velum	E. None of the above.

56. Monocotyledonous plants do not increase in thickness or girth because they have:
A. Diffused cambium	B. Undifferentiated bundles
C. No cambium	D. Poor root system
E. None of the above.

57. The area between two adjacent vascular bundles is known as:
A. The pith	B. The interfascicular cambium
C. The medullary ray	D. The vascular
E. The cambium

58. What is the difference between the collenchyma and the parenchyma cells?
A. Both the collenchyma and the parenchyma cortical cells
B. The parenchyma cells contain chloroplasts while the collenchyma cells do not
C. Collenchyma cells contain starch grains while the parenchyma cells do not
D. Collenchyma cells have thickened cell walls those of the parenchyma cells are thin
E. The collenchyma cells are nucleated while those of the parenchyma cells are not

59. The principal function of the pith of stems is:
A. To protect the vascular bundles
B. To give strength to the stem
C. To manufacture food
D. To accumulate air
E. To store food for the plant

60. The structure that produces cork in the stem is:
A. The phelloderm	B. Phellogen
C. Periderm	D. Complimentary cells
E. None of the above.

61. The meristematic structure of the stem of the plant is:
A. The cambium	B. Epidermis	C. Phloem
D. Pith	E. None of the above.

62. That part of the leaf that secures it to the stem is:
A. The peduncle   B. The pedicel
C. The petiole   D. The patula   E. The perianth

63. The leaves of the onion are modified into a storage organ known as:
A. The rhizome   B. The corm
C. The tuber   D. The stem
E. None of the above.

64. The food stored by the onion is:
A. Starch   B. Sugar   C. Protein   D. Oils   E. Fats

65. The plant Bryophyllum pinnatum can be propagated easily because:
A. The leaf stores food   B. It does not die
C. The stem is stout   D. It is immortal
E. None of the above.

3.6. AMOEBA AND PARAMECIUM
66. UME 79/2
Amoeba moves by means of
A. cilia   B. flagella   C. pseudopodia
D. swimmerets   E. setae

67. UME 88/9
What is the function of trichocyst in paramecium?
A. Movement   B. Defence
C. Excretion   D. Reproduction

68. UME 90/2
Which of the following characteristics is common to Amoeba and Paramecium
A. Oral groove   B. Trichocyst
C. Contractile vacuole D. Cilia

69. PCE 91/5
The structure of Amoeba which performs a similar function as the mammalian kidney is the
A. food vacuole   B. cytoplasm
C. plasmalemma   D. contractile vascuole

70. The delicate skin covering the amoeba is called:
A. Ectoderm   B. Membrane
C. Pellicle   D. Epidermis   E. Plasmalemma

71. The waste nitrogenous product of amoeba is:
A. Ammonia   B. Urea   C. Urine
D. Uric acid   E. Carbondioxide

72. Excretion in amoeba is effected through:
A. its general body surface   B. Its pseudopodia
C. The contractile vacuole   D. Its food vacuoles
E. Its endoplasm

73. Amoeba feeds in the main on:
A. Bacteria   B. Other weak amoeba
C. Microscopic plants   D. Water fleas
E. Sea weed

74. The purpose of the contractile vacuole is:
A. To prevent desiccation of amoeba
B. To regulate the intake of salts
C. To expel excess water taken in by osmosis
D. To encyst
E. To increase the protoplasmic contents of amoeba

75. It has been discovered that those amoebae found in the sea have no contactile vacuole. This is because:
A. Sea amoebae are different from fresh water types
B. Sea amoebae do not eliminate waste materials
C. Sea amoebae have cytoplasm whose osmotic pressure equals that of the surrounding water
D. The contractile vacuole is not an essential structure
E. None of the above

76. Amoeba encyst for purpose of:
A. Reproduction
B. Distribution and survival
C. Maintaining their body fluid
D. Defense against diseases E. None of the above

77. Amoebae encyst for purposes of:
A. Reproduction
B. Distribution and survival
C. Maintaining their body fluid
D. Defence against diseases
E. None of these

78. Amoeba is economically important to man because:
A. It is difficult to find
B. It is found in ditches
C. It pollutes man's source of water supply
D. It can very easily be destroyed
E. When in man it causes dysentery

79. Which of these characteristics identify Euglena as belonging to the plant kingdom
A. The possession of single nucleus
B. The possession of a pellicle
C. The possession of chloroplasts
D. The possession of granular cytoplasm
E. None of the above

80. Which of these is not true of the Euglena as an animal?
A. It has a contractile vacuole
B. It has a flagellum
C. It manufactures carbohydrates
D. It has a gullet
E. None of the above

81. The surface of the paramecium is covered with:
A. Flagellae   B. Cilia   C. Trichocysts
D. Bristles   E. Spines

82. The tough skin surface of paramecium is known as:
A. Plasmalemma   B. Ectoderm
C. Pellicle   D. Neuroneme   E. Epidermis

83. The paramecium takes its food through:
A. the oral groove   B. the undulating membrane
C. the food vacuole   D. the general body surface
E. None of the above

84. The paramecium eliminates its waste products through:
A. the contractile vacuoles
B. the food vacuoles
C. the oral groove
D. the temporary anus
E. None of the above

85. The contractile vacuoles in the paramecium are used mainly for:
A. the elimination of nitrogenous wastes
B. the elimination of carbon dioxide
C. the control of the body fluid
D. the storage of salts
E. the elimination of undigested food

3.7. TAPEWORM (Taenia)

86. UME 80/5
Which one of the following animals is NEVER a secondary host of tapeworms?
A. cow   B. fish   C. pig   D. man   E. bat

87.   UME 81/23
The two types of human tapeworm can be distinguished by the presence or absence of
A. scolex B. hook C. head D. sucker E. proglottids

88. UME 87/12
In tapeworm, the two structures that run throughout the length of the body are the
A. nerve cord and the excretory duct
B. sperm duct and the nerve cord
C. genital pore and the excretory duct
D. sperm duct and the genital pore.

89. PCE 95/6
The parts labeled I and II are the
A. Head and the neck   B. Testis and the scolex
C. Ovary and the rostellum   D. Hooks and the sucker

90. PCE 95/7
The body of the organism lengthens when new proglottids are produced at the part labeled
A. I   B. II   C. IV   D. V

91. PCE/98/7/TYPE: C
Tapeworm feeds by absorbing
A. blood from the host's intestine
B. tissues of the host's intestine
C. partly digested food in the host's intestine
D. undigested food in the host's intestine

92. The tape worm is usually segmented. Each of the segments is known as:
A. Strobila
B. proglottid
C. Embryophore
D. Cysticerus E. Hydatid

93. The most conspicuous system absent in the tape worm is:
A. The excretory system
B. The nervous system
C. The alimentary system
D. The system responsible for anchorage
E. None of the above

94. The region from where new segments arise in the tape worm is:
A. the posterior end or the worm
B. the head of the worm
C. the anterior region above the suckers
D. some distance behind the suckers
E. the middle portion of the worm

95. The tape worm is anchored in the intestine of its host by means of:
A. the suckers   B. the rostellum
C. the body spines   D. Its teeth
E. The curved hooks

96. The eggs in the mature proglottid of the tape worm are stored in:
A. the receptaculum   B. the vagina
C. the uterus   D. the embryo sac
E. the oviduct

97. The secondary host of the tape worm Taenia solium is:
A. the man   B. the dog
C. the snail   D. the pig
E. the hen

28

98. When the hexacanth embryo of the tape worm gets into its host it bores into
A. the muscles   B. the blood stream
C. the heart   D. the walls of the alimentary canal
E. the liver

99. One of the effects of the attacks of tape worm is:
A. The irritation of the gut   B. Sleepiness
C. Frequent stooling   D. The loss of appetite
E. None of the above

MISCELLANEOUS QUESTIONS(3)
100. In ferns, the adult plant is the
A. capsule   B. rhizome   C. gametophyte
D. sporophyte (PCE/2005/45/Types: P)

101. The vegetative structure of most fungi is
A. mycelium   B. thallus
C. conidium   D. sporangium (PCE/96/2)

102. The lichen is a composite organism that consists of a fungus living symbiotically with
A. a bacterium   B. a protozoan
C. a virus   D. an algae (PCE/96/12)

103. In bryophytes, sex organs are produced in the
A. gametophyte   B. rhizoid
C. protonema   D. sporophyte (UME/90/7)

104. The prothallus of a fern is equivalent to the gametophyte generation of a moss because it
A. is inconspicuous   B. has rhizoids
C. bears sexual organs   D. is multicellular (UME/88/8)

105. The essential structural difference between Hydra and tapeworm is that while Hydra
A. has tentacles, tapeworm is parasitic
B. is diploblastic, tapeworm is triploblastic
C. has mouth, tapeworm feeds by suckers
D.has mesoderm, tapeworm has mesogloea
(UME/92/5)

106. Which of the following groups of plants shows the correct trend from simple to complex forms?
A.schizophyta → bryophyte → thalophytha → pteridophyta
B.thalophytha → Schizophyta → bryophyte → pteridophyta
C.schizophyta → thalophytha → bryophyte → pteridophyta
D. pteridophyta → bryophyte → thalophytha → schizophyta
PCE/94/2

CHAPTER 4

COMMON INSECTS IN OUR ENVIRONMENT

Understand the following after reading this chapter:

> Mosquito
> Cockroach
> Housefly
> Termite

GENERAL CHARACTERISTICS OF INSECTS

1. The body is segmented
2. The body is divisible into three-head, thorax and abdomen
3. The body is covered by chitinous exoskeleton
4. Each thoracic segment bears a pair of jointed.
5. On the head are a pair of antennae and a pair of compound eyes
6. Respiration is by means of a network of open air tubes called trachea; these open outside as spiracles
7. One pair of wings is present on each of the last two thoracic segments. In some insects one or more of these may be absent, reduced or modified.
8. The young develops through a series of changes in form and in structure before maturing into an adult; this is known as **metamorphosis**

Complete metamorphosis has four stages: egg, larva, pupa, adult; the larva is completely different from the adult, e.g. butterfly.

Incomplete metamorphosis has three stages: egg, nymph (larva), adult; the nymph resembles the adult except that it is smaller, wingless and sexually immature e.g Termite.

MOSQUITO

It is slender in shape. It has only a pair of wings for flight and the other is reduced to *balancers.* The head bears one compound eyes, and a mouth modified into a long tube called *proboscis* for piercing and sucking. There are 8 visible abdominal segments.

Economic importance

They pollinate flowers in their search for nectar both male and female. In addition, the female requires a blood meal to stimulate egg production; in doing this they become vector of disease organisms. The female *Anopheles* mosquito transmits *Plasmodium,* a protozoan parasite causing malaria. The female *Culex* mosquito transmits *Wuchereriabancrofti,* a roundworm, causing filariasis (elephantiasis). The female Aedes mosquito transmits a virus causing yellow fever.

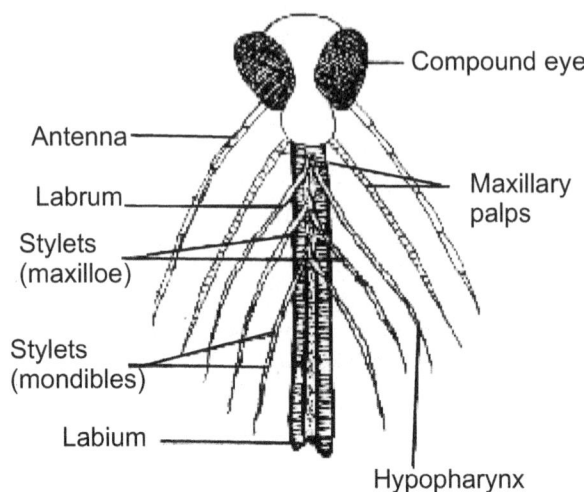

Mouthpart of mosquito

COCKCROACH

It has a dorso-ventrally flattened and oval body. It has a long antenna for taste sensation. Body divided into head, prothorax, thorax and abdomen. It has 2 pairs of wings. Abdomen has 10 segments which 8 are visible. The last segment bears a pair of jointed *Cerci* covered with sensory hairs to detect air movement. In addition, the male bears two unjointed *styles.*

Economic importance

They destroy books, clothing, paper and food.

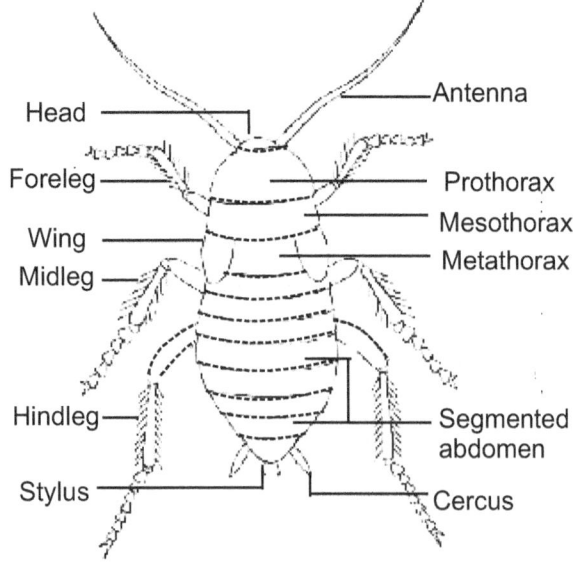

A cockroach

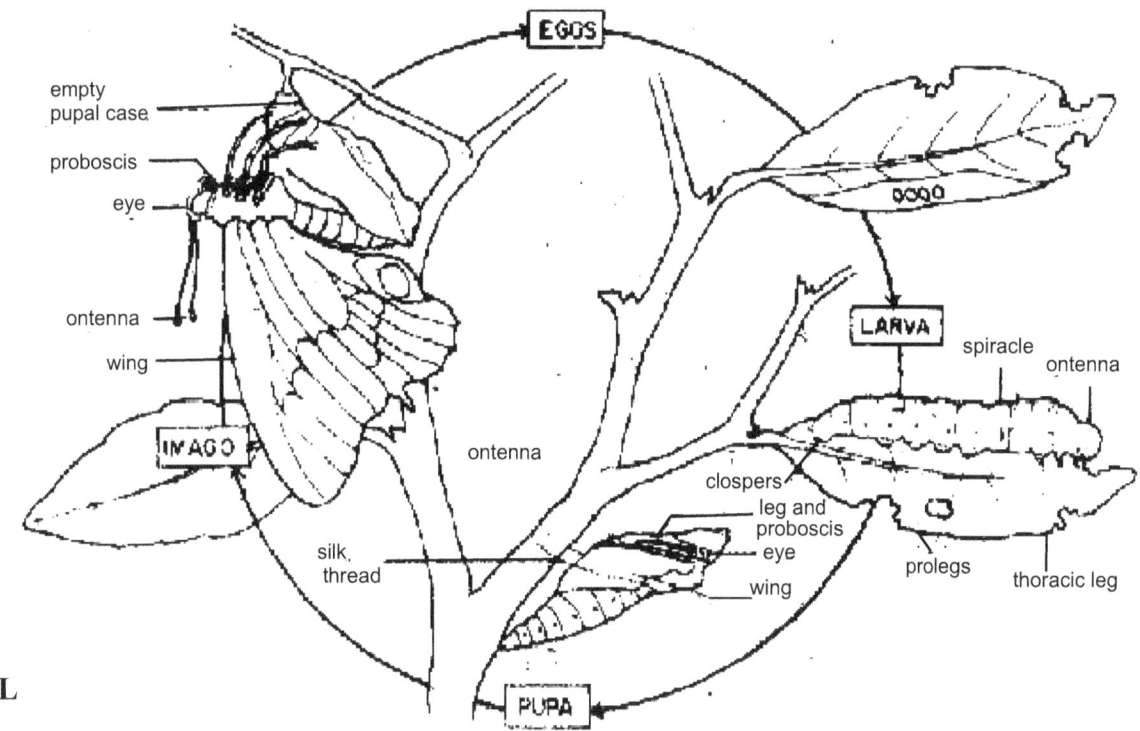

empty
pupal case

proboscis

eye

ontenna

wing

IMAGO

silk
thread

ontenna

L

EGGS

LARVA

spiracle

ontenna

clospers

leg and
proboscis

eye

wing

prolegs

thoracic leg

PUPA

Life cycle of a Butterfly

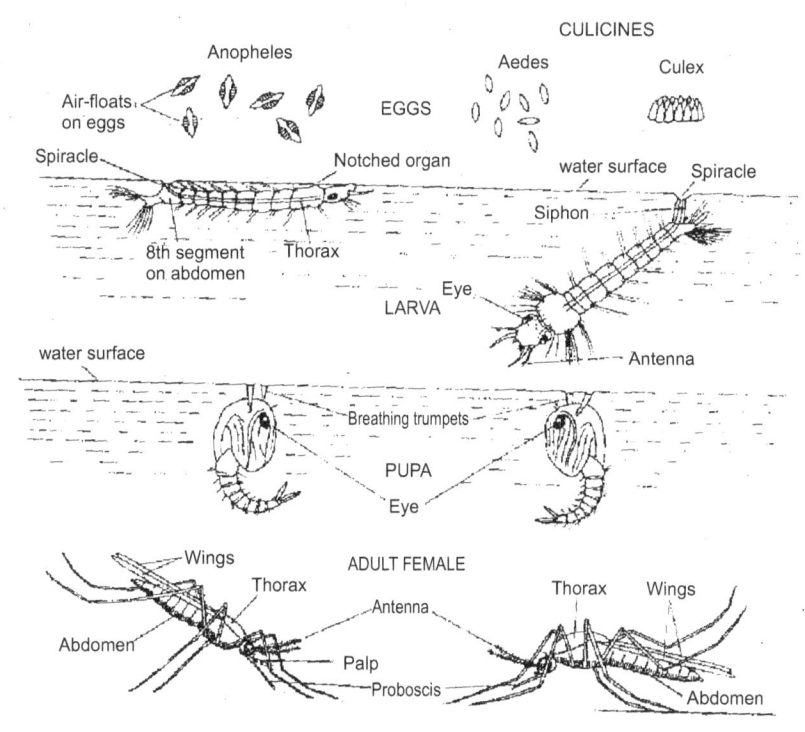

CULICINES

Anopheles

Aedes

Culex

Air-floats
on eggs

EGGS

Spiracle

Notched organ

water surface

Spiracle

8th segment
on abdomen

Thorax

Siphon

Eye

LARVA

Antenna

water surface

Breathing trumpets

PUPA

Eye

Wings

Thorax

ADULT FEMALE

Thorax

Wings

Abdomen

Antenna

Palp

Proboscis

Abdomen

Life cycle of *Anopheles* and *Culex* mosquitoes

31

HOUSEFLY

It has hairy body. The head has a pair of compound eyes and three small simple eyes, a pair of short antennae. There is one pair of membranous wings.

The second is reduced as balancers. There are four visible segments on the abdomen. The mouth is modified as a tubular *proboscis* for sucking up food.

Economic importance

They carry disease-organisms, causing diseases like cholera, dysentery and typhoid. They also transmit eggs of tapeworm and roundworms. They pollinate some foul-smelling flowers. The maggots (larvae stage) speed up decay process of organic matter.

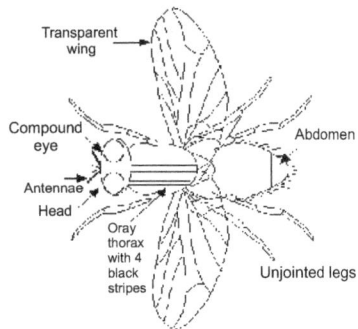

A common housefly

BUTTERFLY

It has a slender body and two pairs of large wings covered with scales. The head has a pair of large compound eyes and antennae which are club-shaped at the tips. The mouth is modified into a long coiled tube called proboscis which can be uncoiled during use.

Economic importance

Caterpillars (larval stage) eat leaves of crops and forest trees thereby causing food losses. Adult butterflies pollinate many flowers.

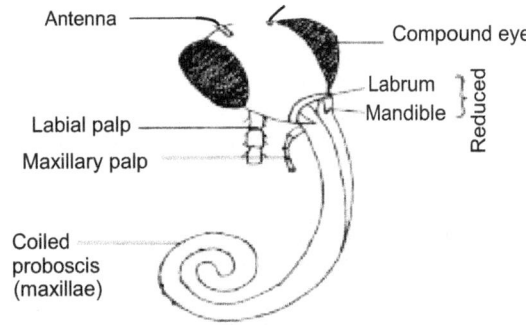

Mouthpart of butterfly

1. PCE 2003/41/Type N
The part labeled I is the
A. maxillary palp   B. proboscis
C. mandible   D. stylet

2. PCE/2003/42/Type N
The mouth part is adapted for
A. cutting
B. biting
C. cutting and sucking
D. piercing and sucking

3. UME/2004/30/Type 4
The larval stage of a mosquito is called
A. caterpillar   B. maggot
C. wriggler   D. grub

4. UME/2002/6/Type Y
Which oil is poured into the breeding site of mosquitoes, it
A. deprives the larvae of water
B. kills the adult
C. suffocates the pupae
D. Slows down egg development

5. UME 95/6
Which of the following is transmitted through mosquito bite?
A. Filariasis   B. Typhus
C. Plague   D. Schistosomiasis

6. PCE/96/7
The saliva injected into the wound during mosquito bite is principally to
A. transmit diseases to the host
B. digests the host's muscle
C. prevent the host's blood from clotting
D. kills germs contained in the host's blood

7. PCE/06/30/TYPE: A
The mosquito that transmits elephantiasis in humans is the
A. male culex   B. female aedes
C. male aedes   D. female culex

8. The anopheles mosquito lays its eggs on water. How are these eggs able to float?
A. the eggs are oily
B. the eggs have 'floats'
C. the eggs are weightless
D. the eggs are laid in rafts
E. the eggs have internal air chambers

9. What is the organ of respiration found in the larvae of mosquito?
A. The trachea   B. The spiracle
C. The siphon   D. Body bristles
E. Mouth brushes

10. How is movement effected by the pupa of the mosquito when disturbed?
A. The pupa sinks by sheer weight of body
B. It moves by means of its tail fins
C. It moves by taking in water through its trumpets
D. The pupa rounds up when disturbed
E. It moves away by the twisting of its abdominal region

11. The name of the malarial parasite transmitted by the mosquito is
A. Plasmania   B. Plasma   C. Filaria
D. Plasmodium   E. Sporozoites

12. In what way is the life cycle of the mosquito different from that of the cockroach?
A. The embryos of the mosquitoes are not protected
B. All adult mosquitoes have wings while some adults of the cockroach have no wings
C. The mosquito has no nymphal stage while the cockroach has no larva pupal stages
D. The cockroach is a domesticated Insect
E. None of the above reasons

4.2. COCKROACH
13. UME/98/7
The ability of the cockroach to live in cracks and crevices is enhanced by the possession of
A. wings and segmented body
B. compound eyes
C. claws on the legs
D. dorso– ventrally flattened body.

14. PCE/91/8
Which of the following can be found in the thoracic region of the cockroach?
A. cerci   B. elytra   C. styles   D. labium

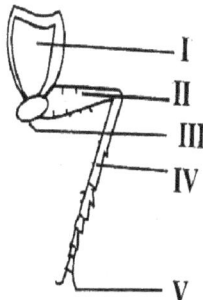

15. PCE 2004/8/Type 9
The structure labeled I is the
A. tibia   B. coxa   C. torsal   D. claw

16. PCE 2004/9/Type 9
The part used for gripping is labeled
A. I   B. III   C. IV   D. V

17. Considering the mouthparts of the cockroach, the organ of mastication is:
A. the maxilla   B. the labrum   C. the mandible
D. the stylets   E. the labium

18. What is the external difference between a male and female cockroach?
A. The female is usually smaller in size
B. The female does not fly but rather prefers to run
C. The female has no styles
D. The male is lighter in colour
E. the female usually carries its egg-case about.

19. Which of these is the most appropriate definition of metamorphosis?
A. It is the development of the insect from the larval into the pupal stage
B. The growth of the insect from the egg to the adult
C. It is the transformation of the insect from the pupa to the adult
D. It is all changes in the development of an insect-from the embryo to the adult
E. None of the above

20. The eggs of the cockroach are usually are usually enclosed in a hard chitinous case called:
A. the puparium   B. the cocoon   C. the otheca
D. the ootheca   E. None of the above

21. What is the name given to the young cockroaches that have just emerged from the egg-case?
A. Grubs   B. Larvae   C. Nymphs
D. Imagines   E. None of the above

21. What is the difference between new emergent and adult cockroaches?
A. The young cockroach has no legs
B. The young cockroach does not feed
C. The young cockroach is wingless
D. The adult cockroach has many more abdominal segments than the emergent one
E. There is no different between the two

22. The cockroach feeds on a number of items which may items which may include meat, paper, wood, etc. It is said to be
A. Carnivorous   B. Insectivorous   C. Omnivorous
D. Herbivorous   E. None of the above
23. The cercus of the cockroach is:
A. a tactile organ   B. a direction-finding hair
C. an organ of defence   D. an auditory organ
E. None of the above

33

24. The cockroach has its two antennae removed. Which of these functions do you think the insect will be unable to perform?
A. it will not see
B. it will not run well
C. It will not detect the direction of wind
D. It will not be able to locate its food
E. None of the above

4.3. HOUSEFLY

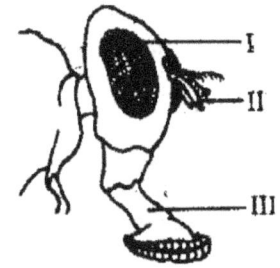

25. UME 99/7 Type B
The structure labeled II is used for
A. tasting   B. feeling   C. biting   D. sucking

26. UME 99/8 Type B
The structure labelleled III represents the
A. mandible   B. palp   C. proboscis D. labium

27. UME/84/1
The mouthparts of the housefly are adapted for
A. lapping and sponging
B. sucking and chewing
C. piercing and sucking
D. chewing and lapping
E. biting and chewing.

28. PCE 97/8
One of the function performed by the hook-like spine at the anterior end of housefly larva is to
A. secrete digestive enzymes
B. attach itself to objects
C. sting its attacker
D. assist the larva in crawling

29. What is the name given to the hind wings of the housefly?
A. Drumstick   B. Halteres
C. Halteers   D. Elytra
E. None of the above

30. What is the function of the balancers in the housefly?
A. To keep it flying
B. To prevent loss of balance
C. To aid a good take-off
D. Serves no useful purpose
E. To prevent rolling from side to side when in flight

31. Which of these diseases cannot be transmitted by the housefly?
A. Diarrhoea   B. Typhoid
C. Dysentry   D. Malaria

4.4. TERMITE
32. PCE/93/7
The insect that has the ability to digest cellulose is the
A. Termite   B. Butterfly
C. Mosquito   D. Housefly

33. UME/86/10
When the original king and queen of termites die, they are replaced by
A. The king and queen of another colony
B. Some adult reproductive from the same colony
C. Some adult which are specially fed to breed
D. Developing nymphs nurtured as secondary reproductive.

34. UME/98/8
The caste of termites that lacks pigmentation is the
A. king   B. worker   C. soldier   D. queen

35. PCE/99/3/Type: 3
The caste of the termite that causes the greatest damage to man is
A. queen B. soldier   C. worker   D. king

36. PCE/92/50
Which of the following should be killed in order to destroy a termite colony?
A. the king   B. the queen
C. the workers   D. the soldiers

37. PCE/93/7
The insect that has the ability to digest cellulose is the
A. termite   B. butterfly
C. mosquito   D. housefly

38. The home of termites is known as
A. the mole hill   B. the habitation   C. the pit
D. the termitarium   E. the sanatorium

39. The termite colony has a number of individuals such as the queen, king, solders and workers. The queen is known by her:
A. Prominent and extended abdomen
B. Sclerotized cuticle   C. Large head
D. Poorly developed brain   E. Reduced legs

40. The termites are important because they:
A. Transmit diseases   B. Destroy wood and crops
C. Eat honey   D. Attack animals
E. None of the above

4.5. BUTTERFLY

41. UME/2003/11/Type A
An insect whose economic importance is both harmful and beneficial is the
A. butterfly   B. mosquito   C. blackfly   D. testsefly

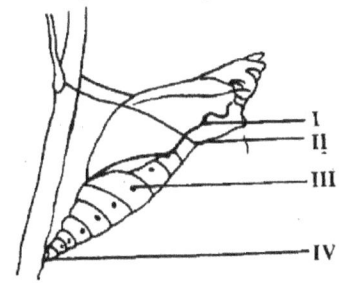

42. UME/2000/1/Type: M
The part that will develop into an organ for feeling is labeled
A. IV   B. III   C. II   D. I

43. UME/2000/2/Type: M
The part lablled II is the
A. silk thread   B. thorax   C. forewing   D. anchor

44. PCE/2002/3/Type:5
The structure labell II is the
A. Head   B. Ear
C. Compound eye   D. Maxillary palp.

45. PCE/2002/4/Type:5
The diagram is the mouth parts of a
A. butterfly   B. cockroach   C. housefly   D. mosquito

46. Which of these characteristics marks out clearly the butterfly?
A. The possession of club-like antennae
B. The possession of two pairs of dissimilar wings
C. The possession of the proboscis
D. The possession of very tiny legs
E. The possession of scales

MISCELLANEOUS QUESTIONS (4)
47. Which of the following spreads malaria in Nigeria?
A. Anopheles Mosquito   B. Culex mosquito
C. Aedes mosquito   D. Tsetse fly
E. Housefly (UME/80/3)

48. The jointed structures in insects that bears organs which are sensitive to touch, smell and vibration is the
A. maxilla   B. labium
C. antenna   D. abdomen (UME/2001//4/Type: T)

49. Which of the following is a typical filter feeder?
A. Mosquito larva   B. Mosquito adult
C. Housefly larva   D. Housefly adult (PCE/92/10)

50. The structures found in male adult cockroaches but absent in females are the
A. anal styles   B. antennae
C. anal cerci   D. proboscis (PCE/2005/47/Type:P)

51. Which of the following insects lays its eggs in horny, purse-like cases?
A. Mosquito   B. Butterfly
C. Termite   D. Cockroach (UME/95/7)

52. The group of insects that undergoes complete metamorphosis is
A. houseflies beetles and cockroaches
B. cockroaches, grass hoppers and bees
C. house flies, beetles and butterflies
D. aphids, grasshoppers and butterflies (UME/97/9)

53. The larval stage of a mosquito is called
A. Catarpillar   B. Maggot
C. Wriggler   D. Grub
(UME/2004/30/Type:4)

54. Which of the following is destructive at the developmental and adult stages?
A. Grass hopper   B. Mosquito
C. Housefly   D. Termite (PCE/2000/9/Type: A)

55. The only caste in the termite colony whose members can feed themselves are
A. Reproductive   B. Workers
C. Nymphs   D. Soldiers
(UME/2005/49/Type: D)

56. The stage in the history of a moth responsible for the destruction of agricultural crops is the
A. nymph   B. imago
C. pupa   D. caterpillar (UME/2005/10/Type:D)

57. The process of shedding the exoskeleton of an arthropod is known as
A. metamorphosis   B. instars formation
C. ecdysis   D. tagmosis (UME/2007/42)

58. Which of the following is common to the mosquito, housefly and butterfly?
A. they undergo complete metamorphosis
B. they are parasites of man
C. their immature stages are aquatic
D. their adults have two pairs of wings (UME/2007/43)

59. The dorsal part of insect is covered by
A. pleurum   B. tegmen
C. sternum   D. tergum (PCE/96/6)

60. The correct sequence that describes the insect segmented leg is
A. coxa → trochanter → femur → tibia → tarsus
B. coxa → stemum → pleuron → femur → tarsus
C. coxa → femur → trochanter → tibia → tarsus
D. coxa → pleuron → trochanter → femur → tarsus (PCE/97/7)

61. A specie with a body divided into three regions, two large compound eyes, two short antennae and a pair of wings is a
A. spider   B. cockroach
C. housefly   D. butterfly (PCE/98/6)

62. An example of a nocturnal insects is the
A. termite   B. blackfly
C. cockroach   D. housefly(PCE 06/28 TYPE: A)

63. The skin covering the dorsal aspect of the insect is called:
A. the pleuron   B. the tergum
C. the steron   D. the ternum E. the sternum

64. The cuticle of the insect is made up of a substance known as:
A. Wax   B. Chitin   C. Cutin
D. Sclerotin   E. None of the above

65. On which of these are the tracheal openings of insects found?
A. the sternum   B. the abdominal walls
C. the thoracic walls   D. the pleuron
E. None of these

66. The thorax of insects is divided into segments namely: prothorax, meso-thorax and metathorax. Which of these bears the first pair of wings or the elytra?
A. the prothorax   B. the mesothorax
C. the metathorax   D. no segment in particular
E. None of the above

67. The abdominal region of insects differs from the thoracic region in this respect:
A. it is always bigger than the thorax
B. it is segmented   C. it has spiracles
D. it has no locomotory structures
E. It is laterally flattened

68. The metamorphosis of insects is either complete or incomplete. Which of these insects shows an incomplete form of metamorphosis?
A. the bee   B. the beetle   C. the tsetse fly
D. the mosquito   E. None of the above

69. What are the trumpets?
A. A sound producing apparatus found on the thorax of certain insects
B. Respiratory structures of the housefly
C. Respiratory structures found only in the pupa of an Insect
D. A sound – producing apparatus of the mosquito
E. None of the above

70. What is the difference between the adult cockroach and mosquito?
A. The mosquito is a harmful insect while the cockroach is not
B. The cockroach is protected by the cuticle while the mosquito is not
C. The mosquito has no hind wings
D. The cockroach cannot eat blood
E. None of these reasons

71. What is the main function of the adult insect?
A. To fly about   B. To distribute the insect
C. To eat up predators
D. To provide food for other animals
E. To reproduce

72. The larvae of all beetles resemble the adults in one respect and this is:
A. Both have biting mouthparts
B. Both possess the anal cerci
C. Both possess abdominal
D. Both possess an equal number of abdominal segments
E. None of the above reasons

73. Moulting in insects is
A. the change is colour of insects
B. the emergence of young larvae from eggs
C. the casting away of old skin in insects
D. the emergence of the adult from the pupa
E. none of the above

74. Insects shed their skin:
A. to remove waste materials in the body
B. To prevent the attack of diseases
C. To transform the insect
D. To allow for growth   E. None of the above reasons

75. It is discovered that a once – active insect cannot fly early in the morning. But as the sun rises in the sky it takes to wings. This implies that:
A. Insects do not fly in the morning
B. In the mornings the muscles that control the wings of insects are non-functional
C. High temperature is a vital factor in the activity of insects
D. The blood of insects freeze up at night
E. None of the above reasons

CHAPTER 5

VERTEBRATES

Understand the following after reading this chapter:

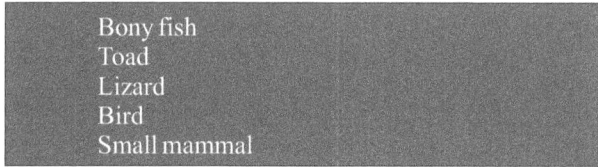

Bony fish
Toad
Lizard
Bird
Small mammal

Vertebrates are animals with vertebral column or backbones, internal skeleton, bilateral symmetry, brain in cranium, paired fore and hind limbs, well developed sense organs and 'closed' circulatory system. They include: Pisces (Fishes), Amphibians, Reptiles, Aves (Birds) and Mammals.

CHARACTERISTICS AND ADAPTATION
1. Pisces (Fishes): They are all aquatic, variable body temperature (poikilothermic), overlapping scales pointing backwards, endoskeleton, gills used for respiration. They have paired and unpaired fins for swimming.
They reproduce by laying eggs (oviparous) and external fertilization. e.g. Tilapia, Salmon, Shark etc.
Adaptation of Tilapia (bony fish) to:

a. Movement: Streamlined shape. Paired and unpaired fins for fast movement and stability.
Muscles in tail region used for movement. Absence of neck. Slimy body for locomotion and escape.

b. Protection: Greyish in colour blending with surrounding to give protective camouflage. Body covered with scales for protection.
Lateral line to detect vibrations or changes in water pressure, lidless eye to give a wide range of vision. Nostrils to detect chemical substances (smell). Swim bladder for buoyancy. Vascularized gills for respiration in water.

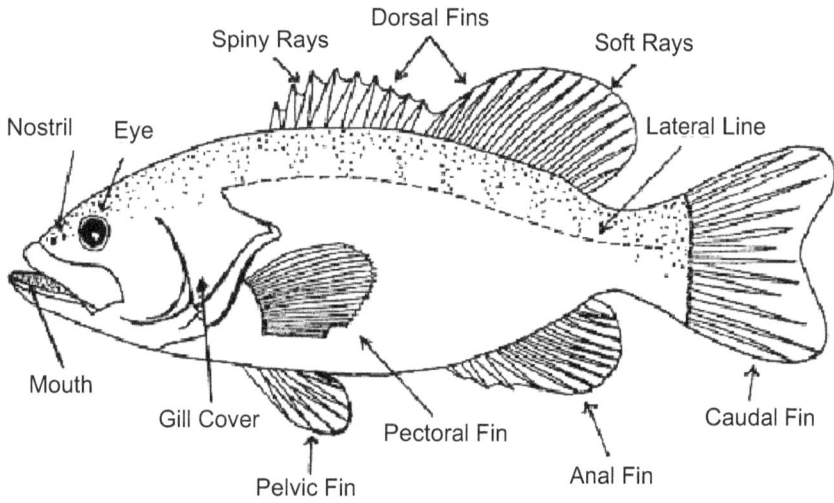

External structure of a bony fish

2. Amphibians: They are both terrestrial and aquatic. They spend most of their life time on land but usually return to water during breeding season. They are poikilothermic. Immovable eyelids, oviparous, gills used for respiration during larval stage and lungs, skin and mouth for respiration at adult stage. Their bodies are divided into head and trunk, e.g. toad, frog, salamander etc.

Adaptation of Toad (*Buforegularis*)
A. Movement: Stream lined shape, absence of neck and tail suitable for hopping, fore limbs for landing, long and powerful hind limbs for propulsion.

b. Feeding: Carnivorous- absence of teeth, anteriorly fixed and sticky tongue with mucous. Ability to swallow prey alive.

c. Protection (tadpole stage): Jelly coat provides buoyancy to eggs for absorption of light, protection from predators, prevention from injury, attack of bacteria and fungi and also for oxygen circulation. Tail fin with muscles or movement in water.

(Adult stage): Brown, rough and warty skin for protective camouflage and to reduce water loss on land. Ability to fill lungs with air to scare away enemies.Poisonous secretion from gland under skin making it unpalatable for predators like snakes. Well developed and positioned eyes and tympanum to give a wide range of vision and detect vibrations of preys and predators.

Webbed hind limbs for swimming in water during breathing.Ability to breathe through skin, lungs and mouth cavity.Can carry out anaerobic respiration.

3. Reptiles: They are mostly terrestrial. Poikilothermic. Dry sharp pointed overlapping scales, dry skin, long tail, fore and hind limbs with five digits ending in claws. Well developed eyes with movable eyelids, have

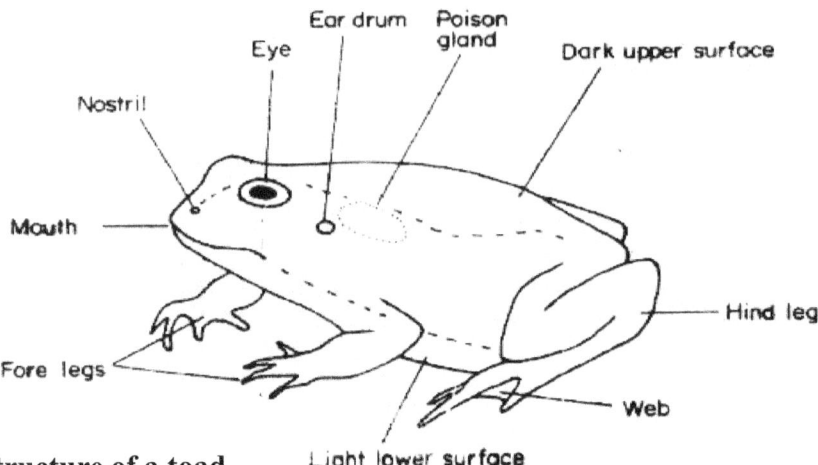

External structure of a toad

two pairs of walking limbs. Their bodies are divided into head, neck and trunk. Homodont dentition.

Internal fertilization and oviparous.e.g. tortoise, turtle, snakes, lizards etc.

Adaptation of Lizard (*Agama agama*)

Stream lined shape. Powerful jaws to hold prey.Homodont dentition to crush preys.Well developed eyes and tympanum to afford a wide range of vision and to detect vibration respectively.

Nostrilsfor smell and respiration, lungs for respiration on land.Flexible neck to turn to all directions.Dry scaly skin to

prevent dessication on land.Clawed digits to hold on to surfaces during movement.Ability to complete life cycle on land and also exhibit internal fertilization.

Ability to blend with environment, power of regeneration of tail, leathery shells can absorb oxygen and water and also protect eggs.

4. Aves (Birds): They are mostly terrestrial and their bodies are covered with feathers. They are homoeothermic i.e. maintain constant body temperature. Periodic moulting. Skull extended to form beak. Stream lined shape.

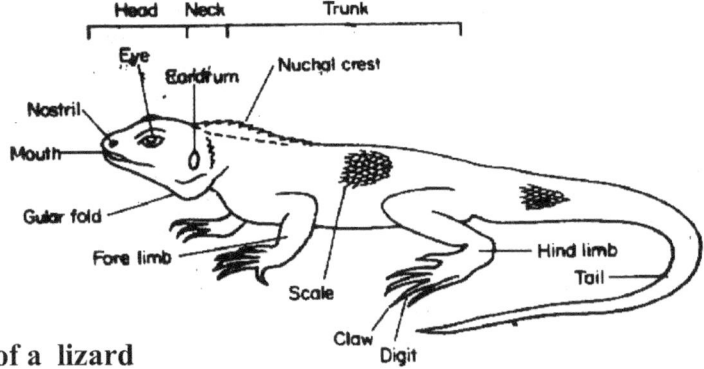

External structure of a lizard

Hind limbs covered with scales, fore limbs modified to form wings, feet adapted for feeding and to habitat; movable eyelids with a nictitating membrane. Body divided into head, neck, trunk, and a short tail. Internal fertilization.e.g. ostrich, kiwi, fowl, duck etc.

Adaptation of Birds for flight

Dry skin without sweat glands, body covered with feathers, wing and tail feathers for flight. Feathers act as insulators, fusion of skeleton to provide strength. Light hollow bone to reduce weight, sternum modified into keel for attachment of muscles. Streamlined shape, large air sacs in lungs, fusion of pelvic girdle with backbone. Flexible neck for easy movement.

Types of feathers: Contour (covering body and giving shape). Contour made up of **quill** and **convert** feathers. Convert to cover body and quill found on wing and tail for flight.

Down feather gives insulation. Short shafts and loose barbs. No barbules.

Filoplume feather has slender shaft with tuft of loose barbs at end. They are specializes sensory structures that aid in the operation or movement of other feathers for flight, insulation or display. Others include: **Semiplumes** and **Bristles.**

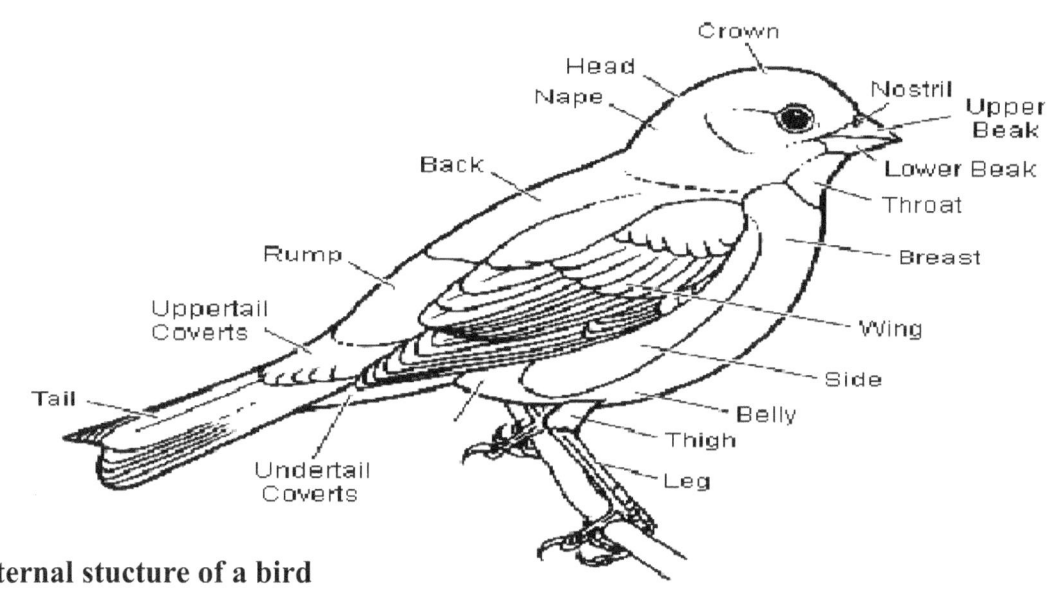

External stucture of a bird

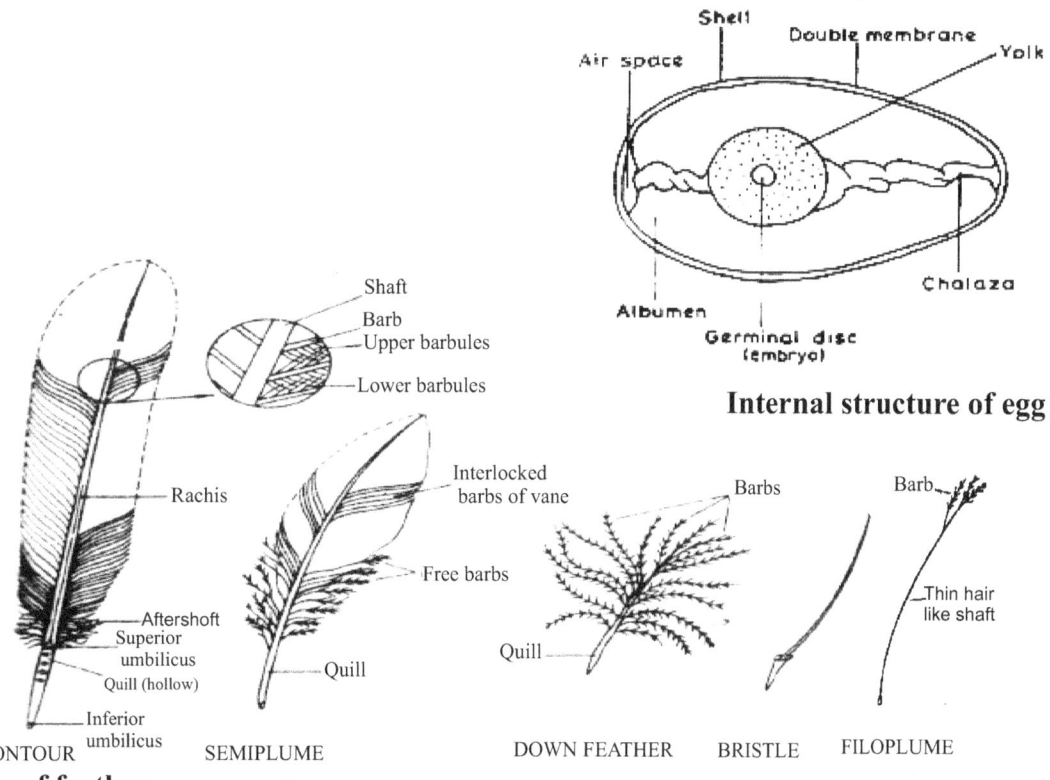

Internal structure of egg

Types of feather

5.Mammals: Their bodies are covered with hairs. The have mammary glands and give birth to their young ones alive (viviparous). They are homoeothermic. They have sweat gland, heterodont dentition, external ears (pinnae), large and complicated brain. Coelom is divided by diaphragm into thorax and abdomen. e.g. man, lion, bat, rat etc.

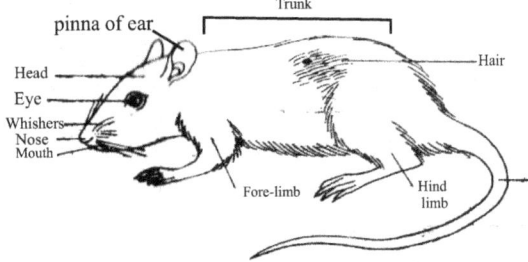

External stucture of a rat

Revision Questions

5.1. BONYFISH

1. UME 86/9
Fishes are cold-blooded because their body temperature is
A. constantly low
B. constantly high
C. dependent on that on their surroundings
D. regulated at will

2. UME 94/9
Which of the following combinations differentiates a bony fish from a cartilaginous fish?
I. Presence of gills
II. Absence of gills slits
III. Possession of bony skeleton
IV. Possession of lateral compressed body
V. Possession of dorso – ventrally compressed body
A. I, II and III   B. I, II and IV
C. II, III and IV   D. II, III and V

3. UME/94/10
Which is the most important adaptation of a bony fish in water?
A. the possession of a streamlined shape
B. The presence of overlapping scale
C. The covering of the body by thin film of slime
D. The possession of a caudal fin

4. UME/82/17
The main function of swim bladder is for
A. swimming   B. detecting sound
C. buoyancy   D. breathing
E. reproduction

5. UME/84/14
The dorsal and anal fins of fish are used for
A. upward movements
B. controlling rolling movements
C. downward movements
D. steering   E. buoyancy

6. UME/85/5
In fish, the sense organs which detects movement in water are located within the
A. gills   B. operculum
C. nostrils   D. median fins
E. lateral lines

7. UME/86/12
The fins making up the limbs of the bony fish are
A. caudal and ventral   B. ventral and pelvic
C. pelvic and pectoral   D. pectoral and dorsal

8. UME/91/15
The gills rakers of fishes take part in
A. feeding   B. respiration
C. swimming   D. diffusion

9. UME/2008/10
Which of the following is used mainly for balancing in fish?
A. the caudal fin   B. the pectoral fin
C. the anal fin   D. The dorsal fin

10. PCE/96/9
Which pair of adaptive features enables the lung fish to live in different environment?
A. Gills and fins   B. Gills and swim bladders
C. Barbells and fins   D. Barbells and nostrils

11. PCE/97/9
In tilapia, the fins lying just behind the gill covers are
A. caudal   B. pectoral
C. pelvic   D. anal

12. The osmotic pressure of sea water is higher than that of the body fluid of the fish living in it. The cartilaginous fish, despite this are at home in the sea. This is because they retain a large quantity of urea in their body fluids. The purpose of the urea is:
A. To prevent the cartilaginous fish from desiccation
B. To assist the cartilaginous fish in excretion
C. To raise the osmotic pressure of the body fluid of the cartilaginous fish to that of the surrounding water
D. To neutralize the effect of salt drunk by the cartilaginous fish
E. None of the above

13. The cartilaginous fish differ from the bony fish
A. in having no anal fin
B. in having internal fertilization
C. in having no claspers
D. in having no lateral system
E. in having no vertebrae

14. The operculum is possessed by:
A. the deep-sea fish   B. the bony fish
C. all types of fish   D. Fishes with the swim bladder
E. All fishes other than cartilaginous ones

15. The fish will roll from side to side if:
A. the pectoral fins are cut off
B. the caudal fin is cut off
C. the ventral fin is cut off
D. the dorsal, ventral and anal fins are cut off

16. The propulsive forward movement of the fish is due to:
A. the flapping of the pelvic fin
B. the movement of the operculum
C. The side to side movement of the tail
D. The up and down movement of the pectoral fin
E. None of the above

17. Many fish have swim bladders which are used for varied purposes. For which of these do the fish not employ the swim bladder?
A. Respiratory purposes
B. Maintenance of body position in water
C. Production of gases D. Sound detection in water
E. Storage of excretory material

18. Why are the olfactory lobes of the brain of the fish much more developed than any other part?
A. They control the life of the fish
B. They correlate the activities of the fish
C. The fish depends much on its sense of smell for existence
D. The olfactory lobes are anteriorly placed and hence come into new environment before any other part of the body
E. None of the above reasons

19. With reference to the fish, the cerebellum controls:
A. Their movement   B. Their muscular activities
C. Their escape reaction   D. Reason and thought
E. None of the above

20. Which of these is not true of the fish living in fast moving water?
A. they are robust
B. they have adhesive structures
C. they are strong swimmers
D. they are laterally flattened
E. None of the above

5.2. TOAD
21. UME/83/28
In what order do the following structures develop during mentamorphosis of the toad?
1. External gills   2. Internal gills   3. Forelimbs
4. Hindlimbs   5. Mouth
A. 1, 2, 3, 4, 5   B. 1, 5, 2, 4, 3
C. 2, 1, 3, 4, 5,   D. 5, 3, 4, 1, 2   E. 5, 4, 3, 2, 1

22. UME/78/29
Which of these is not associated with the tadpole stages of the toad or frog?
A. V-shaped gland   B. operculum
C. external gills   D. jelly
E. shell

23. UME/88/12
The long and coiled intestine of a young tadpole is an adaptation to its.
A. herbivorous diet
B. carnivorous diet
C. aquatic diet
D. insectivorous habit

24. UME/90/11
In which of the following does external fertilization takes place?
A. Toad   B. Lizard
C. Bird   D. Cockroach.

25. PCE/91/9
The function of the layer of jelly which surrounds the eggs of frogs when laid is to
A. provide mineral salts for the developing eggs
B. provide food for the developing eggs
C. protect the eggs from being eaten by aquatic animals
D. disallow gaseous exchange

26. PCE/99/1/Type: E
The poison glands found in the toad produce a substance which
A. kills its predators
B. Protects it against predators
C. keeps the body surface shinny and moist
D. blends it with the surrounding

27. PCE/92/9
What function does the nictitating membrane perform?
A. it gives the body a streamline shape and enables the animal to move easily through water
B. it is pulled upwards to cover the eyes of the animal when there is danger
C. it secretes materials that moisten the body of the animal
D. it help reduce water loss from the body of the animal

28. Which of these characteristics may NOT be regarded as amphibian?
A. Smooth and moist skin
B. Presence of scales
C. Permeable skin
D. A four – digit fore limb
E. None of the above

29. A frog such as Xenopus is placed in a basin of salt water. It dies a moment later and analysis shows that it died of desiccation. This means:
A. Xenopus cannot live long in water
B. Xenopus has a permeable skin
C. Xenopus prefers land to water
D. Xenopus has no mucous glands in its skin
E. None of the above

30. The moveable thin film of skin that moves over the eyes of the toad is:
A. the conjuctiva
B. the eyelash
C. the cornea
D. the membrane ocule
E. the nictitating membrane

31. That gland in the skin of the toad that produces a milky white poisonous fluid is
A. the paraxoid gland   B. the mucus gland
C. the parotid gland   D. the parotoid gland
E. the milk gland

32. The amphibians have different respiratory surfaces. Which of these is not a respiratory surface?
A. the skin B. the lining of the cloaca
C. the lungs D. the lining of the buccal cavity
E. the webs of the feet.

33. Other than its slight build the male toad is distinguished from the female by:
A.The possession of the five digits in the fore limb
B.The possession of a black breeding pad
C.The possession of a penis
D.The possession of a dark skin beneath the chin
E. None of the above

34. With regard to the toad, which of these may not be considered functions of the jelly to the eggs in them?
A. It protects the eggs from injuries
B. It prevents the attack of bacteria
C. It prevents overcrowding
D. It protects eggs from attack by fleas
E. It provides the eggs and hence the developing tadpoles with mineral salts

35. Looking at the early life of the toad in water, which of these structures is not essential for a successful life?
A. Ventral sucker
B. External and internal gills
C. Ciliated skin
D. Elongated tail
E. None of the above

36. With reference to the toad, which of these is true of metamorphosis?
A. The change from the tad-pole to the adult toad
B. The development of the lungs
C. The process involving the internal rearrangements of the blood vessels in the toad
D. The total development of the toad from the eggs to the adult
E. None of the above

5.3. LIZARD
37. UME/80/6
Some of the features of an animal are scales, teeth, nares and backbone. The animal is likely to a
A. Toad   B. Bird
C. Lizard   D. Rat   E. Bat

38. UME/83/6
Which of the following structures is NOT found in the female Agama Lizard?
A. Pre-analpads   B. Eardrum
C. Gularfold   D. Nasal Scale
E. Nuchal crest.

39. UME/2004/46/Type 4
One adaptation of reptiles to water loss is the presence of
A. long nails   B. long sticky tongues
C. keratinous scales   D. claws on limbs

40. PCE/96/8
Lizards are able to breed on land because they produce eggs covered with a
A. thin leathery shell   B. thick leathery shell
C. thick slimy mucus   D. thin slimy mucous

41 UME/2007/15
In lizards, the lowering of the guler fold is used to
A. catch insects   B. attract mates
C. defend their territory   D. frighten enemies

42. PCE/91/10
A major characteristic of reptiles is that they
A. are four-legged
B. have homodont dentition
C. have long narrow bodies
D. have hair under their scales

43. Which of these is not true of the reptiles?
A. Their skins have hard and horny epidermal scales
B. They have no external ears
C. They undergo no metamorphosis
D. They lay large yolky eggs which are protected by shells
E. Fertilization is external

44. Which of these can be used to separate the reptile from the amphibians?
A. the presence of claws on digits
B. the presence of a neck
C. Viviparity
D. The presence of a tail
E. none of the above

5.4. BIRD
45. UME/95/9
Birds maintain their body temperature with the help of their
A. blood which maintains constant temperature
B. feathers which cover the body
C. skin which conserves moisture
D. veins which transport fluid of constant temperature to all body tissue

46. UME/84/32
Which of the following adaptations is NOT concerned with the flight of birds?
A. Stream lined shape
B. Presence of powerful muscles
B. Reduced body weight
D. Broad sternum
E. Webbed feet.

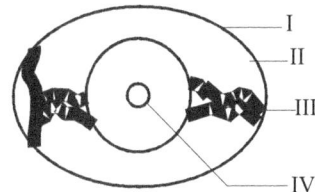

47. UME/2001/14/Type T
The structure that holds the yolk in position is labeled
A. II   B. II
C. IV   D. III

48. UME/2001/15/type: T
The part labeled IV is the
A. albumen   B. germinal disc
C. yolk   D. air space.

49. PCE/96/10
Which of the following feathers of a bird make a warm light covering?
A. Down feathers   B. Covert feathers
C. Quill feathers   D. Filoplumes

50. UME/78/31
Which of these is NOT a part of the feather of a bird?
A. Wing   B. Aftershaft
C. Inferior umbilicus   D. Vane E. Barbule

51. UME/2008/11
The beak of a duck is structurally adapted for
A. scooping and sieving food
B. catching and grasping food
C. picking and cracking food
D. boring and sucking food

52. UME/2010/5/TYPE: D
Parental care is exhibited by
A. snails   B. earthworms
C. birds   D. toads

53. Which of these is not true of the birds?
A. They have air sacs
B. Their bodies may be covered with epidermal scales
C. They have no teeth
D. The number of digits of the fore limbs are reduced
E. None of the above

54. Unlike the toad, the skin of the bird is:
A. Non-glandular   B. without blood vessels
C. without stratified epithelia
loose E. None of the above

55. The feathers on the skin of birds are arranged on rows. Those tracts without feathers on the body of birds are called:
A. the pterium   B. the apteridia
C. the apteria   D. the pterylea
E. the tract

56. Which of these is not connected with the feather system of the bird?
A. Filoplumes
B. Contour feathers
C. Model feathers
D. Down feathers
E. Plumules.

57. The feathers serve the birds in:
A. Providing a protective and a water-repellent covering
B. Providing a good device for flight
C. Insulation and hence a good device for body temperature regulation
D. Colouration and sex display
E. All of the above

58. The only true skin gland of the bird is:
A. the sebaceous gland
B. the sweat gland
C. the avecian gland
D. the lipid gland
E. the uropygial gland

59. The adaptations of the fish-eating birds are:
A. The possession of beaks with serrated edges
B. The possession of hook like structures on the palate and tongue of birds to hold prey
C. The possession of long necks
D. The possession of long beaks
E. None of the above

60. The adaptation of the birds associated with water are:
A. Possession of long legs
B. Possession of webbed feet
C. Possession of oily feathers
D. Possession of flipper like wings
E. Possession of sharp claws

5.5. SMALL MAMMALS

61. UME/78/8
A traveler on a fishing trip landed a type of animal with hairs on its body. This animal could have been
A. an amphibian   B. a reptile
C. shark   D. a mammal
E. seagull

62. UME/2004/28/Type: 4
A peculiar characteristics of mammals is that they
A. have sebaceous glands   B. have teeth
C. are warm blooded   D. have lungs

63. UME/2004/38/Type:4
Rodents gnaw on food with their
A. strong jaws
B. flat – ridged teeth
C. chisel-like front teeth
D. molar teeth

64. UME/84/10
The mammalian organ through which nourishment and oxygen diffuse into a developing embryo is called
A. amnion   B chorion
C. umbilical cord   D. oviduct
E. placenta

65. UME 2002/7/Type: Y
The correct evolutionary sequence of the organisms represented is
A. I → III → II → IV
B. II → III → IV → I
C. III → II → I → IV
D. IV → II → III → I

66. UME/2002/8/Type : Y
Ovoviviparity is the type of fertilization exhibited by the organism labeled
A. I   B. II   C. III   D. IV

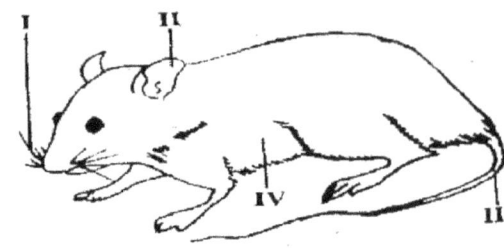

67. UME/2007/11
The type of protective adaptation exhibited by the animal is
A. disruptive colouration
B. flash colouration
C. countershading colouration
D. warning colouration

68. UME/2007/12
The structure labelled I is
A. photosensitive   B. radiosensitive
C. chemosensitive   D. tactile

69. UME/2006/39
A characteristic that best exemplifies the evolutionary advancement of mammals over other vertebrates is the
A. viviparous mode of reproduction
B. possession of paired limbs
C. terrestrial mode of life
D. possession of heart

MISCELLANEOUS QUESTIONS (5)
70. The set of fins that controls steering, balancing and change of direction and pitch in fish is
A. dorsal and anal
B. pectoral and pelvic
C. candual and dorsal
D. anal and pelvic (UME/2000/4/Type: M)

71. The main function of the caudal fin is tilapia is to
A. propel it forward in water
B. steer it while changing direction
C. balance it in water
D. enable it to float in water (UME/2005/13/Type:D)

72. The thick, short and muscular fore limbs of toads are an adaptation for
A. path-finding during swimming
B. capturing prey
C. absorbing shock on landing
D. take-off during jumping (PCE/2003/17/Type:N)

73. An animal that combines cutaneous, bucal and pulmonary gaseous exchange is the
A. Dog
B. Grasshopper
C. Tilapia
D. Toad (PCE/95/11)

74. An example of cold-blooded animal is
A. man
B. lizard
C. bird
D. rat (PCE/2003/14/Type N)

75. Which of the following animals has homodont dentition:
A. Rat   B. Man
C. Lizard   D. Pigeon (UME/90/10)

76. The possession of scales, laying of eggs with shells and bony structure of the head are characteristics shared by
A. birds and reptiles
B. fishes and birds
C. reptiles and fishes
D. birds and molluscs (UME 99/12/Type: D)

77. The muscles that are used for flight in birds are
A. pelvic muscles
B. pectoral muscles
C. dorsal muscles
D. lateral muscles
(PCE/2000/10/Type: A)

78. A feature which adapts birds to flight is the possession of
A. scaly legs
B. light bones
C. two walking legs
D. a pointed beak
(UME/90/13)

79. The correct sequence that represents vertebrate evolutionary trend is
A. fish→ amphibians→ birds → reptiles→ mammals
B. fish→ reptiles→ amphibians→ birds→ mammals
C. fish→ birds→ reptiles→ amphibians→ mammals
D. fish→ amphibians→ reptiles→ birds→ mammals
(PCE/95/3)

80. One of the main social characteristics of the pigeon is that it is
A. polygamous
B. monogamous
C. polyandrous
D. both polygamous and monogamous
(PCE/01/9/TYPE4)

CHAPTER 6

INTERNAL STRUCTURE OF ORGANISMS

Understand the following after reading this chapter:

> Flowering plants (roots, stems and leaves)
> Mammals

INTERNAL STRUCTURE OF ROOTS

The piliferous layer is a single row of cells, which, in young roots, may project to form root hairs used for absorption of water and salts. The cortex is wide and consists of loose parenchyma cells. The innermost cortex is a layer of tightly arranged rectangular cells, the endodermis.

The lateral walls of the endodermal cells are sealed with surberin thereby forcing water and solutes to pass through the cell's protoplasm where the movement of dissolved material is regulated.

Inside the endodermis is the pericycle consisting of fibres and a central stele of vascular tissues. The xylem is in the centre and has many radiatory arms (protoxylem points). The phloem occurs between these arms. During secondary growth, cambium is developed where the xylem and phloem meet. There is no pith.

In the monocot root, there is a wide pith around which is a layer of xylem and phloem tissues alternately arranged.

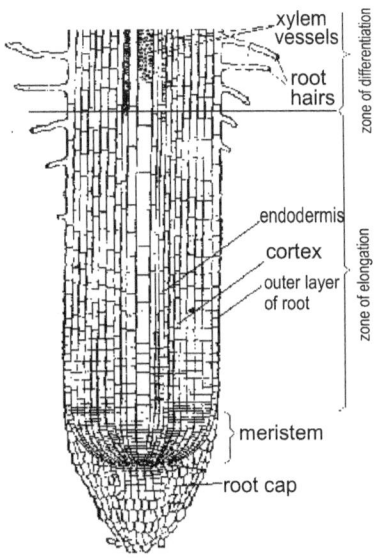

Longitudinal section of a dicot root

INTERNAL STRUCTURE OF STEMS

The outcome layer is the epidermis. Within this is a wide cortex predominatlyparenchymatous. Some of the cells contain chloroplasrs and are photosynthetic. In some plants, an outer hypodermis layer made up of collenchyma is dermarcated; it provides rigidity to the stem to withstand swaying in the wind.

The innermost layer of the cortex is the endodermis; it is a single layer of rectangular-shaped cells containing starch granules. The vascular tissues are organized into separate bundles arranged in a radial pattern. Each consists of an inner xylem, an outer phloem and a cambium in the middle. On the outer side of the phloem is a pericycle made up of thickened fibres. The innermost region is the pith, it is parenchymatous.

In monocot stems the cortex is narrow. The vascular bundles are scattered in the parenchymatous ground tissue. The bundles are smaller and more numerous towards the outside, while those towards the centre are fewer and bigger. Each vascular bundle is surrounded by a sheath of strengthening sclerenchyma cells. There is no cambium.

During secondary growth in the dicot stem, inter-fascicular cambium is formed in the pith ray, linking all the cambium cells into a continuous ring. The cambium then divides and produces new xylem cells in the inside and new phloem cells on the outside. Simultaneously a cork cambium is developed just below the epidermis. This divides to produce layers of water proof cork cells which inter form the bark. Perforations called lenticles are developed on the bark to allow movement of air in the stem.

INTERNAL STRUCTURE OF LEAVES

The leaf surface is covered by the upper and lower epidermis. The epidermis is perforated by stomatal pores; they are more numerous on the lower epidermis. Stomata allow exchange and gases for respiration and photosynthesis, as well as movement of water vapour during transpiration. In general, the stomata open in the presence of light and close in the dark.

The mesophyll is the photosynthetic tissue between the upper and lower epidermis. It consists of an upper palisade parenchyma and spongy parenchyma. The palisade cells are cylindrical and are arrange in one or more rows perpendicular to the epidermis. The spongy mesophyll consists of loosely arranged cells with large air spaces. All mesophyll cells contain chloroplasts.

Small vascular bundles called veins, arise from the vascular bundle in the mid-rid and form a network in the leaf tissue. Veins conduct water, mineral salts, and food, and also mechanically support the mesophyll tissue.

In the leaves of monocots, the mesophyll not differentiated into distinct pallisade and spongy parenchyma layers.

MAMMALS

1. The body cavity is divided into the chest and abdomen by muscular sheet – the diaphragm
2. Organs in the chest include: Lungs, heart (covered by pericardium). Organs in the abdomen include: Stomach, liver, kidney intestines.
3. Embryonic development occurs in the uterus of mother and the young are born alive. No eggs are laid.
4. The parents care for their young; the mothers feed the young on milk from mammary gland

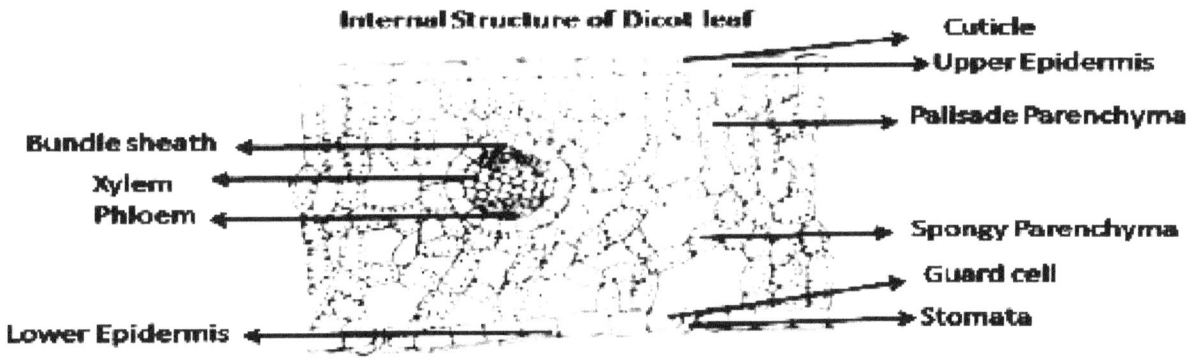

Internal Structure of Dicot leaf

Cuticle
Upper Epidermis
Palisade Parenchyma

Bundle sheath
Xylem
Phloem

Spongy Parenchyma
Guard cell
Stomata

Lower Epidermis

Internal Structure of Monocot leaf

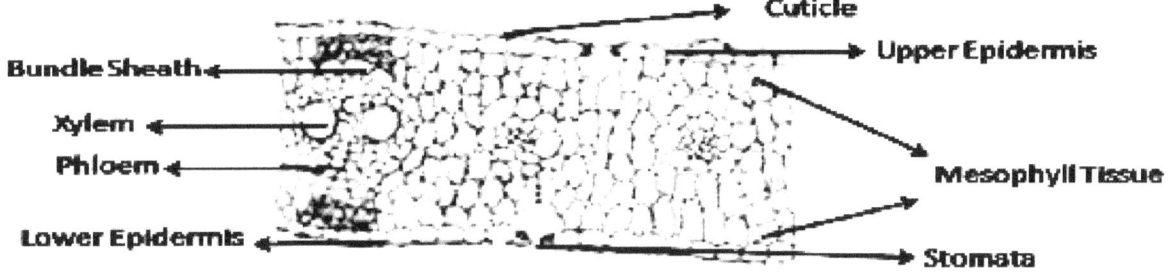

Cuticle
Upper Epidermis

Bundle Sheath
Xylem
Phloem

Mesophyll Tissue

Lower Epidermis
Stomata

Internal structure of dicot and monocot leaf

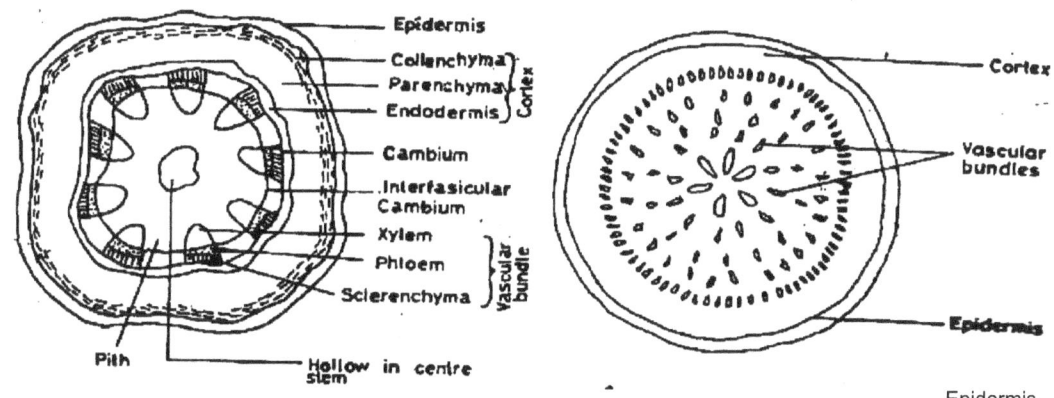

Epidermis
Collenchyma
Parenchyma — Cortex
Endodermis
Cambium
Interfasicular Cambium
Xylem
Phloem — Vascular bundle
Sclerenchyma
Pith
Hollow in centre stem

Cortex
Vascular bundles
Epidermis
Epidermis

Monocot and dicot stem

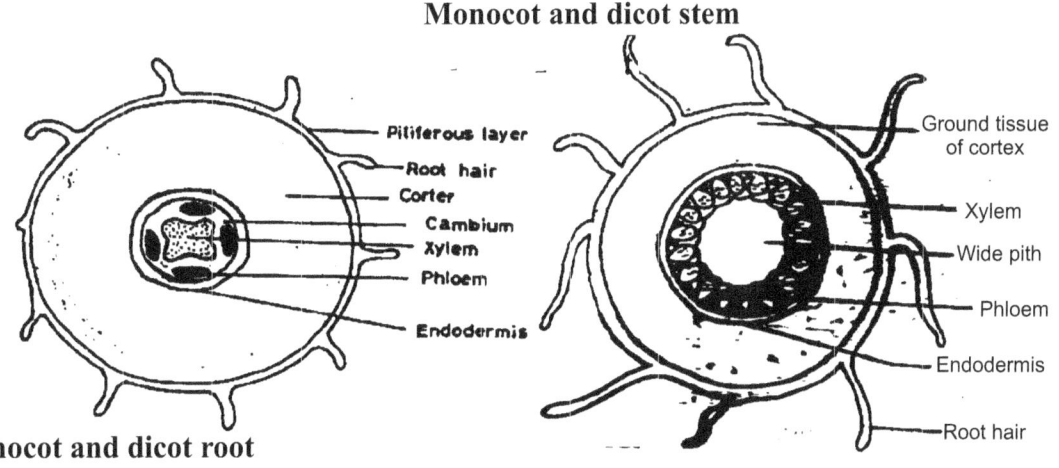

Pitiferous layer
Root hair
Corter
Cambium
Xylem
Phloem
Endodermis

Ground tissue of cortex
Xylem
Wide pith
Phloem
Endodermis
Root hair

Monocot and dicot root

6.1.1. ROOT, STEM AND LEAVE

1. UME/84/27
Which of the following statements is NOT true of the piliferous layer of a root? It
A. has a very thin cuticle
B. is the outermost layer of the cortex
C. may bear root hairs
D. breaks down as root ages
E. is replaced by cork in old roots.

2. UME/85/13
Which of the following is common to a dicotyledonous stem and a monocotyledonous root?
A. medullary ray   B. central pith
C. wide cortex   D. narrow cortex
E. pericycle fibers

3. UME/86/13
The stem differs from the root in having the xylem and phloem stands
A. on the same radii   B. scattered
C. on alternate radii   D. towards the pith

4. UME/87/15
The correct sequence of tissues in the anatomy of a young dicotyledonous stem from the inside to the outside is
A. pith, phloem, cambium, xylem, parenchyma, collenchyma and epidermis
B. xylem, phloem, cambium, cortex, endodermis, collenchyma, and epidermis
C. pith, xylem, cambium, phloem, collecnchyma, parenchyma, and epidermis
D. phloem, xylem, cambium, cortex, endodermis, collenchyma and epidermis

5. UME/87/16
Secondary thickening is initiated in dicotyledonous stem by the
A. Xylem parenchyma   B. Secondary phloem
C. Endodermis   D. Cambium

6. UME/88/17
In the tranverse section of the leaf of maize vascular bundles are arranged in
A. a row   B. one circle
C. alternate positions   D. two circles

7. UME/89/15
In a dicotyledenous stem, each companion cell is found beside the
A. Endodermal cell   B. Xylem vessel
C. Sieve tube   D. Pericyclic fibre.

8. UME/90/16
Which of the following cells is thin-walled and living at maturity?
A. collenchyma   B. sieve tubes
C. xylem vessel   D. sclerenchyma

9. UME/92/11
The flow of air and water in or out of the mesophyll layer of a leaf is controlled by the
A. stomata   B. lenticels   C. air spaces   D. guard cells

10. UME/92/13
The major site of photosynthesis in the leaf is the
A. palisade parenchyma
B. mesophyll parenchyma
C. upper epidermis
D. lower epidermis

11. UME/98/11
Which of the following structures is capable of producing more tissues in the stem of a herbaceous flowering plant?
A. Epidermis   B. Pericycle
C. Xylem   D. Cambium

12. UME/2001/21/Type T
The veins of the leaf are formed by the
A. vascular bundles   B. cambium cells
C. palisade tissue   D. spongy mesophyll

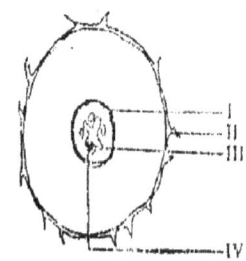

13. UME 2002/41 Type: Y
The function of absorption is performed by the structure labeled
A. I   B. II
C. III   D. IV

14. UME/2002/42/ Type: Y
The structure labeled I represents the
A. phloem   B. Xylem
C. cortex   D. pericycle.

15. UME/2004/6/Type: A
In the internal structure of plants, a wide pith in the centre of common to
A. dicot roots and monocot stems
B. dicot stems and monocot stems
C. dicot stems and monocot roots
D. dicot roots and monocot roots

16. UME/2007/23
Monocot stems differ from dicot stems in that monocots have
A. no cambium
B. no pith
C. fewer vascular bundles
D. phloems with parenchyma

17. UME/2006/17
In the transverse section of a dicot stem, the region lying between the endodermis and the vascular bundle is the
A. parenchyma   B. pericycle
C. phloem   D. hypodermis

18. UME/2008/13
In a dicotyledonous stem, the zone between the epidermis and the pericycle is the
A. cortex   B. stele   C. xylem   D. phloem

19. UME/2009/6
In plants, the structures that play roles similar to the arteries and veins of animals are the
A. xylem and phloem   B. root hairs and xylem
C. lenticels and phloem   D. roots and stems

20. UME/2010//11/TYPE: D
Which of the following is true of the transverse section of a dicot stem?
A. the xylem is more interiorly located than the phloem
B. the cambium lies between the cortex and the vascular bundles
C. the vascular bundles are randomly scattered within the cortex
D. the epidermis is completely encircled by the cortex

6.2. MAMMAL
21. UME/84/10
The mammalian organ though which nourishment and oxygen diffuses into a developing embryo is called
A. amnion   B. chorion   C. umbilical cord
D. oviduct E. placenta

22. UME/2002/35/Type Y
Which of the following pairs of organs is located in the anterior half of the mammaliam body cavity?
A. Kidneys and lungs   B. Heart and ovary
C. Lungs and heart   D. Kidneys and heart

23. UME/2006/7
The organ situated in the pericardial cavity of a mammal is the
A. heart   B. Liver
C. Stomach   D. spleen

24. UME/2006/9
In rabbits, the chamber of the heart that receives oxygenated blood from the lungs is the
A. left ventricle   B. left auricle
C. right ventricle   D. right auricle

25. UME/2006/16
The mammalian vein which starts with and ends in a capillary network is the
A. pulmonary vein   B. mesenteric vein
C. renal vein   D. hepatic portal vein

26. UME/2010/20/TYPE: D
The sheet of muscle that separates the thoracic and the abdominal cavities is the
A. intercostals muscle
B. pleural membrane
C. pericardium
D. diaphragm

27. The mammals are distinguished from the amphibians by
A. the possession of scales
B. the possession of a tail
C. the possession of the mammary glands
D. the possession of five digits in the limbs
E. the possession of glands in the skin

28. The teeth of the mammal are
A. Monodont   B. Bonodont
C. Heterodont   D. Diplodant
E. None of the above

29. With reference to the mammal, what is the perineum?
A. it is a mesentery in the body
B. it is a fold in the abdominal cavity
C. it is the membrane covering the heart
D. it is a partition separating the anus from the urinogenital aperture
E. It is an organ

30. The ear of the mammal is almost like the ear of the bird except for:
A. the presence of auditory meatus
B. the absence of the tympanic membrane
C. the absence of the utriculus
D. the absence of the cochlea
E. the presence of a fold of skin called the pinna

31. Hair regulates the body temperature of the mammal by:
A. Causing the goose flesh
B. Preventing heat radiation from the body
C. Acting as an insulator
D. Preventing the loss of sweat from the body
E. None of the above

32. A rabbit went through on operation. On recovery it is discovered that it could not digest fat. It is also seen that one of the ducts in its body had been tied up. This duct is:
A. the dorsal aorta   B. the bile duct
C. the oesophageal duct   D. the celiac duct
E. none of the above

33. The pancreatic duct opens into this region of the intestines
A. the ileum   B. the colon
C. the duodenum   D. the pyloric region
E. the caecum

34. This structure has a blind ending:
A. the wind pipe   B. the caecum
C. the colon   D. the spleen
E. the gall bladder

35. This structure produces enzymes from digestion. It is also a ductless gland
A. the walls of the stomach   B. the liver
C. the gall bladder   D. the pancreas
E. None of the above

36. The is the biggest organ in the body
A. the kidney   B. the stomach
C. the liver   D. the heart
E. the spleen

37. This structure is responsible for the formation of urea;
A. the kidney   B. the spleen
C. the gall bladder   D. the liver
E. the bladder

38. When there is a great muscular activity in the animal, which is these structures will be called upon to supply more blood so the muscles
A. it is the heart   B. it is the spleen
C. it is the lungs   D. it is the liver
E. None of the above.

39. Hydrochloric acid is produced in:
A. the pyloric region of the stomach
B. the spleen
C. the small intestine
D. the cardiac region of the stomach
E. the rectum

40. The thyroid gland is located on this structure
A. the oesophagus
B. the auricle of the heart
C. the trachea
D. the aortic arch
E. the gall bladder

41. Which muscle structure in the body is tranversed only by the oesophagus and the blood vessels?
A. the mesentery
B. the liver
C. the diaphragm
D. the lungs
E. None of the above

42. The blood cells have a certain amount of iron. When they are destroyed, the iron is stored in
A. the liver   B. the gall bladder
C. the spleen   D. the pancreas
E. None of the above

43. Considering the alimentary canal which of these regions is solely for the absorption of water?
A. the duodenum   B. the ileum
C. the colon   D. the stomach
E. the oesophagus

44. The bladder stores urine which is carried into it from the kidney through
A. the urethra
B. the urethrum
C. the ureter
D. the spermatic duct
E. the vas deferens

MISCELLANEOUS QUESTIONS (6)

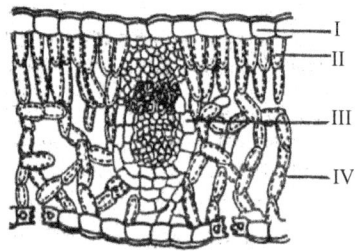

45. The major site for photosynthesis is labeled?
A. I   B. II
C. III   D. IV (PCE/2000/11/Type:A)

46. The function of the structure labeled II is to
A. allow gaseous exchange between the tissues of the leaf
B. allow light penetrate the leaf
C. prevent bacterial and fungal infections of the leaf
D. transport water and mineral salts to the leaf (PCE/2000/12/Type:A)

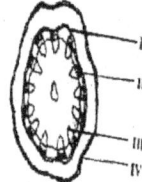

47. The structure labeled I represent the
A. parenchyma
B. endodermis
C. cortex
D. collenchyma (PCE/2003/35/Type:N)

48. Water and mineral salts are transported in the structure labeled
A. II
B. I
C. IV
D. III (PCE/2003/36/Type:N)

49. In the root vascular system, the stele is directly surrounded by the
A. Pericycle
B. cortex
C. Endodermis
D. parenchyma (UME/2005/14/Type: D)

50. In which part of a leaf are stomata found
A. in the palisade layer
B. in the spongy mesophyll layer
C. between two guard cells on the epidermis
D. guard cells and air spaces (PCE/91/12)

51. The components of the photosynthetic tissues of a leaf are the
A. palisade and spongy mesophyll
B. upper and lower epidermis
C. sclerenchyma and collenchyma tissues
D. guard cells and air spaces (PCE/2001/10/Type:4)

52. The part of the stem between the epidermis and the vascular bundles is called the
A. cortex
B. collenchyma
C. parenchyma
D. endodermis (PCE/98/11/type C)

53. In plants, the cell-type that die off at maturity is
A. sclerenchyma
B. parenchyma
C. collenchyma
D. cambium (PCE/94/11)

54. In the leaf, each stoma is usually enclosed by two kidney-shaped cells called
A. endodermis cells
B. border parenchyma cells
C. guard cells
D. phloem cell (PCE/95/12)

55. The part of the internal structure of the leaf where photosynthesis occurs most is the
A. spongy mesophyl II
B. palisade mesophyl II
C. lower epidermis
D. upper epidermis (PCE/96/11)

CHAPTER 7

NUTRITION

Understand the following after reading this chapter:

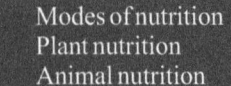

Modes of nutrition
Plant nutrition
Animal nutrition

NUTRITION
Processes by which organisms obtain and use the nutrients required for maintaining life.

Nutrients - Substances that are needed for
Sources of chemical potential energy (ATP)
Sources of molecules and atoms which are used for building up cells

Inorganic Nutrients - An inorganic substance in which the molecules of ions are relatively small and which does not contain carbon. E.g. H_2O, O_2, CO_2 and nitrate ions.

Organic Nutrients – In an organic substance the molecules are usually relatively large and contain carbon. E.g. carbohydrates, fats and protein

TYPES OF NUTRITION
Different types of organism will use different types of nutrition. There are two types of nutrition used.

Autotrophic
The use of inorganic nutrients
Carried out by all plants, algae and some bacteria
Organisms which use this type of nutrition are known as AUTOTROPHS
They need an energy source which is usually the sun but can be chemical energy

Heterotrophic
The use of a mixture of organic and inorganic nutrients
Organisms which use this form of nutrition are known as heterotrophs.

Holozoic Nutrition
This is a form of heterotrophic nutrition when the nutrients that are taken into the body are passed down some form of alimentary canal. Holozoic organisms can be divided into herbivores, carnivores, omnivores and scavengers.

Saprophytic or Saprobiontic Nutrition
Organisms that use this type of nutrition are know as saprophytes or saprobionts
They don't invade living tissue
They feed on dead or rotting matter
They secrete digestive enzymes directly onto the food this called extracellular digestion and occurs outside the cell. They then reabsorb the digested products back into their cells.
These organisms are involved in decomposing and decay. They are really important in the recycling of nutrients.

Mutualistic Nutrition
These organisms are involved in symbiotic relationships
An example is the bacteria Rhizobium with legumes
The bacteria are nitrogen fixing, they convert nitrogen and hydrogen to ammonia.
They are found in the soil and in the root nodules of legumes.
The plants use the ammonia produced by the bacteria is used by the plants to make amino acids.
The bacteria benefits from this relationship because they gain carbohydrate and ATP from the plant.
The plant benefits because the ammonia produced by the bacteria allows them to flourish in nitrogen deficient soil.

Parasitic Nutrition
This is a relationship between parasite and host. The parasite is dependent on the host for nutrition and protection. The parasite will generally do the host harm.
Parasites are highly adapted for this type of nutrition. There are two main types of parasite ectoparasites which lived outside the organism and endoparasites which live inside the organism.

Carnivorous Plants
Carnivorous plants are plants that derive some or most of their nutrients (but not energy) from trapping and consuming animals or protozoans, typically insects and other arthropods. Carnivorous plants have adapted to grow in places where the soil is thin or poor in nutrients, especially nitrogen, such as acidic bogs and rock out croppings. Examples of carnivorous plants include: Venus Flytraps, Sundews, Pitcher Plants a.k.a. Monkey Cups, North American Pitcher Plants, and Bladderworts.

PHOTOSYNTHESIS

Plants use light energy from the sun as their external energy source, this type of autotrophic nutrition is **photosynthesis.**

During photosynthesis carbon dioxide and water are used to manufacture carbohydrates. Light energy is trapped by chlorophyll and used to generate ATP (**cyclic photophosphorylation**) and also to split water (**the light reaction**); the hydroxyl (-OH group) is recombined to make oxygen (which is **excreted**); the hydrogen is later used to fix carbon dioxide and reduce it to form carbohydrates. Plants manufacture all the other organic molecules (DNA, vitamins, lipids etc. etc.) that they need from these carbohydrates and very small quantities of mineral ions obtained from the soil.

Carbon dioxide + water $\longrightarrow$(light + chlorophyll) $\longrightarrow$glucose + oxygen

$$6CO_2 + 6H_2O \longrightarrow C_6H_{12}O_6 + 6O_2$$

The biochemistry of photosynthesis.
There are two basic stages to photosynthesis.

1. The Light Dependent Reaction
This takes place in the **thylakoids** of a chloroplast.

During this stage light energy is trapped by chlorophyll molecules in the light harvesting stage. This raises the energy level of electrons in the chlorophyll molecules. In the process of non-cyclic photophosphorylation this energy is used to convert NADP to NADPH + H^+ and ADP to ATP.
Non-cyclic photophosphorylation involves photosystems II and I.

During this process water is split into oxygen - a waste gas (most of which will be released from the plant) - and into hydrogen ions and electrons.
These are used to generate **NADPH + H⁻and** ATP. In cyclic photophosphorylation **only** photosystem I is involved and **only** ATP is produced.

$$NADP + 2H^+ + 2e^- \longrightarrow NADPH + H$$

2. The Light Independent Reaction
This takes place in the **stroma** of a chloroplast. The process is known as **The Calvin Cycle.**

Carbon dioxide fixation takes place.
Initially the carbon dioxide combines with ribulosebisphosphate (RuBP), a compound containing five carbon atoms.
The enzyme RuBisCO catalyses this reaction. Thus life on Earth largely depends on this one chemical!

This reaction can also be called a carboxylation, since carbon dioxide is being added: it gives a six carbon intermediate compound, which **instantly** splits into 2 molecules of glycerate-3-phosphate (**GP**) a 3-carbon compound.
The GP is then phosphorylated (phosphate is added), using the ATP generated in the Light-dependent reaction **and then reduced** to glyceraldehyde-3-phosphate (**GALP**)
These reactions use up the NADPH + H⁺ and ATP which were produced in the Light Dependent reaction. NADP and ADP are regenerated.

Most of the GALP is used to regenerate RuBP, but some is used to produce 6 carbon sugars (monosaccharides), which can then be used to produce a range of organic molecules.

These include: -
1. Starch, a polysaccharide, stored temporarily in the leaf, or in organs such as potato tubers for more long-term storage.
2. Sucrose, a disaccharide transported to other parts of the plant in the phloem.
3. Lipids, for energy storage, in seeds, for example.
4. Amino acids and proteins - the manufacture of these requires nitrogen, the source of this is normally nitrate ions from the soil.

The factors that affect the rate of photosynthesis and their interactions:

a. *Temperature* – higher temperatures raise the speed of the light-independent reactions, up to about 35-40°C (= gross photosynthetic rate); it has little effect on the speed of the light reactions. **However,** they also raise the speed of respiration in the plant and so net photosynthesis **slows down at higher temperatures.** Peak crop yields are at about 25°C

b. *Carbon Dioxide* –normally 360 ppm, growers raise this to about 1000 ppm in greenhouses to raise yields. Peak photosynthesis is at even higher concentrations, but it is not economical to produce this. N.B. Higher world CO_2 levels = better plant growth.

c. *Light -* the more the better (**brightness**); 16-20 hours a day (**duration**); blue and red must be in balance (i.e. filament lamps are no good!)

d. *Water* – often a limiting factor – plants shut their stomata long before they wilt. Shut stomata = no gas exchange and no photosynthesis!

How leaves are adapted to carry out photosynthesis effectively:

stomata for gas exchange (largely on the underside to maximise palisade cells)

thin to allow gas to reach mesophyll cells quickly *moist* to allow gasses to dissolve *waterproof cuticle* to reduce transpiration and so water stress *good transport system*
xylem - to bring water and minerals
phloem – to carry away the products of photosynthesis
large surface area – to trap light
palisade mesophyll – near top to trap maximum light
spongy mesophyll to aid in light absorption
leaves arranged in space so as not to shade each other.
Transparent cuticle/upper epidermis to allow maximum light penetration

MINERAL NUTRIENTS

Mineral nutrients are the inorganic matters or chemicals in the soil released during weathering of rocks. They are grouped into two, based on the quantitites required of them by plants;

1. Macro-nutrients: These are minerals required in large quantities e.g. sodium, potassium, phosphorus, nitrogen, calcium, sulphur, magnesium.

2. Micro-nutrients or trace elements: These are minerals required in small quantities e.g. iron, manganese, molybdenum, zinc, copper, boron, cobalt. Animals obtain minerals indirectly by feeding on plants or their products.

Name	Function in plant	Deficiency symptoms
Nitrogen	Proteins, amino acids	Light green to yellow appearance of leaves, especially older leaves; stunted growth; poor fruit development.
Phosphorus	Nucleic acids, ATP	Leaves may develop purple coloration; stunted plant growth and delay in plant development.
Potassium	Catalyst, ion transport	Older leaves turn yellow initially around margins and die; irregular fruit development.
Calcium	Cell wall component	Reduced growth or death of growing tips; blossom-end rot of tomato; poor fruit development and appearance.
Magnesium	Part of chlorophyll	Initial yellowing of older leaves between leaf veins spreading to younger leaves; poor fruit development and production.
Sulphur	Amino acids	Initial yellowing of young leaves spreading to whole plant; similar symptoms to nitrogen deficiency but occurs on new growth.
Iron	Chlorophyll synthesis	Initial distinct yellow or white areas between veins of young leaves leading to spots of dead leaf tissue.
Copper	Component of enzymes	Poor growth.
Manganese	Activates enzymes	Interveinal yellowing or mottling of young leaves.
Zinc	Activates enzymes	Interveinal yellowing on young leaves; reduced leaf size.
Boron	Cell wall component	Leaf tips become yellow followed by necrosis. Leaves get a scorched appearance and later fall off.
Molybdenum	Involved in N fixation	Poor growth.
Chlorine	Photosynthesis reactions	May affect growth.

ANIMAL NUTRITION

ENZYMES

Enzymes are biological catalysts, made of protein, that speed up the rate of reactions. They can be breakdown (degradation) or build up (synthesis) reactions.

Substrate	⇒	Enzyme	⇒	End Product
(what you start with)		(biological catalysts)		(what you end up with)

Enzyme rate is affected by temperature and pH. All enzymes work best at 37^0C.
Pepsin works best at pH 2.5 and catalase works best at pH 9.
Enzymes are therefore said to be "**specific**" and work best at their "**optimum**".
Enzymes work like a "**lock and key**". The substrate is the key and the enzyme is the lock.
The enzyme remains unchanged at the end of the experiment and can be reused.

Substrate	Enzyme	End Product	Synthesis (S) / Degradation (D)
Starch	Amylase	Maltose	D
Hydrogen Peroxide	Catalase	Water and oxygen	D
Protein	Pepsin	Peptides	D
Fat	Lipase	Fatty acids	D
Glucose – 1 – phosphate	Phosphorylase	Starch	S

Amylases, **proteases** and **lipases** are group of enzymes that digest **carbohydrates**, **proteins** and **lipids** respectively.

FOOD SUBSTANCES
The main classes of food are as follows:

Carbohydrates:
This nutrient is an organic compound composed of carbon, hydrogen and oxygen.

Function:
It is used as an energy resource, essential in respiration to release energy.
It is used in creating the cellulose, the substance forming cell walls of plant cells.
Carbohydrates are 3 types:

Monosaccharides:·
The smallest and simplest form·
Water soluble·
Chemical formula $C_6H_{12}O_6$·
Examples: Glucose-Fructose-Galactose·
Sources: Fruits, Honey.

Disaccharides:·
Each molecule consists of two Monosaccharide
j oined together·
Water soluble·
Examples: Lactose, Sucrose, Maltose·
Sources: Table sugar, Milk.

Polysaccharides:·
Each molecule has many joined monosaccharide forming a long chain.·
Insoluble in water·
Examples: Starch-Glycogen-Cellulose·
Sources: Bread, Potatoes, Pasta, Cellulose in plant cells and Glycogen in livers.

Monosaccharide and Disaccharides are sugars, they are reducing for Benedict's reagent, except for the disaccharide sucrose, it is non-reducing.
Polysaccharides are not considered as sugars and don't have a sweet taste. Excess polysaccharides are stored in the liver and muscles.

Lipids (Fats):
These are composed of carbon, hydrogen and oxygen. But their ratios are different than that of carbohydrates. One fat molecule is made of a glycerol unit and three molecules of fatty acids.

Fats are essential in a diet because they are needed to:
Release high amounts of energy
Make cell membranes
Store them under the skin to insulate heat.
Forming a layer of fats around organs to protect them from damage
Storing energy (better than glycogen)
When fats are respired, they produce about twice as much energy as carbohydrates.

Proteins:
These are also organic compounds; they contain the elements Carbon, Hydrogen, Oxygen, Nitrogen and sometimes Phosphorus or Sulfur.
A molecule of protein is a long chain of simpler units called amino acids.
These amino acids are linked together by what's called "peptide bond".

Types of protein:
Animal Protein: It contains the most biological value because it contains all essential amino acids (Meat, Milk, Fish, Eggs etc).
Plant Protein: It contains a lower biological value to humans because it contains fewer essential amino acids (Cereals, Peas, Beans etc).

Needs of proteins:
Making and new body cells
Growth and repair
Making enzymes (they are proteins in nature)
Build up hormones
Making antibodies
Although proteins are needed in high amounts, the body will only absorb as much as needed, so excess protein is delaminated in the liver and excreted as urea.

Vitamins: These are organic, soluble substances that should be present in small amount in our diets, they are very important though. Most of the amount of vitamins in our bodies was taken in as nutrients, the body its self can only make few Vitamins, so we have to have to get them from organisms that make them, such as plants. Each type of Vitamin helps in chemical reactions that take place in our cells.

Types of Vitamins:
Vitamin C: This is present in most fruits and vegetables specially citrus fruits like lemon and oranges, however, it is damaged by heating so it these foods have no value of Vitamin C if they are eaten cooked. Vitamin C is essential for the formation of Collagen, a protein that functions as cementing layer between cells, Vitamin C also increases immunity.

Vitamin D: This is present in fish oils, egg yolk, milk and liver. Unlike Vitamin C, Vitamin D is made by animals as well as plants, this occurs when the skin is exposed to the Ultra Violet Rays of the sun. Vitamin D plays a big role in absorbing Calcium from the small intestine and depositing it in bones. So it is responsible for having healthy bones.

Minerals (Inorganic Ions):
These are a lot of types, each needed in small quantities. Iron and Calcium are the most important minerals, and they are needed in higher amounts.

Types of Minerals:
Calcium: This mineral is needed for the formation of bones and teeth as they are made of calcium salts, it also helps in blood clotting and transmission of nerve impulses. Good sources of the mineral Calcium are milk, dairy products and hard water.

Iron: This mineral is needed for the formation of the red pigment haemoglobin which is essential for the transport of oxygen around the body in red blood cells. Good sources of Iron include red meat specially liver and green leafy vegetables.

Roughages (Fibre):
Although roughages are not even absorbed by the body, they are a very important nutrient in our diet. Roughages are mostly cellulose, which is the substance that makes up the cell walls of plants we eat. We humans, have no enzyme that could digest cellulose, that means that roughages enter the body from the mouth, go through the digestive system, and out through the anus unchanged. But as it goes through the digestive system, roughages take space in the gut to give the gut muscles something to push against, this process of pushing the food through the gut is called peristalsis, without roughages peristalsis is very slow and weak. Quick and strong peristalsis means that food stays in the alimentary canal for a shorter period, this prevents harmful chemicals of certain foods from changing the DNA of cells of the alimentary canal causing cancer, so roughages also helps stay away from cancer. Roughages are found in leafy vegetables.

Water:
About 70% of your weight is water. Water is perhaps a very essential nutrient we should take in. The functions of water include:·

As a solvent which reactants of metabolic reactions are dissolved in.
• It makes up most of the blood plasma which red blood cells,

nutrients, hormones and other materials are carried in.
• It helps in lowering the body temperature in hot conditions by secreting it as sweat on the skin, the sweat evaporates using heat energy from the body, thus lowering the temperature.

THE MAMMALIAN TOOTH
Structure of tooth: -**enamel**: the strongest tissue in the body made from calcium salts -**cement**: helps to anchor tooth -**pulp cavity**: contains tooth-producing cells, blood vessels, and nerve endings which detect pain. -**dentine**: calcium salts deposited on a framework of collagen fibres -**neck**: in between crown and root, it is the gums

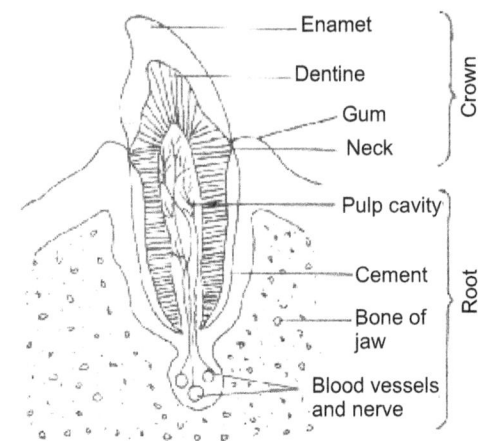

Section through a tooth

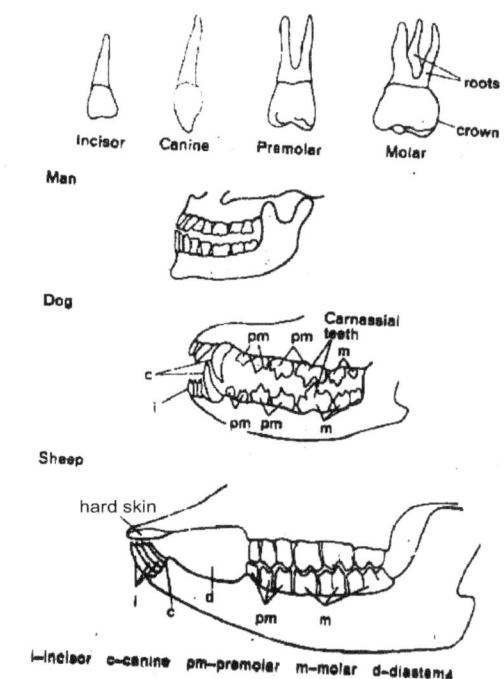

Teeth and dentition in mammals

56

DENTITION AND TYPES OF TEETH

Dentition is the number and arrangement of teeth in the mouth. The teeth are used for holding, cutting, biting and grinding food. A tooth is made of three parts namely; crown, neck and root. A longitudinal section of a tooth shows the enamel, dentine pulp cavity and cement. The types of teeth in mammals are:

1. Incisors: these have been sharp flat edges suitable for cutting and bitting food. They are found at the centre in front of the jaw.

2. Canines: These are pointed and are strong teeth for tearing food.

3. Pre-Molars: These have broad crown sharp edges and flat tops for cutting and grinding.

4. Molar: These have broad surface with projections for crushing bones and flesh.

Carnassial teeth: found in advanced carnivores, it is the last upper premolar and the first lower molar. Dental formula is the figurative and numerical expression of each half of the upper and lower jaws. Examples:

Man   - **I**2/2' **C**1/1' **Pm**2/2' **M**3/3   = 32
Dog   - **I**3/3' **C**1/1' **Pm**4/4' **M**2/3   = 42
Rabbit - **I**2/2' **C**0/0' **Pm**3/2' **M**3/3   = 28

DIGESTIVE SYSTEM

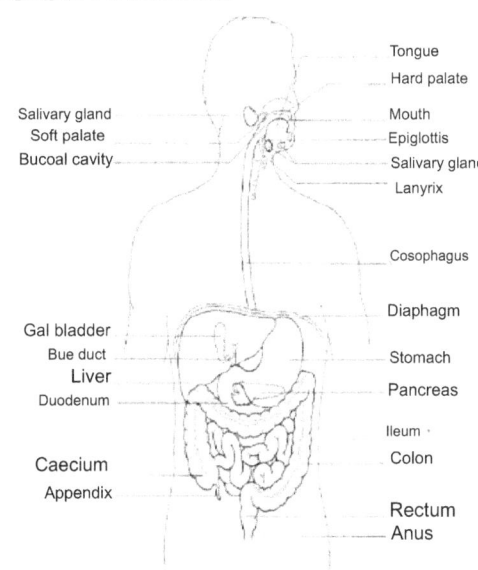

Alimentary canal of man

The digestive system consists of a tube which leads from mouth to anus. The digestive system is made up of the oesophagus, the stomach, small intestine and large intestine. Each of these parts has a different function and a different structure. Food and fluids which are ingested are pushed through the digestive system by a process of peristalsis.

Parts of the Digestive System
Stage 1 - The Mouth
Food first enters the digestive system in the mouth. Here is it initially broken down both mechanically (through chewing) and chemically through the enzymes in saliva, salivary amylase (ptylin) which breaks down starch.
Starch   (ptylin)→ Maltose

Stage 2 - The Oesophagus
This is a tube which connects the mouth to the stomach, food travels down this tube by peristalsis. There are a few mucus secreting glands but only small villi.

Stage 3 - The Stomach
The stomach is effectively a bag of muscle which can expand there are three layers of muscle which churns the contents of the stomach. Food is stored in the stomach for a few hours during which time it becomes a liquid known as chyme. The mucosa of the stomach wall has no villi but does have structures which are known as gastric pits which are associated with gastric glands. The gastric glands secrete hydrochloric acid to kill bacteria. The enzymes pepsin and rennin are secreted to digest proteins. The chyme is gradually released into the small intestine gradually through a large muscular sphincter known as the pyloric sphincter.
Protein   (pepsin)→ Peptide
Milk   (renin) → Curdled milk

Stage 4 - The Small Intestine
The small intestine is made up of three parts the duodenum the jejunum and the ileum. In the duodenum digestion continues and two separate secretions assist this. The pancreas secretes pancreatic juices which contain carbohydrases (amylases), proteases and lipases. Bile is secreted from the liver and contains bile salts which aid lipid digestion and sodium hydrogen carbonate which neutralize the stomach acid. The jejunum and the ileum are very similar they are where final digestion occurs, glands in the mucosa and sub mucosa secrete enzymes, mucus and sodium hydrogen carbonate.
Duodenum:
Starch (amylase)→ Maltose
Protein (peptidase)→ Polypeptide
Lipid   (lipase)→ Fatty acid and glycerol

Stage 5 - The Large Intestine
The large intestine is made up of the caecum, appendix, colon and rectum. There are numerous villi and mucus secreting glands. Water is reabsorbed from the large intestine and semi solid faeces are formed.

ABSORPTION
Absorption is the movement of digested food molecules through the wall of the intestine into the blood or lymph.

The small intestine is the region for the absorption of digested food.

The small intestine is folded into many **villi** which increase the surface area for absorption. One villus will have tiny folds on the cells on its outside called **microvilli**. More surface area means more absorption can happen.

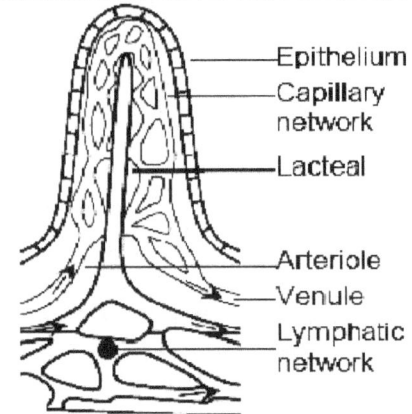

Structure of a villus

THE LIVER

The liver is a reddish-brown organ lying against the posterior surface of the diaphragm and partly covering the stomach.

Functions of the Liver

1. Regulation of blood sugar
2. Deamiantion of proteins
3. Detoxication of harmful drugs and chemicals
4. Manufacture of blood plasm proteins
5. Storage of irons
6. Storage of vitamins A and D
7. Production of bile
8. Reservoir for blood production

EXPERIMENTS ON NUTRITION

Experiment: 1 Test for starch (cooked starch)

Materials required. Cooked starch, iodine solution, a white tile, a dropper.

Method. A few drops of the test material (cooked starch) are placed on different spots of a white tile with the aid of a dropper. Iodine solution is added to each drop of the starch solution.

Observation. A bluish black colour appears.

Conclusion. A bluish black colour with the iodine solution shows that starch is present.

Experiment :2 Tests for sugars
(a) reducing sugars (e.g. glucose)

Materials required. Fehling's solution, Benedict's solution, glucose (a reducing sugar), test tube, Bunsen burner.

Method. Some glucose is placed in a test tube and a few drops of Fehling's solution are added to the glucose. The mixture is then heated to boiling point; OR Into glucose in a test tube, some drops of Benedict's solution are added and the mixture is warmed for some time.

Observation. There is orange (brick) red precipitate with Fehling's solution or Benedict's solution.

Conclusion. The orange red precipitate indicates the presence of a reducing sugar.

(b) non-reducing sugars (e.g. sucrose)

Materials required. Sucrose, water, HCl, test tube, NaOH, Fehling's solution.

Method. Some sucrose is put in attest tube and a few drops of water and HCl are added to the sucrose. Next, some drops of NaOH are added to neutralise the acid. Finally, Fehling's solution is added.

Observation. There is an orange red precipitate.
Conclusion. The test material is sucrose.
Experiment :3(a) Test for fats and oils.

Materials required. Test material (e.g. groundnut seeds), filter papers.

Method. The test material is placed in between folds of filter papers. The test material is the pressed. Next, the paper is held before light.

Observation. The spots with fats become translucent.

Conclusion. The test material contains fats or oils. If the translucent spots are washed with benzol or ether, the spots clears off.

Experiment : 3(b) test for fats and oils using Sudan (III) solution.

Materials required. Test material, Sudan (III) solution, test tube.

Method. The test material is put in a test tube and a few drops of Sudan (III) solution is added. Heat the mixture.

Observation. A red colouration appears before boiling. A black precipitate is formed on boiling.

Conclusion. The test material contains fats or oils.

Experiment :4 Test for protein (e.g. milk, egg albumen)

Materials required. Egg albumen (or milk), test tube, NaOH solution, CuSO4 solution.

Method.
1. **Millon's test.** Some egg albumen or milk is put in attest tube and Millon's reagent is added to it. Heat the mixture. OR
2. **Biuret test.** Some milk or egg is put in a test tube. Drops of NaOH and CuSO, are added to the test material. Both solutions are made dilute.
3. **Xanthoproteic test.** Some milk or egg albumen is put in a test tube and concentrated HNO, is added and warmed. The mixture is allowed to cool under a running tap and concentrated NH, is added.

Observation.
Millon's test. There is a deep red colouration of the test material.
Biuret's test. There is a violet colouration.
Xantoproteic test. There is a yellow colouration which on cooling and adding NH3 becomes orange.

Conclusion. The test material is protein.

Experiment: 5 To demonstrate that ptyalin is necessary for the digestion of starch.

Materials required. A solution of enzyme, a solution of starch, sodium bicarbonate, small size beaker, 3 test tubes, water bath, iodine solution, a white tile, a dropper and labels.

Method. the enzyme solution is made by taking some water into the mouth and turning it for some time. This is done after the mouth has been thoroughly washed with warm water to remove any food substances. The water in the mouth is poured into a small beaker. The starch is made by adding a little cold water into powder starch and stirring it. Some boiled water is then added. This gives the starch solution. The solution is left to cool. A little starch solution is poured into 3 test tubes, labelled 1, 2, and 3 and they are treated with the enzymes (ptyalin) as follows:-

Test tube 1. Into this tube, some saliva is poured and sodium bicarbonate is added to provide an alkaline medium.

Test tube 2. Into this tube, no enzyme is added.

Test tube 3. Into this tube, boiled saliva is added, i.e. the tube contains starch and killed enzyme.

The three tubes are transferred to a boiling water bath. The temperature of the solution in the test tube is maintained around 37·C which is the body temperature of mammals. The mixture in the test tube is stirred continually. After some time, the solution in each test tube is tested for starch with iodine on white tile. The test tube is repeated every 30 minutes for 1 hour, 30 minutes. Next, the content of the first tube is boiled with Benedict's solution.

Observation. The second test tube which contained only starch, continued to give a blue-black colour with iodine, i.e. No change. The third test tube which contain killed enzyme gives the same result as in the second tube i.e. no change. The first test tube which contained the enzyme and the alkaline, showed gradual changes from blue- black to purple and then to light brown. A brick red colour appeared when the first test tube content was boiled with Benedict's solution.

Conclusion. This first test tube shows that saliva contains an enzyme (ptyalin) which acts on starch. It converts starch into complex sugar (maltose) which on boiling with Benedict's solution gave a brick red colour. In other tube (2 and 3) where there were no changes, means that either the enzyme was absence or destroyed.

Experiment: 6 To test for leaf for starch.

Material required. A leaf, a beaker, 70% alcohol, test tube, water bath, iodine solution, a white tile.

Method. A leaf (e.g. cassava leaf) is poured at the end of a sunny day. The cells of the leaf are killed by boiling for five minutes in a beaker. Some alcohol (70%) is poured into a test tube and is warmed in a water bath the killed leaf is then transferred into the test tube. Heat is continually applied to the test tube until the alcohol is close to boiling and the green colour of the leaf will gradually disappear and the whole leaf will have become decolourised. The leaf is removed from the test tube and washed with hot water for about five minutes to soften it. The leaf is then tested for starch with iodine solution. The test is done on a white tile.

Observation. There is a blue-black colour.

Conclusion. The blue-black colour shows the presence of starch.

Experiment: 7 To demonstrate that carbon (IV) oxide is necessary for photosynthesis.

Material required. Two young potted plant, two bell jars, a beaker, sodium hydroxide solution, and candle sticks.

Method. Two young potted plants are kept in the dark such as inside a cupboard. This is to remove the starch from the leaves. After this period the plant are brought from the cupboard and the bell jar is placed over each plant as shown in the above figure. A small beaker which contains sodium hydroxide (NaOH) is placed beside one of the plants while a burning candle in the other. The NaOH is to remove carbon (IV) oxide under the first bell jar and the burning candle is to remove oxygen under the second bell jar but to leave carbon (IV) oxide. Both experiments are then exposed to light for some time. After this period a leaf from each plant is plucked and tested for starch as in experiment 7.

Observation. The leaf from the second experiment gave a bluish black colour with iodine. There was no change in the leaf from the first experiment.

Conclusion. The bluish black colour indicates the present of starch grains. This shows that carbon (IV) oxide is necessary for photosynthesis.

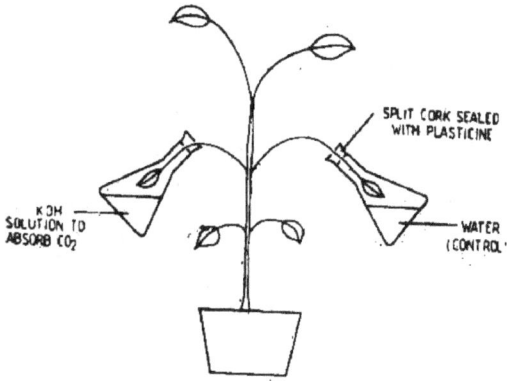

Experiment: 8 To demonstrate that light is essential for photosynthesis.

Material required. A young potted plant, paper clips, strips of brown paper, iodine solution, and white tile.

Method. A young potted plant is lift in the dark for about 48 hours to remove all starch in the leaves. To ensure that starch is not present, some leaves are tested two strips of brown paper are taken and a hole of letter S is made in each. Next, a leaf is selected from the plant (without plucking it). Carefully, the brown papers are placed over the leaf, one on each side of the lamina. The papers are then clipped together by means of paper clip as shown in fig. 40.6. The potted plant is then exposed to adequate sunlight for a period of 4 hours. After this period, the leaf is plucked and tested for starch as in experiment 7.

Observation. The area of the leaf which received sunlight gave a bluish black colour. Letter S is printed on the leaf. Other areas of the leaf were dark.

Conclusion. The areas of the leaf which gave a bluish black colour contained starch grains and hence, a letter S with starch grains was printed on the leaf. This show that light is necessary for photosynthesis.

Experiment :9 To demonstrate that only green areas of a leaf can photosynthesize.

Materials required. A variegated leaf, iodine solution, a white tile, water bath, test tube, 70% alcohol.

Method. A variegated leaf is selected from a plant around the laboratory at the end of a bright day. All the area of the leaf which bear green colour are carefully mapped out. The leaf is treated and tested for starch with iodine as in experiment 7.

Observation. Only the area that is green gave blue-black with iodine.

Conclusion. The blue-black colour shows the present of starch grain in the green areas. This shows that only the green area of the leaf can photosynthesize.

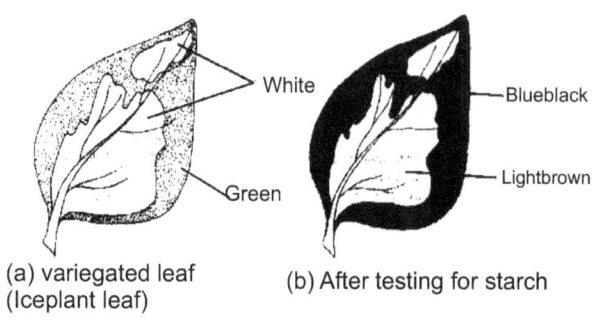

(a) variegated leaf (Iceplant leaf) (b) After testing for starch

Experiment :10 To demonstrate that oxygen is given out during photosynthesis.

Material required. A beaker, funnel, test tube, water weed (e.g. Elodea).

Method. A beaker is filled with water, almost mid-way. Carbon (IV) oxide is bubble into this water until it is clearly saturated. Some water weeds (e.g. Elodea) are place in water and a funnel is place over the weeds. The rim of the funnel is raised from the bottom of the beaker to allow for free circulation of air. Next, medium sized test tube which nearly filled with water is used to cover the stem of the funnel as shown in fig. 40.8. After this, the setup is kept in a place where it can receive sufficient sunlight.

Observations: Bubbles of a gas are noticed to rise up the funnel. The gas displaces the water in the test tube and it collects at the upper end. When the test tube was carefully removed (after sufficient gas has collected in the tube) and tested with a glowing splint, the splint is rekindled.

Conclusion: The gas in the tube is oxygen which rekindled the glowing splint. This shows that oxygen is liberated during photosynthesis.

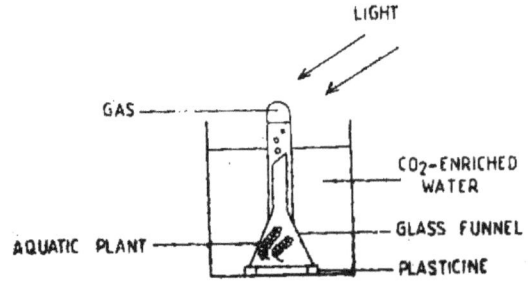

7.1. MODES OF NUTRITION

1. UME/83/30
Which of the following statements is NOT true of insectivorous plants?
A. They obtaining part of their food by trapping and feeding on insects
B. They attract insect simply because of pollination
C. They grow in soil poor in nitrogenous salts
D. They can supplement their nitrogen supply by feeding on insects
E. Examples include butterworts, sundews and pitcher plants.

2. UME/90/17
The mode of nutrition in which digestion is extracellular is
A. holophytic   B. parasitic
C. Holozoic   D. saprophytic

3. UME/92/12
Fungi are heterotrophic because they
A. have no leaves   B. lack roots
C. are filamentous   D. lack chlorophyll

4. UME/99/14/Type: D
The modes of nutrition in Nitrobacter, *Ascaris* and *Homo sapiens* respectively are
A. Photosynthetic, parasitic and holozoic
B. Chemosynthetic, parasitic and holozoic
C. Photosynthetic, parastic and heterophytic
D. Chemosynthetic, holophytic a nd holozoic

5. UME/2002/28/Type :Y
Examples of organisms in which extracellular digestion occurs are
A. fungus, loranthus and housefly
B. rhizopus, sponges and earthworm
C. roundworm, tapeworm and hydra
D. rhizopus, housefly and hydra

6. UME/2002/31/Type:Y
The capture and digestion of insects by a pitcher plant is a special form of nutrition termed
A. autotrophic   B. heterotrophic
C. chemosynthetic   D. saprophytic

7. UME./2002/36/Type:Y
The mode of nutrition exhibited by a tapeworm is
A. symbiotic   B. saprophytic
C. parasitic   D. Holozoic

8. PCE/2004/12/Type: 9
The possession of haustoria by some plants is an adaptation for
A. parasitic life   B. saprophytic life
C. carnivorous plant   D. symbiotic life

9. UME/2008/15
Insectivorous plants trap and kill their prey to derive
A. phosphorus   B. calcium
C. nitrogen   D. zinc

10. UME/2009/15
Which of the following is an example of a carnivorous plant?
A. Hydra   B. Bladderwort
C. Yeast   D. Spirogyra

11. UME/2010/15/TYPE: D
The mode of nutrition of sundew and bladderwort can be described as
A. saprophytic   B. holozoic
C. chemosynthetic   D. autotrophic

12. UME/2010/25/TYPE: D
Chemosynthetic organisms are capable of manufacturing their food from single inorganic substances through the process of
A. denitrification   B. reduction
C. phosphorylation   D. oxidation

7.2. PLANT NUTRITION

7.2.1. PHOTOSYNTHESIS
Use the equation below to answer examples 13 and 14
$$6CO_2 + 6H_2O \xrightarrow[\text{Chlorophyll}]{\text{Sunlight}} C_6H_{12}O_6 + 6O_2$$

13. UME/78/2
The process represented by the above equation is
A. proteinsynthesis   B. respiration
C. photosynthesis   D. transpiration
E. translocation

14. UME 78/3
The oxygen given off during the process is derived from:
A. sunlight   B. chlorophyll
C. carbon dioxide   D. atmosphere
E. water

15. UME/90/18
The first step in the process of photosynthesis is the
A. activation of the chlorophyll
B. photolysis of water
C. reduction of carbondioxide
D. formation of sugar

16. In photosynthesis, the following processes are part of the light reaction except
A. Transfer of radiant energy which reduces a co-factor
B. absorption of radiant energy by chlorophyll
C. utilization of energy in the electron transfer chain from ATP
D. formation of glucose using energy from NADPH
E. splitting of water into its component parts.

17. UME/87/20
In standard experiment to show that oxygen is given off during photosynthesis, sodium bicarbonate is used to
A. neutralize the acid in water
B. supply mineral salt to water plant
C. supply carbon dioxide for photosynthesis
D. kill micro-organisms in water.

18. UME/78/34
Which of these is a direct photosynthetic product?
A. Glucose   B. Starch
C. Protein   D. Fats
E. Lactose

19. UME/89/18
If a healthy potted plant is continuously kept in dim light?
A. the rate of respiration may equal that of photosynthesis
B. more carbondioxide and water are taken in
C. respiration may be halted
D. the volume of oxygen released increases

20. UME/95/11
$2H_2O \quad 2H_2 + O_2(g)$.
The equation above represents a part of the light stage of photosynthesis. Which of the following must be present for this reaction to occur?
A. Enzyme and light energy
B. carbodioxide and light energy
C. light energy and chlorophyll
D. chlorophyll and enzyme

21. UME/98/12
The manufacture of carbohydrates by plants takes place only
A. the leaves   B. the green stems
C. chlorophyllous parts D. flowering plants

22. UME/99/15/Type D
The dark reaction of photosynthesis involves
A. Fixation of carbon (iv) oxide to give a six-carbon sugar
B. Fixation of carbon (iv) oxide with the help of oxygen
C. Use of carbon (iv) oxide to produce glucose using ATP
D. Fixation of carbon (iv) oxide on chlorophyll using Hydrogen

23. PCE/97/14
The dark stage of photosynthesis takes place in the
A. stoma   B. stroma   C. guard cell   D. lamellae.

24. UME/2007/20
Photosynthetic pigments include
A. chlorophyll and carotenoids
B. chloroplasts and cytochromes
C. melanin and haemoglobin
D. carotenoids and haemoglobin

25. Chlorophyll absorbs from sunlight more of:
A. Infra-red rays   B. Violet rays   C. Red rays
D. Yellow rays   E. Green rays

26. The metallic ion present in chlorophyll is:
A. Iron   B. Copper
C. Magnesium   D. Carotone
E. Sodium

27. One of the yellow pigments of chlorophyll is:
A. Carotene   B. Lycopene   C. Xanthopyll
D. Hydrocarbon   E. Chlorophyll

28. Carbon dioxide gets to the chloroplast of cells in:
A. Liquid form
B. In gaseous form
C. In bicarbonate form
D. As carbonic acid
E. None of the above

7.2.2. MINERAL NUTRIENTS

9. UME 79/47
The element Nitrogen is utilized in
A. formation of ATP
B. formation of glucose
C. formation of amino acids
D. photosynthesis E. None of the above

30. UME/79/48
Magnesium is utilized in the formation of
A. ATP   B. glucose   C. amino acids
D. chlorophyll   E. fats

31. UME/85/36
A young plant showing yellowing leaves is likely to be deficient in
A. calcium
B. magnesium
C. potassium
D. boron
E. molybdenum

32. UME/82/4
Some of the major elements required by plant are
A. Potassium, Nitrogen, Phosphorus, Sodium, Calcium
B. Nitrogen, Phosphorus, Molybdenum, Sodium, Calcium
C. Potassium, Phosporus, Molybdenum, Sodium, Calcium
D. Potassium, Nitrogen, Iron, Sodium, Calcium
E. Potassium, Nitrogen, Phosphorus, Silicon, Calcium

33. UME/99/16/Type D
Yellowing of the leaves in a symptom associated with deficiency of
A. iron, calcium and magnesium
B. nitrogen, sulphur, potassium
C. sulphur, phosphorus and iron
D. magnesium, nitrogen and iron.

34. UME/89/21
The seedling in a rice field were found to have thin lanky growth with reddish leaves and poor root development. This is because; the soil lacks
A. sulphur   B. phosphorus   C. potassium   D. iron

35. UME/97/13
Trace elements are required by plants mainly for the
A. formation of pigments and enzymes
B. production of energy and hormones
C. manufacture of carbohydrates
D. manufacture of proteins.

36. UME/98/13
In a water culture experiment, a plant showed poor growth and yellowing of the leaves. These may be due to deficiency of
A. Copper   B. Iron   C. Magnesium   D. Calcium

37. UME/2007/25
Stunted growth and poor root development are a result of a deficiency in
A. sulphur   B. phosphorus
C. calcium   D. iron

38. In the synthesis of protein, the essential substance added to simple sugars is:
A. Magnesium   B. Iron   C. Nitrogen
D. Hydrogen   E. Boron

7.3. ANIMAL NUTRITION
7.3.1. FOOD SUBSTANCES

39. Food is defined as :
A. Substance that contain starch and protein
B. Those substances which when eaten supply fats and oils
C. The material requirements of organisms
D. Substance containing potential energy
E. None of the above

40. UME/79/13
Which of these is not an enzyme?
A. Pepsin   B. Gastrin
C. Amylase   D. Chemotrypsin
E. Trypsin

41. UME/80/8
Human beings require vitamins in their diet because vitamins
A. contain carbohydrates and fats
B. prevent kwashiorkor
C. stimulate the alimentary canal
D. digest proteins in the body
E. influence many important chemical processes in the body

42. UME/81/41
Which of these substances is likely to be deficient in the diet of a person having goiter?
A. potassium   B. calcium   C. iodine
D. sodium   E. phosphorus

43. UME/82/6
Kwashiorkor is caused by severe deficiency of
A. water   B. oil   C. drugs   D. proteins   E. sugars.

44. UME/83/9
If an organic compound has its Hydrogen: Oxygen ratio as 2:1, it is likely to be
A. a protein   B. a carbohydrate
C. a fat   D. a fatty acid and glycerol
E. an amino acid.

45. UME/85/16
If calcium is deficient in food, this may cause
A. anaemia   B. retarded growth
C. sterility   D. goiter E. beri-beri

46. UME/86/16
The main organic substances found in the human body are
A. carbohydrates, proteins and salt
B. salts, fats and proteins
C. fats, carbohydrates and proteins
D. salts, fats and carbohydrate

47. UME/87/18
The vitamin which is important in the formation of the retina pigment is
A. vitamin A   B. vitamin B.
C. vitamin C.   D. vitamin D.

48. UME/95/13
Which vitamin plays an important role in blood clotting?
A. Vitamin A   B. Vitamin K
C. Vitamin B$_{12}$ D. Vitamin C

49. UME/2000/22/Type: M
The greatest amount of energy will be obtained by the oxidation of 100kg of
A. meat   B. butter   C. sugar   D. biscuit

50. UME/2010/14/TYPE: D
Which of the following is lacking in the diet of a person with kwashiorkor?
A. proteins   B. carbohydrates
C. minerals   D. vitamins

51. Animals depend for their existence on carbohydrates, proteins, fats and oils, mineral salts, water and vitamins. Vitamins are:
A. Deficiency diseases caused by insufficient food
B. Some kind of food
C. Substances essential for life
D. Substance present in food and which are responsible for good health and growth
E. Substance necessary for body metabolism.

52. Vitamin A, a fat soluble substance, found in milk, liver oils and green vegetables when absent in food not only cause night blindness, but also:
A. Blindness   B. Loss of blood
C. Weak bones   D. Dry and scaly skin
E. Loss of appetite

53. The anti-sterility vitamin is:
A. Vitamin K   B. Vitamin E   C. Vitamin C
D. Vitamin D   E. None of the above

54. Mineral salts are essential for the maintenance of:
A. Growth of the tissues
B. Healthy growth of the muscles
C. Osmotic pressure of the body fluids
D. Efficient excretory systems
E. Body temperature

55. Milk sugar is:
A. Simple sugar   B. Maltose   C. Sucrose
D. Cane sugar   E. Lactose

56. The glucose of the blood is known as:
A. Simple sugar   B. Liver sugar
C. Glycogen   D. Blood sugar
E. Blood carbohyhrates

7.3.2. FOOD TESTS

57. UME/79/22
A sugar solution was boiled with Fehling's solutions A and B and the colour remained blue. The sugar tested was
A. glucose   B. maltose
C. fructose   D. sucrose
E. lactose.

58. UME/80/18
In testing for glucose the necessary reagent and the condition under which the reagent reacts best are
A. Fehlin's reagent, in the cold
B. Millon's reagent, boiled
C. Fehling's reagent in acid medium, heated
D. Iodine solution, boiled
E. Fehling's reagent in neutral or alkaline medium, boiled.

59. UME/85/15
Food substance which produces red colouration with sudan III contains
A. protein   B. sugar
C. starch   D. cellulose
E. fat
60. UME/86/20
The extract from a food substance reacting with sodium hydroxide and copper sulphate solutions will produce violet to purple coloration if
A. fats are present
B. carbohydrate is present
C. protein is present
D. reducing sugar is present

61. UME/2000/21/Type:M
The production of violet colouration, when dilute NaOH solution is added to a solution of food substance, followed by drops of 1% $CuSO_4$ solution while making indicates the presence of
A. protein   B. carbohydrates
C. fats   D. reducing sugar

62. UME/2001/24/Type:T
When specimen X is mixed with few drops of iodine solution, the appearance of a blue black colour confirms that X is
A. galactose   B. starch
C. sucrose   D. glucose

63. UME/2010//16/TYPE: D
When a mixture of a food substance and Benedict is solution was warmed, the solution changed from blue to brick-red. This indicates the presence of
A. fatty acid   B. sucrose
C. amino acid   D. reducing sugar

7.3.3. THE MAMMALIAN TOOTH

64. Which of the following is the dental formula of a man?
A. i. 2/1, c0/0 pm 3/2, m3/3
B. i. 0/3, c0/1, pm3/3, m3/3
C. i 2/2, c1/1, pm2/2, m3/3
D. i 3/3, c1/1, pm4/4, m2/3
E. i 2/2, c1/1, pm2/3, m3/3

65. UME/81/8
The kind of teeth used for tearing food materials is the A.
fang   B. canine   C. premolar
D. carnassial   E. incisor

66. UME/83/46
Which of the following statements about the dentition of
man is INCORRECT? Man has
A. more molars than incisors
B. no diastema
C. the same number of teeth on upper and lower jaws
D. a total of thirty-two teeth
E. a total of sixteen molars.

67. UME/84/36
The presence of a diastema in the jaw bone indicates that
the mammal lacks the teeth suitable for
A. tearing
B. chewing
C. cutting
D. chewing and tearing
E. cutting and grinding

68. UME/86/22
Evidence that a tooth is a living part of the mammalam
body can be found within the
A. gum
B. pulp cavity
C. cement
D. enamel

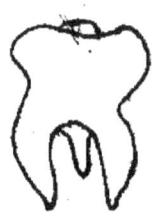 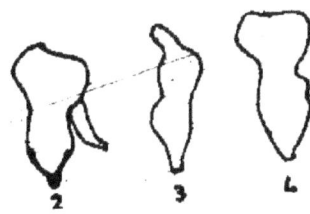

69. UME/89/16
Which of the structures is the molar
A. 1   B. 2   C. 3   D. 4

70. UME/89/17
What is the function of 3?
A. cutting of large pieces of food
B. seizure and tearing of prey
C. grinding of food
D. tearing of flesh only

71. UME/91/22
The crown of the mammalian tooth is covered with
A. cement
B. dentine
C. caries
D. enamel

72. UME/2002/23/TYPE:Y
The carnassial teeth of a carnivorous animal consists of the
A. last upper premolar and the first lower molar
B. last upper molar and the last lower molar
C. first upper premolar and the first lower molar
D. first upper molar and the first lower molar

7.3.4. MAMMALIAN ALIMENTARY CANAL

73. UME/78/40
The mammalian stomach can carry out a number of
processes except
A. absorption of glucose
B. secretion of hydrochloric acid
C. secretion of gastric juice
D. churning of food
E. production of trypsin

74. The first part of the stomach to receive food from the
oesophagus is:
A. The duodenum
B. The jejunum
C. The pyloric zone
D. The ileum
E. The cardiac zone

75. UME/79/4
Which of these is the terminal portion of the alimentary
canal of a mammal?
A. Oesophagus
B. Stomach
C. Rectum
D. Colon
E. Appendix

76. UME/81/15
Which of the following is NOT a function of the liver? A.
Regulation of blood sugar
B. Storage of iron
C. Formation of bile
D. Breakdown of excess amino acids
E. Excretion of urea from the blood

77. UME/81/49
The correct route for a piece of bread eaten by a mammal is
A. Oral cavity→ oesophagus→ stomach → duodenum
          → ileum→ rectum
B. Oral cavity→ stomach → oesophagus → duodenum
          → ileum → rectum
C. Oral cavity→ oesophagus→ deodonum → stomach→ ileum
          → rectum
D. Oral cavity→ oesophagus→ ileum→ stomach→ duodenum
          → rectum
E. Oral cavity→ oesophagus→ stomach→ ileum → duodenum
          → rectum

78. That portion of the alimentary system into which the pancreatic juice is poured is:
A. The duodenum   B. The pancreatic zone
C. The small intestine   D. The pyloric zone
E. The colon

79. UME/82/39
If the gall bladder of a man is removed by surgery, which of the following process will be most seriously affected?
A. digestion of fats and oils
B. formation of urea
C. digestion of starch
D. conversion of proteins
E. storage or release of urine.

80. UME/86/21
The three important organs that are situated close to the stomach are
A. kidney, liver and gall bladder
B. pancreas, liver and kidney
C. liver, kidney and spleen
D. gall bladder, pancreas and spleen

81. UME/87/19
Which of the following lists of organs is directly involved in nutrition?
A. oesophagus, bronchus, stomach, pancreas and anus
B. spleen, pharynx, duodenum, jejunum, and rectum
C. teeth, oesophagus, ileum, lungs and large intestine
D. salivary gland, liver, stomach, villi and colon

82. UME/91/20
One of the accessory organs of the digestive system is the
A. kidney   B. spleen
C. liver   D. lung

83. UME/97/12
The part of the stomach nearer the gullet is called the
A. epiglottis   B. cardiac sphincter
C. duodenum   D. pyloric sphincter

84. UME/80/15
Which of the following food substances is digested in the stomach?
A. carbohydrates
B. fats and oils
C. fats and proteins
D. proteins
E. carbohydrates and fats

85. UME/2006/14
Which of the following produces both hormones and enzymes?
A. ileum   B. pancreas
C. gall bladder   D. kidney

86. UME/2008/14
The order of passage of food in the digestive system is A. ileum → caecum → large intestine → rectum
B. ileum → colon → caecum → rectum
C. large intestine → ileum → caecum → rectum
D. colon → caecum → ileum → rectum

87. UME/2010/26/TYPE: D
The part of the human gut that has an acidic content is the
A. duodenum   B. ileum
C. colon   D.   stomach

88. That flap of tissue that prevents food from straying into the larynx is:
A. The glottis   B. The arytenoids
C. The cricoid   D. The epiglottis
E. The palate

7.3.5. NUTRITIONAL PROCESS
89. UME/79/34
The digestive enzyme that coagulates proteins in milk is
A. ptyalin   B. pepsin
C. rennin   D. trypsin
E. amylase

90. UME/79/50
The arrangements below are steps in protein digestion. Which is the correct sequence?
a-polypeptides, b-protein,
c-amino acids, d-peptones
A. a→b→c→d
B. c→d→a→b
C. b→c→a→d
D. b→d→a→c
E. b→a→c→d

91. UME/82/22
In the enzymic reactions: starch à sugar. Starch is referred to as the
A. substrate   B. product
C. enzyme   D. enzyme – substrate complex
E. reaction mixture.

92. UME/84/13
Which of the following statement is NOT true of enzymes? They
A. are proteins
B. need co-factors to activate them
C. are sensitive to hydrogen ion concentration
D. are specific in their action
E. can withstand high temperature.
93. UME/85/17
Partially digested food ready to leave the stomach is referred to as
A. chyme   B. curd   C. glycogen
D. paste   E. roughage

94. UME/88/20
Which of the following will be digested first, if ingested at the same time?
A. cooked beans   B. cooked rice
C. cod liver oil   D. roasted beef

95. UME/89/19
The pancreatic juice contains the enzymes amylopsin,
A. pepsin and trypsinogen   B. rennin and steapsin
C. steapsin and trypsinogen   D. steapsin and ptyalin.

96. UME/90/20
The organ which secretes digestive enzymes as well as hormone is the
A. liver   B. salivary glands
C. pancreas   D. spleen.

97. UME/97/21
The process of deamination is essential for the
A. digestion of protein   B. secretion of bile
C. formation of urea   D. formation of antibody

98. UME/2007/26
The pancreas secretes enzymes for the digestion of
A. fats, proteins and carbohydrate
B. fats, vitamins and cellulose
C. fats, carbohydrates and vitamins
D. proteins, cellulose and minerals

MISCELLANEOUS QUESTIONS (7)
99. The mode of feeding in Amoeba and Hydra is
A. heterotrophic   B. holophytic
C. autotrophic   D. symbiotic
(UME/2001/8/Type 7)

100. The gall bladder of a mammal has a duct connected to the
A. duodenum   B. liver
C. pancreas   D. small intestine
(UME/2004/1/type: 4)

101. The part of the mammalian digestive system where absorption of materials takes place is the
A. duodenum   B. colon
C. ileum   D. oesophagus (UME/2003/19/type: A)

102. Metabolic production of urea is carried out in the
A. urinary bladder and kidney   B. pancreas
C. kidney and malphigian tubule
D. liver (UME/2003/22/Type:A)

103. The element common to protein, carbohydrate and lipid is
A. Hydrogen   B. Sulphur
C. Nitrogen   D. Phosphorus (UME/91/21)

104. The crown of the mammalian tooth is covered with
A. cement   B. dentine
C. caries   D. enamel (UME/91/22)

105. The blood vessel which carries blood from the alimentary canal to the liver is the
A. hepatic artery   B. hepatic vein
C. hepatic portal vein   D. mensenteric artery
(UME/92/15)

106. Which enzymes are contained in the pancreatic juice?
A. ptyalin, lipase and pepsin
B. maltase, erepsin and trypsin
C. renin, sucrase and lipase
D. amylase, lipase and trypsin (UME/92/22)

107. Interveinal chlorosis is normally associated with the deficiency of
A. magnesium   B. potassium
C. iron   D. calcium (UME/93/14)

108. Osmic acid boiled with a solution of food substances gave a black precipitate. This indicates the presence of
A. fats and oils   B. proteins
C. aminoacids   D. starch (UME/93/15)

109. Which of the following is an autotrophic mode of nutrition?
A. saprophytism   B. chemosynthesis
C. parasitism   D. commensalisms
(PCE/2005/7/type:P)

110. In millon test, when the reagent is added to a protein food item, a white precipitate is produced which turns
A. blue on heating   B yellow on heating
C. green on heating   D. red on heating (UME/98/14)

111. The modes of nutrition in Nitrobacter, *Ascaris* and *Homo sapiens*, respectively are
A. photosynthetic, parasitic and holozoic
B. chemosynthetic, parasitic and holozoic
C. photosynthetic, parasitic and heterotrophic
D. chemosynthetic, holophytic and holozoic
(UME/99/14/Type: D)

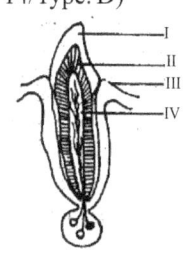

112. The section that is extremely sensitive to heat and cold is labeled
A. IV   B. III   C. I   D. II (PCE/2005/16/Type:P)

113. The part labeled II is the
A. dentine   B. cement   C. enamel   D. gum
(PCE/2005/17/Type:P)

114. Autotrophic nutrition involves
A. only unicellular animals
B. the use of complex organic substances as starting materials
C. the production of energy rich organic compounds
D. deactivation of green pigments
(PCE/91/13)

115. A cleaning agent containing lipase will be most effective in removing stains caused by
A. proteins   B. fats
C. mud   D. pollen grains (PCE/92/17)

116. The sulphur bacteria which use the energy from acidation of substances to fix carbon are called
A. holozoic heterotrophs
B. chemosynthetic autotrophs
C. parasite heterotrophs
D. photosynthetic autotrophs (PCE/92/15)

117. The dark reaction of photosynthetis involve the
A. reduction of carbon dioxide
B. absorption of water and carbon dioxide
C. photolysis of water
D. conversion of starch into sugar
(PCE/92/13)

118. Sterility in man may be caused by a deficiency of
A. vitamin A   B. vitamin C
C. vitamin D   D. vitamin E (PCE/93/16)

119. Osteomalacia is a symptom of the deficiency of
A. vitamin A   B. vitamin C
C. vitamin D   D. vitamin K (PCE/96/14)

120. The disease that results from eating mainly polished rice is
A. cancer   B. cholera
C. beri-beri   D. dysentery (PCE/98/12/TYPE: C)

121. Deamination in the liver involves the
A. storage of iron from the haemoglobin
B. conversion of poisonous substances to harmless ones
C. conversion of excess amino acids to glucose and urea
D. storage of excess fat and sugar (PCE/06/TYPE: A)

CHAPTER 8

TRANSPORT

Understand the following after reading this chapter:

> Need for transportation
> Mammalian circulation system
> Plant vascular system
> Media, process and mechanism of transport

NEED FOR TRANSPORTATION
Surface Area: Volume Ratio
Ratio of SA: Vol. ratio in a small organism is greater than in a large organism

Smaller organisms have the advantage of a relatively large absorbing surface

Larger organisms have a low SA: Vol. ratio, their skin not very permeable to gases

Thus large organisms need special systems (e.g. circulatory systems) to help them transport the substances and special absorbing substances such as lungs or gills to increase the surface area for absorption

The ratio decreases with increasing size of the organism

The Mammalian Circulatory System
The circulatory system is made up of blood, blood vessels, lymph and the heart.

Composition of Blood
Plasma: 1. Approximately 55% of blood is made up of plasma, the straw-coloured liquid portion of blood; it is 90% water and 10% dissolved molecules (mainly plasma proteins).

2. These can be divided into three types:
a) Albumins - these help to regulate water potential, by maintaining normal blood volume and pressure. They are the most common plasma protein.
b) Immunoglobins (antibodies) – These are very large proteins that target infection and so cause infected or foreign cells to be attacked by white blood cells (WBC's). Together with the WBC's they form the immune system.
c) Fibrinogen – these are tightly coiled proteins that unwind to form a blood clot.

Other molecules include: glucose, amino acids, vitamins, minerals (mainly NaCl), urea, CO_2, hormones.

BLOOD CELLS
Red Blood Cells
Tiny (8μ) biconcave disc-shaped cells (thus **large SA).**

Do not have nucleus, mitochondria, ribosomes.

Cell full of haemoglobin – binds O_2 (and CO).

Made in the bone marrow – live about 120 days. Destroyed and recycled by the liver.

White Blood Cells (leucocytes)
These are colourless cells and possess a nucleus. They function in defending the body against pathogens.

Phagocytes -**'granulocytes'** 'feed' on pathogens by phagocytosis.

Monocytes are one form of phagocytes.

Lymphocytes – **'agranulocytes'** - produce antibodies, the specific defence proteins.

Made in bone marrow and lymphatic tissue.

Platelets
Responsible for **clotting of the blood**

Responsible for repair of damaged tissue – releasing the hormone **platelet growth factor**.

Short life – under 7 days.

Made in bone marrow

Functions of Blood System
Transport: to and from tissue cells

Nutrients to cells: amino acids, glucose, vitamins, minerals, lipids (as lipoproteins).

Oxygen: by red blood corpuscles (as *oxyhaemoglobin)*

Wastes from cells: urea, CO_2

Temperature Regulation: by altering the blood flow through the skin.

Immunity: protection against pathogens — blood clotting; phagocytes, lymphocytes and antibodies distributed in blood.

Communication: hormones distributed to all parts of the body in the blood.

Defence: clotting following a wound

BLOOD VESSELS
These are channels or tubes which convey blood to all part of the body. They include the arteries, veins and capillaries.

Arteries – carry oxygen rich blood (except pulmonary artery) away from the heart at a high pressure. They have thick walls and no valves.

Veins – carry blood without oxygen (except pulmonary vein) towards the heart at a low pressure. They have thinner walls and valves.

Capillaries

Capillaries are the link between arteries and veins – where exchange with tissues occurs.

The capillary wall is one cell thick and somewhat porous — ideal to allow materials to pass in and out.

All tissue cells very close to a capillary so exchange is very efficient.

Exchange at the capillaries is by diffusion, mass flow and active transport.

Blood flow in capillaries is slow giving enough time for effective exchange.

One type of cell, thus a **tissue.**

Differences Between Artery and Vain

Artery	Vein
I. It has thick/muscular wall	It has thin/less muscular wall.
ii. It has elastic wall	Its wall is non-elastic
iii. It carries blood away from the heart	It carries blood to the heart
iv. It carries oxygenated blood, except the pulmonary artery	It carries deoxygenated blood, blood except the pulmonary vein
v. Blood in it is pink or bright red in colour	Blood in it is dark-re in colour
vi. it is situated deep in the muscles	it is superficially located or situated
vii. it has small lumen	it has large lumen
viii. pressure is high	pressure is low
ix. pulse is readily detectable	pulse is not readily detectable
x. it has no valve except semiluma valves	it has valves

Closed System of Blood Vessels

The blood does not make direct contact with the tissue cells.

The blood is retained in the blood vessels.

A closed system is very responsive to the change needs of the organs and is highly efficient.

Double Circulation

The double circuit prevents mixing of oxygenated and deoxygenated blood. Therefore oxygen supply is highly efficient.

Pulmonary Circulation: deoxygenated blood flows from the heart to the lungs

oxygen is taken on and carbon dioxide is excreted, oxygenated blood flows from the lungs back to the heart.

Systemic Circulation: oxygenated blood flows from the heart to the organ systems of the body.

oxygen is delivered and carbon dioxide is taken on, deoxygenated blood flow from the organs systems back to the heart.

Portal System

A portal blood vessel has a set of capillaries at each end.

The **hepatic portal vein** carries blood rich in absorbed nutrients from the capillaries in the gut to capillaries in the liver.

Hepatic artery supplies liver with oxygen

Hepatic vein takes **all the blood away** from the liver, back to heart.

The Heart

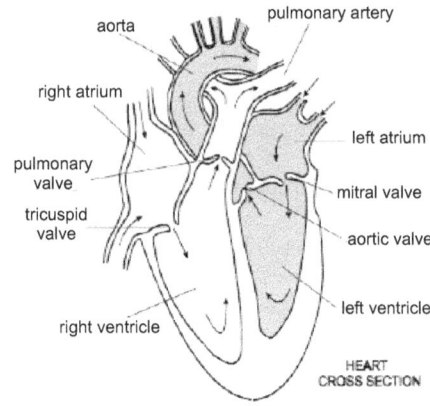

Internal structure of the heart

The heart is a muscular pumping organ situated between lungs in the thoracic cavity. It is made up of specialized muscles called cardiac muscles and protected by the pericardium. The heart consists of four chambers. The two upper chambers are called auricles while the two lower chambers are known as ventricles and are seperated by septum into left and right. The left ventricle has a thicker wall in order to generate enough pressure on contraction needed to pump blood through the aorta. Each pumping action of the heart is known as a heart-beat.

Summary of circulation in the heart

Superior vena cava→right auricle→ through tricuspid valve →right ventricle→ pulmonary artery → lungs→ pulmonary vein→left auricle→ through bicuspid valve→left ventricle aorta→ all parts of the body.

Action of heart Two phases

Diastole: auricles contract and ventricles relax. The heart fills with blood under low pressure from the veins.

Systole: auricles relax and ventricles contract. The chambers of the heart are emptying of blood.

THE LYMPHATIC SYSTEM

A collection of special drainage vessels receiving excess tissue fluid.

Once the tissue fluid enters the lymphatic capillaries it

is called **lymph**.

Lymph nodes (e.g. tonsils) filter the lymph and produce lymphocytes.

The lymph vessels have many valves, but low pressure.

The lymph is moved along by the squeezing action of:

the skeletal muscles,

pressure changes in the thorax during breathing and

by the rhythmic contraction of the lymph vessel walls.

Lymph re-enters the blood just before the right atrium.

Functions of the Lymphatic System:

Circulatory role

Return the excess tissue fluid to the blood: this maintains blood volume, pressure and concentration.

Collect and deliver the absorbed lipids from the small intestine to the blood

Defence role

The lymph nodes filter out pathogens in the lymph. Production and 'export' of lymphocytes to the blood system for general distribution. Detection of antigens and production of specific antibodies.

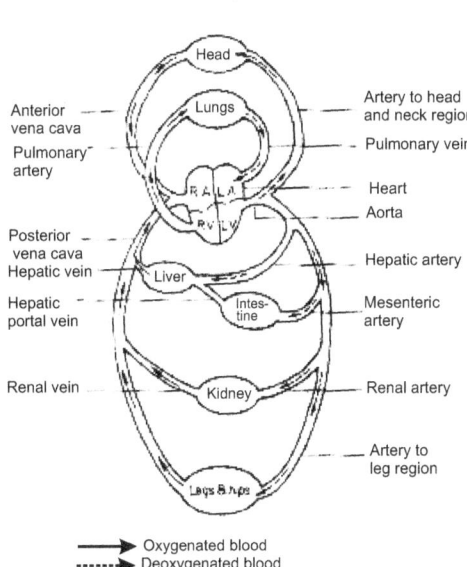

Diagram of blood circulation in mammals

Transport Systems in Plants

The plants have two transport systems:

1. Xylem which carries water and minerals, and

2. Phloem which carries the food materials which the plant makes (Phloem also carries the hormones made by the plants in their root and shoot tips).

The transport of materials in a plant can be divided into two parts:

(i) Transport of water and minerals in the plant, and

(ii) Transport of food and other substances (like hormones) in the plant.

Transport of Water and Minerals:
The root hair cell:

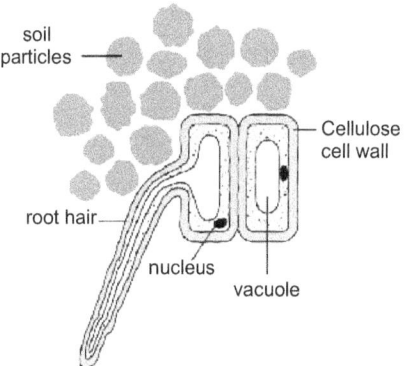

Path of water from root hair to xylem vessels

Function: to absorb water and minerals from the soil. They have an elongated shape for more surface area. The path of water in plants is as follows:
1. Water enters the root hair cell for the moist soil because the water potential is higher in the soil, than in the cytoplasm. 2. Water passes through the cortex cells by osmosis but mostly by "suction". 3. Water (and dissolved substances) is forced to cross the endodermis. 4. Water enters the xylem then leaves when it gets to the mesophyll cells. The uptake of water is caused by water loss in leaves through the stoma lowering the water potential in leaves, then water moves from xylem to enter leaf tissues down water potential gradient, then water moves up the stem in the xylem due to the tension (because of cohesion / sticking of water molecules to each other) caused by water loss from the leaves, and ends with the gain through roots. The upward flow of water is called the transpiration stream. The minerals present in soil dissolve in water to form an aqueous solution. So, when water is transported by the root of the plant to its leaves, then the minerals dissolved in it also get transported along with it.

Transport of Solutes:

The phloem contains a very concentrated solution of dissolved solutes, mainly sucrose, but also other sugars, amino acids, and other metabolites. This solution is called the <u>sap</u>, and the transport of solutes in the phloem is called <u>translocation</u>.

Unlike the water in the xylem, the contents of the phloem can move both up or down a plant stem, often simultaneously. It helps to identify where the sugar is being transported from (<u>the source</u>), and where to (<u>the sink</u>).

- During the summer sugar is mostly transported from the leaves, where it is made by photosynthesis (the source) to the roots, where it is stored (the sink).

- During the spring, sugar is often transported from the underground root store (the source) to the growing leaf buds (the sink).

- Flowers and young buds are not photosynthetic, so sugars can also be transported from leaves or roots (the source) to flowers or buds (sinks).

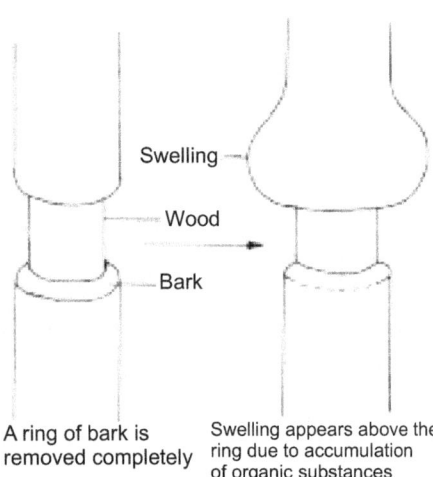

Swelling

Wood

Bark

A ring of bark is removed completely

Swelling appears above the ring due to accumulation of organic substances which moved down the phloem from the leaves

Wood and bark

Wood and some bark

No swelling appears above the ring since food can still pass through the phloem from the leaves to the roots

To show that the phloem tissue transport food subtances downwards from the leaves

Factors affecting Transpiration

The rate of transpiration can be measured in the lab using a <u>potometer</u> ("drinking meter"):

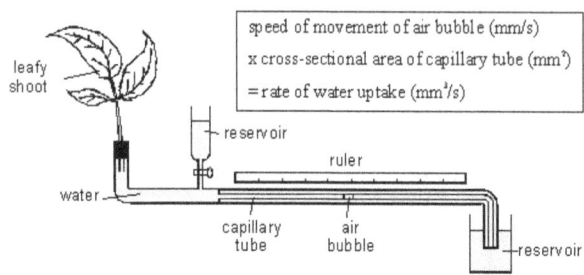

speed of movement of air bubble (mm/s)
x cross-sectional area of capillary tube (mm²)
= rate of water uptake (mm³/s)

leafy shoot

reservoir

ruler

water

capillary tube

air bubble

reservoir

A potometer actually measures the rate of water uptake by the cut stem, not the rate of transpiration; and these two are not always the same. During the day plants often transpire more water than they take up (i.e. they lose water and may wilt), and during the night plants may take up more water than they transpire (i.e. they store water and become turgid). The difference can be important for a large tree, but for a small shoot in a potometer the difference is usually trivial and can be ignored.

The potometer can be used to investigate how various environmental factors affect the rate of transpiration.

- <u>Light</u>. Light stimulates the stomata to open allowing gas exchange for photosynthesis, and as a side effect this also increases transpiration. This is a problem for some plants as they may lose water during the day and wilt.

- <u>Temperature</u>. High temperature increases the rate of evaporation of water from the spongy cells, and reduces air humidity, so transpiration increases.

- <u>Humidity</u>. High humidity means a higher water potential in the air, so a lower water potential gradient between the leaf and the air, so less evaporation.

- <u>Air movements</u>. Wind blows away saturated air from around stomata, replacing it with drier air, so increasing the water potential gradient and increasing transpiration.

Many plants are able to control their stomata, and if they are losing too much water and their cells are wilting, they can close their stomata, reducing transpiration and water loss. So long periods of light, heat, or dry air could result in a decrease in transpiration when the stomata close.

Movement across Cell Membranes

Cell membranes are a barrier to most substances, and this property allows materials to be concentrated inside cells, excluded from cells, or simply separated from the outside environment. Obviously materials need to be able to enter and leave cells, and there are five main methods by which substances can move across a cell membrane: Diffusion, Osmosis, Passive Transport, Active Transport; and Vesicles.

Diffusion (or Simple Diffusion)

Diffusion is the movement of molecules of a substance from a region of higher concentration to a region of lower concentration till equilibrium is reached.

For example, the movement of oxygen from lungs to bloodstream and absorption of nutrients from small intestine into bloodstream etc. When there is no net movement, equilibrium is achieved.

Factors affecting the rate of diffusion:

1. Amount of substances on either side (i.e. the concentration)
2. Permeability of membrane
3. Temperature (movement of particles, increased temperature usually increases the speed of the particles, according to the kinetic theory of collision)
4. Distance to be travelled
5. Size of the particles

Biological significance of diffusion

1. It is responsible for the exchange of gases between living organisms and their environment.
2. It is responsible for the exchange of materials between the mother and foetus through the placenta.
3. It is responsible for movement of materials from cell to cell along the phloem during translocation.
4. It is responsible for even distribution of substances admitted into the body.

Osmosis

Osmosis is a process which involves the movement of water molecules or any solvent across a semi-permeable membrane from a weaker to a stronger solution. A semi-permeable membrane is a membrane which allows certain solvents to pass through while disallowing others. E.g. cellophane paper, fish bladder egg membrane and ectoplasm of plant cells. The contents of cells are essentially solutions of numerous different solutes, and the more concentrated the solution, the more solute molecules there are in a given volume, so the fewer water molecules there are. Water molecules can diffuse freely across a membrane, but always down their concentration gradient, so water therefore diffuses <u>from a dilute to a concentrated solution</u>.

Osmotic Pressure (OP). This is an older term used to describe osmosis. The more concentrated a solution, the higher the osmotic pressure. It therefore means the opposite to water potential, and so water move from a low to a high OP. Always use ? rather than OP.

Cells and Osmosis. The concentration (or OP) of the solution that surrounds a cell will affect the state of the cell, due to osmosis. There are three possible concentrations of solution to consider:

- <u>Isotonic</u> solution   a solution of equal OP (or concentration) to a cell
- <u>Hypertonic</u> solution   a solution of higher OP (or concentration) than a cell
- <u>Hypotonic</u> solution   a solution of lower OP (or concentration) than a cell

The effects of these solutions on cells are shown in this diagram:

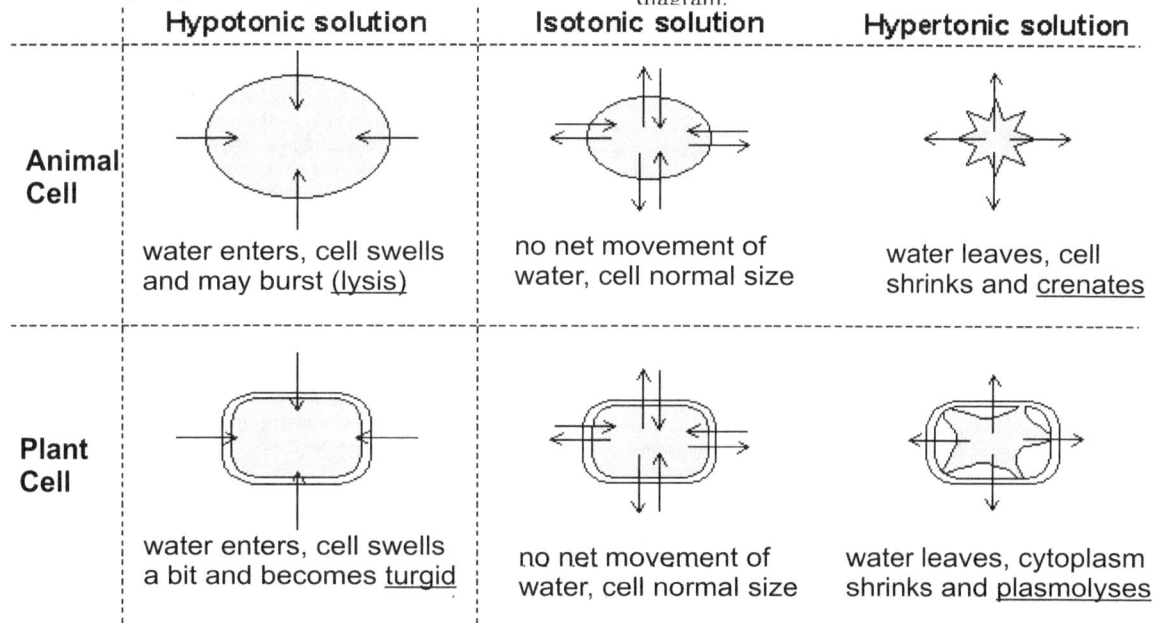

73

These are problems that living cells face all the time. For example:

- Simple animal cells (protozoans) in fresh water habitats are surrounded by a hypotonic solution and constantly need to expel water using <u>contractile vacuoles</u> to prevent swelling and lysis.
- Cells in marine environments are surrounded by a hypertonic solution, and must actively pump ions into their cells to reduce their water potential and so reduce water loss by osmosis.
- Young non-woody plants rely on cell turgor for their support, and without enough water they wilt. Plants take up water through their root hair cells by osmosis, and must actively pump ions into their cells to keep them hypertonic compared to the soil. This is particularly difficult for plants rooted in salt water.

Biological Significance of Osmosis
1. It is responsible for the absorption of water from the soil by the root hairs
2. It is responsible for the movement of water from one cell to another
3. It is responsible for the turgidity of cells
4. It is responsible for the re-absorption of water in the kidney tubules.

Passive Transport (or Facilitated Diffusion).
It is the movement of particles across a membrane with the help of a molecule in the membrane (globular protein in bilayer of cell membrane).
The energy comes from the kinetic energy of the molecules involved.
For example, the movement of ADP into mitochrondria

Active Transport (or Pumping).
It is the process by which molecules of a substance move from areas of lower concentration of that substance to areas of higher concentration
Against a concentration gradient
For example: Absorption of mineral salts by the root hair cells, absorption of glucose and amino acids by cells in the small intestine of humans

Bulk Movement via vesicles
Entocytosis:
- Substances are being imported into cells via vesicles
- Phagocytosis = cell eating
- Pinocytosis = bulk transportation of liquid (when taking in of nutrients)

Exocytosis:
- Substances being exported out of cells

Summary of Membrane Transport

Method	Uses energy	Uses proteins	Specific	Controllable
Diffusion	N	N	N	N
Osmosis	N	N	Y	N
Passive Transport	N	Y	Y	Y
Active Transport	Y	Y	Y	Y
Vesicles	Y	N	Y	Y

EXPERIMENTS ON TRANSPORT

Experiment :1 To demonstrate that leaves give out water vapour (by using cobalt chloride paper).

Materials required. A young potted plant, cobalt chloride solution, glass slides, filter papers.

Method.two filter papers are dipped in a solution of cobalt chloride. As these papers become soaked, they gradually turn blue. They are then dried and put aside for the experiment. A leaf, preferably the one at the lower part of the stem of a potted plant is selected. The two cobalt chloride papers are placed on the leaf, one on the papers and rubber binds are used to tie the glass slides together. By placing a glass on the leaf, vapour from outside is cut off. After this, the potted plant is placed near a window to receive sunlight. Both surfaces of the leaf are then checked from time to time.

Observations. The cobalt chloride paper on each surface of the leaf turns pink.

Conclusion. Anhydrous cobalt chloride paper is blue when dry but turns pink when wet. Since the paper turns pink, it means that the vapour from the leaf must have made them so. Therefore, leaves give off water vapour.

NOTE: the degree of pinkness on both leaf surfaces is either same or different. This is due to the distribution of stomata on the different surfaces.

Experiment :2 To demonstrate that leaves give out water vapour (with a potted plant).

Materials required. A young potted plant, polythene bag, a bell jar, glass plate, Vaseline, anhydrous copper (II) tetraoxosulphate (VI).

Method. A young potted plant is collected and a polythene is used to cover the pot and the base of the stem of the plant as shown in Fig. 40.9. A bell jar is placed over the pot. The bell jar is stood on a glass plate whose surface has been rubbed with Vaseline. A second experiment with no plant is set up to serve as a control. Both experiments are then exposed to sunlight, for some hours. If there are drops of liquid on the iner side of the bell jar, the liquid is tested with an anhydrous copper (II) tetraoxosulphate (VI).

Observations. Drops of colourless liquid are noticed on the inner side of the bell jar with a plant. No liquid in the control. When the liquid is tested with an anhydrous copper(II) tetraoxosulphate (VI), it is noticed that the white anhydrous copper (II) tetraoxosulphate (VI) turned blue.

Conclusion. The white anhydrous copper (II) tetraoxosulphate (VI) turning blue shows that the colourless liquid is water.

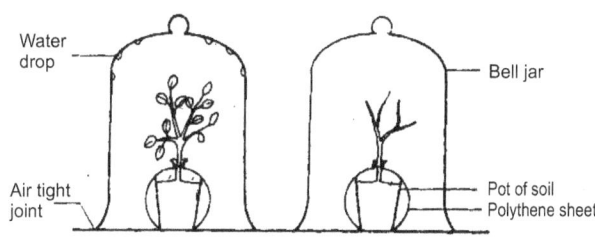

Experiment :3 To demonstrate the loss in weight of a potted plant.

Materials required. A young potted plant, polythene bag, weighing instrument, note book.

Method. A young potted plant is obtained and its soil is properly moistened. A polythene bag is used to cover the pot and part of the stem as shown in Fig. 40.9. After this, the set-up is weighed and the initial weight is recorded. Next, the potted plant is placed close to a window. More weighing is made and recorded every 30 minutes for a day. The weights are then carefully studied.

Observations. There is a loss in weight of the potted plant.
Conclusion. The loss in weight is due to water vapour given off from the plant. The water vapour must have come from the parts of the plant which were not covered. This shows that there is loss in weight of a potted plant.

Experiment :4 To demonstrate the rate of water loss during transpiration.

Potometer is the name of the apparatus which is used for this experiment. The apparatus is of various designs. One type is the one shown in Fig. 40.10. the apparatus is used mainly to measure the rate at which water is absorbed by a plant. However, this rate approximately equals the rate at which water is lost during transpiration. Because of this, one can conveniently, by means of a photometer determine the rate of transpiration in a plant shoot.

A healthy plant shoot is cut under water. It is fitted into a rubber stopper and the rubber stopper is inserted in **C.** The apparatus is filled with water by turning the tap of the funnel. After this, the setup is left for some time for the plant to absorb water. Air will enter the apparatus at **A,** once absorption of water begins. The air is noticed as a bubble in the graduated tube. The bubble moves along and the quantity of water absorbed is then measured. When the bubble gets to somewhere close to **B,** the tap of the funnel is then turned so that water can again fill the graduated tube. In this way the experiment can be repeated many times. With this apparatus, the rates of transpiration in different external conditions, e.g. in a dark room, in laboratory, out of door in the field, etc. are compared.

Experiment :5(a) To demonstrate osmosis with a non-living membrane.

Materials required. A cellophane or parchment paper or a pig's bladder, thistle funnel, sugar solution, a beaker.

Method. A semi-permeable membrane such as cellophane or parchment paper or a pig's bladder is tied around the mouth of a thistle funnel. Some sugar solution is poured into the bulb of the funnel. The level of the sugar solution is then noted. A beaker is collected and water is poured into it until it is about half way filled. Next, the thistle funnel is immersed in the water inside the beaker as shown in Fig. 40.13A. A control experiment is also prepared but the bulb of the funnel contains distilled water instead of sugar solution. Both experiments are then left for some hours.
Observations. There is rise in liquid up the stem of the thistle funnel of the first experiment. No rise in liquid in the control experiment.

Conclusion. The rise in the liquid up the stem must have been due to water molecules which moved from the beaker into the thistle funnel through the non-living membrane. This shows that osmosis has taken place.

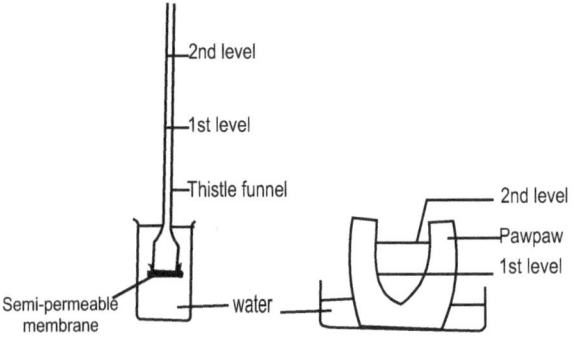

2nd level
1st level
Thistle funnel
2nd level
Pawpaw
1st level
Semi-permeable membrane
water

Experiment :5(b) To demonstrate osmosis with a living membrane.

Materials required. A yam or pawpaw, sugar solution, a trough of pure water.

Method. A yam tuber or pawpaw is taken. A lump is cut out of the tuber. The lump is cut in such a way that there is a base for it. Part of the base is peeled, say about an inch down. A cavity or hollow is made in the yam. The cavity is made deep down to the base. Next, the tissues are washed with tap water to remove any debris. Some sugar solution is added into the cavity which has been made out of the yam or pawpaw. The yam or pawpaw is then stood in a petri dish of pure water. A control experiment is also prepared but the cavity contains distilled water instead of sugar solution. Both experiments are left for some hours.

Observations. There is a rise in liquid up the cavity in the first experiment. No rise in the control experiment.

Conclusion. The same conclusion as with a non-living membrane.

Experiment :6 To demonstrate plasmolysis.

Materials required. Filaments of *Spirogyra* plant, microscope, plain slides, 30% of sodium chloride solution.

Method. A few cells of *Spirogyra* plant are placed on a plain slide. They are examined under a compound microscope. The structure of the cells are then noted. A few drops of 30% sodium chloride solution are added to the cells and they are further examined.

Observations. The protoplasm shrinks and separate from the cell wall. A tiny space appears between the protoplasm and the cell wall. Further shrinkage occurs and the protoplasm finally rounds up.

Conclusion. The changes as observed above are due to fact that the surrounding medium has a higher osmotic concentration than that of the cell. Because of this, water is lost to the surrounding medium from the cell. This shows that plasmolysis has taken place.

8.1. NEED FOR TRANSPORT
1. UME/98/16
Unicellular organisms transport essential nutrients directly to all parts of their bodies by the process of diffusion because, they have
A. a large volume of surface area ratio
B. a large surface area to volume ratio
C. their bodies immersed in the nutrients
D. their outer memberane made of cellulose.

2. UME/2003/16/Type A.
Organism I, II, III and IV have survace area/volume ratios of 1:2, 1:3, 1:4 and 1:5 respectively. The organism that is likely to have the most complex transport system is
A. III   B. IV   C. II   D. I

8.2. MAMMALIAN CIRCULATORY SYSTEM
(heart, arteries, veins, and capillaries)

3. UME/83/36
Which of the following statements is the mammalian circulatory system is NOT true?
A. Blood in the pulmonary artery is richer in oxygen content than blood in the pulmonary vein
B. The blood in the hepatic portal vein in the richest in food substances
C. Blood flow is controlled by valves in veins
D. Arteries are generally thicker and larger than veins
E. Fibrin helps in the formation of blood clot

4.UME/84/37
Normally the flow of blood is never from
A. artery to arterioles
B. arterioles to capillaries
C. capillaries to venules
D. arterioles to the arteries

5. UME/86/23
Blood circulation in a mammal is said to be double because
A. it passes twice through the heart in a complete cycle
B. it moves in both arteries and veins
C. it circulates in both the heart and other organs
D. the heart contains auricles and ventricles

6. UME/99/22
The lymphatic system of mammals rejoins the blood circulatory system at the
A. hepatic vein   B. subclavian vein
C. renal vein   D. common iliac vein

7. UME/83/50
The path taken by glucose from the ileum to the heart is
A. ileum→ hepatic portal vein→hepatic artery→ vena cava →heart
B. ileum→ hepatic portal artery→ hepatic artery→vena cava →heart
C. Ileum→ hepatic portal vein→ vena cava→ heart
D. Ileum→hepatic vein→ vena cava→ heart
E. Ileum→ hepatic portal vein → hepatic vein→ vena cava →heart

8. UME/78/43
A carbohydrate molecule in the right ventricle of the heart is pumped into the cell of the toe of man. Which of these structures is it unlikely to pass through en-route?
A. pulmonary artery
B. lungs   C. Heart
D. Liver   E. Aorta

9. UME/81/46
The hepatic portal vein is unique because it
A. carries deoxygenated blood
B. begins and ends with capillaries
C. is the longest vein in mammals
D. carries digested food
E. is the shortest vein in mammals.

10. UME/91/28
Deoxygenated blood flows into the right and left lungs through the
A. pulmonary vein   B. vena cava
C. pulmonary artery   D. subclavian vein

11.   UME/85/19
The vein which returns blood from the head and arms to the heart is called
A. aorta   B. inferior vena cava
C. superior vena cava   D. pulmonary vein
E. pulmonary artery.

12. UME/98/17/Type :C
The mammalian heart chamber that pumps blood through the longest distance is the
A. right auricle   B. right ventricle
C. left ventricle   D. left auricle

13. UME/2006/16
The mammalian vein which starts with and ends in a capillary network is the
A. pulmonary vein
B. mesenteric vein
C. renal vein
D. hepatic portal vein

14. UME/2009/19
Which of the following describes the sequence of blood flow from the heart to a tissue?
A. heart→artery→ arteriole→ tissue
B. heart→ vein→ venule→ tissue
C. heart→ venule → vein→ tissue
D. heart→ arteriole→ artery→ tissue

15. UME/2010/17/TYPE: D
The primary structure responsible for pumping blood for circulation through the mammalian circulatory systems is the
A. right auricle   B. arteries
C. left ventricle   D. veins

16. UME/2010/18/TYPE: D
Circulation of blood to all parts of the body except the lungs is through
A. systemic circulation
B. the lymphatic system
C. pulmonary circulation
D. the pulmonary artery

17. UME/2010/31/TYPE: D
The blood component that has the greatest affinity for oxygen is the
A. leucocytes
B. erythrocytes
C. thrombocytes
D. lymphocytes

8.3. PLANT VASCULAR SYSTEM
18. PCE/91/18
The path of translocation of mineral ions in the plant is through the
A. cortex   B. phloem
C. pith   D. xylem

19. UME/99/20/Type D
Substances manufactured by the leaves are transported to other parts of the plant through the
A. xylem
B. companion cells
C. sieve tubes
D. cambium.

20. UME/2001/27/Type: T
Salts and water are absorbed in the roots and transported to the leaves by
A. diffusion through the xylem tissues
B. osmosis through the phloem tissues
C. diffusion through the phloem tissues
D. osmosis through the xylem tissues.

21. UME/2003/14/Type: A
If water that has been coloured red is poured at the base of wilting plant, it will appear as a red stain in the cells of the
A. phloem
B. parenchyma
C. xylem
D. epidermis

8.4. MEDIA, PROCESS AND THE MECHANISM OF TRANSPORT

22. UME/83/11
Which of the following statements is NOT true of mammalian erythrocytes?
A. They have haemoglobin
B. They appear yellow when looked at singly
C. They are disc-shaped
D. The cells are more numerous than leucocytes
E. They have nuclei at maturity.

23. UME/84/40
Which of the following sequences represents the process of blood clotting?
1. Fibrin forms a network of threads
2. Red blood cells are caught and a clot is formed
3. Fibrinogen in plasma changes into soluble fibrin
4. Blood is exposed to air

A. 4, 3, 2, 1
B. 4, 3, 1, 2
C. 2, 1, 4, 2
D. 1, 2, 3, 4
E. 3. 1, 2, 4

24. UME/85/18
The function of lymph nodes is to
A. supply oxygen
B. filter out bacteria
C. form red blood cells
D. supply amino acids
E. supply simple sugars

25. UME/87/21
One cubic centimeter of lymph is richer than an equal volume of blood in
A. erythrocytes   B. leucocytes
C. amino acid   D. glucose.

26. UME/93/21
The main function of blood in mammals is to transport
A. excretory materials from tissues
B. carbondioxide from lungs to tissues
C. digested food from cell the body tissues
D. oxygen to the lungs

27. UME/97/11
In mammals, the exchange of nutrients and metabolic products occur in
A. lymph   B. lungs   C. heart   D. liver

28. PCE/95/19
Which of the following materials is NOT carried by the blood plasma?
A. dissolved food   B. oxygen
C. waste products   D. protein

29. PCE/2001/17/Type:4
The medium of transportation in plants is the
A. ectoplasm   B. cell sap
C. lenticel   D. lymph

30. UME/80/25
A potometer is used to determine the rate at which a shoot
A. respires   B. sucks air bubbles
C. loses weight   D. absorbs water
E. transpires.

31. UME/80/30
If the bark and phloem tissues of a woody shoot one peacked off by ringing, the whole plant will eventually die because:
A. water does not reach the leaves
B. water and salts remain below the rough portion
C. there is a withdrawal of water from the root by soil
D. manufactured food does not reach the roots.
E. the roots store too much water.

32. UME/81/21
The rate of transpiration of a leafy shoot would be highest under
A. damp, cold, still air
B. damp, warm, moving air.
C. dry, warm, moving air
D. dry, cold, moving air
E. dry warm, still air.

33. UME/83/35
Which of the following will NOT allow osmosis to take place?
A. pig's bladder   B. cellophene
C. parchment paper   D. transparent polythene
E. cow's bladder

34. UME/86/24
Which is the correct order of water loss from the leaf? 1. Mesophyll 2. Veins 3. Substomatal cavity
4. Stomata
A. 3→ 2→ 1→ 4
B. 2→ 3→ 1→ 4
C. 2→ 1→ 3→ 4
D. 1→ 2 → 3→ 4

35. UME/89/9
Mineral salts can be absorbed into the roots by
A. osmosis only
B. osmosis and diffusion
C. diffusion and active transport
D. imbibition only

36. UME/90/22
If a ring of bark and phloem is removed from a stem, the
A. plant dies immediately
B. plant dies after two days
C. movement of food is not affected
D. movement of mineral salts is hardly affected.

37. UME/97/17
Oozing out of water from the leaves of plant in a humid environment is known as
A. Transpiration   B. Osmosis
C. Pinocytosis   D. Guttation.

38. PCE/95/20
What is the correct equation of water movement in plants?
A. Root hairs → cortex → phloem vessels → leaves
B. Root hairs → xylem vessels → cortex → leaves
C. Root hairs → phloem vessels → cortex → leaves
D. Root hairs → cortex → xylem vessels → leaves

39. PCE/2002/41/Type:5
In higher plants, the mass flow of food and water within the vascular tissues to all parts of the plant is known as
A. translocation   B. diffusion
C. absorption   D. conduction

40. UME/2008/19
What will happen when two equal sized pieces of unripe pawpaw labelled x and y are dropped into equal volumes of concentrated salt solution and distilled water respectively?
A. pawpaw x will become turgid
B. both will increase in size
C. pawpaw y will become turgid
D. both will decrease in size

MISCELLANEOUS QUESTIONS (8)
41. The main function of blood in mammals is to transport
A. excretory materials from tissues
B. carbon dioxide from lungs to tissues
C. digested food from all the body tissues
D. oxygen to the lungs (UME/93/21)

42. Salts and waters are absorbed in the roots and transported to the leaves by
A. diffusion through the xylem tissues
B. osmosis through the phloem tissues
C. diffusion through the phloem tissues
D. osmosis through the xylem tissues
(UME/2001/27/Type: T)

43. Serum differs from blood plasma because it
A. contains blood cells and fibrinogen
B. contains soluble food and mineral salts
C. lacks blood protein, fibrinogen
D. lacks blood cells and albumin
(UME/2000/24/Type:M)

44. The function of the fluid – filled pericardium is to
A. reduce the friction caused by the pumping movements of the heart
B. supply the heart with oxygen and nutrients
C. prevent disease organisms from attacking the heart
D. reduce the intensity of the pumping action of the heart
(UME/2005/18/Type:D)

45. Excess water in plants is excreted as water vapour and as droplets respectively through
A. respiration and guttation
B. transpiration and guttation
C. photosynthesis and guttation
D. guttation and condensation
(UME/2005/22/Type:D)

46. The dark-red colour of blood is determined by it's
A. plasma   B. protein
C. pigment   D. cells (PCE/2002/29/Type:5)

47. An organism with double circulation is the
A. duck   B. rat
C. toad   D. lizard (PCE/2003/46/Type:N)

48. Substances manufactured by the cells are transported to other parts of the plant through the
A. xylem
B. companion cells
C. sieve tubes
D. cambium (UME/99/20/Type:D)

49. The opening and closing of the stoma are regulated by
A. transpiration   B. respiration
C. diffusion   D. osmosis (UME/2007/29)

50. Poison from a snake bite on a man's leg is carried to his heart through the
A. anterior vena cava   B. posterior vena cava
C. hepatic portal vein   D. hepatic artery (PCE/92/19)

51. The blood cells responsible for body defense against disease are the
A. erythrocytes   B platelets
C. fibrinogens   D. leucocytes (PCE/92/20)

52. The total force by which the cells of the root absorb water from its surrounding is known as
A. osmotic pressure   B. suction pressure
C. turgor pressure   D. water pressure
(PCE/93/19)

53. The tube that carries blood from the heart to the liver is called
A. pulmonary vein   B. hepatic vein
C. hepatic artery   D. pulmonary artery
(PCE/93/21)

54. The return of blood to the heart from the main arteries is prevented by the
A. semi-lunar valve   B. bicuspid valves
C. tricuspid valve   D. mitral valves (PCE/96/15)

55. The primary protein that keeps the blood plasma in osmotic equilibrium with the cells of the body is the
A. albumin   B. Myoglobin
C. myosin   D. actin (PCE/97/16)

56. The mammalian heart chamber that pumps blood through the longest distance is the
A. right auricle   B. right ventricle
C. left ventricle   D. left auricle (PCE/98/17)

57. An organism with an open transport system is
A. grasshopper   B. earthworm
C. rat   D. lizard (PCE/07/14)

58. That is turgor pressure?
A. It is the force that forces water into a cell
B. It is the force of equilibrium in a cell
C. It is the force set up by the cell to prevent entry of water into it
D. It is the force exerted on the wall of cells
E. None of the above

59. What is transpiration?
A. it is the loss of water by plants
B. it is a process in which water escapes through some parts of plants
C. it is the evaporation of water from the leaves of plants
D. it is the loss of water from plants in vapour form
E. None of the above

60. What is guttation?
A. This is the loss of water by leaves
B. it is the water found on plants early in the morning
C. It is the water lost by young coleoptiles of maize
D. it is the exudation of water in liquid form from plants in water – saturated atmosphere
E. The exudation of water from leaf tips

61. A young plant is uprooted from the garden by mistake. It is immediately replanted. The plant wilts but its healthy state is restored three days later. What was wrong with the plant?
A. Transpiration had stopped in the plant
B. The leaves were scorched by the sun
C. The xylem vessels responsible for water uplift in plants were defective
D. It lacked an efficient root system
E. The rate of water absorption from the soil was slow due to the destruction of the root hairs

62. A cassava plant is ringed by the removal of a complete circle of its bark about one inch wide. It is seen that at the end of the season no tuberous roots have formed. This shows that:
A. The phloem cells do not transport food materials
B. The xylem vessels can transport manufactured food
C. Xylem vessels are not as important as phloem cells
D. Manufactured food of plants are transported to areas of storage in the roots through the bark cells
E. None of the above

63. The experiment in question 62 shows that in transpiration water gets from the roots to the leaves through the
A. Xylem vessels   B. Wood fibres   C. The cambium
D. Pith cells   E. Some bark cells

64. With regard to defence, which of these is untrue of the blood?
A. blood, by forming a mesh-work of fibrin, can prevent loss of blood
B. Blood induces the red blood cells to fight bacteria
C. Phagocytes of the blood eliminate unwanted matter in the body
D. Blood can produce antibiotic effect on the body
E. None of the above

65. How does the blood regulate the body temperature of an animal?
A. It stores the heat for the body
B. As the origin of heat, it releases it into the body at regular intervals
C. It dissipates the heat round the body
D. It lowers the amount of heat in the body
E. None of the above

66. The liquid part of blood is called
A. The matrix   B. The serum
C. The lymph   D. The plasma
E. None of the above

67. Which of these cells in the blood is responsible for the storage of haemoglobin?
A. the white blood corpuscles
B. the red blood corpuscles
C. the blood plasma D. the blood platelets
E. The blood corpuscles

68. What protein structure in the blood is responsible for blood clotting?
A. Lecithin
B. Fibrinogen
C. Blood platelets
D. Heparin
E. Thrombin

69. What is the membranous structure that encloses the heart?
A. the pleural cavity
B. the diaphragm
C. the pericardium
D. the mesentery

70. Why are the ventricles of the heart more muscular than the auricles?
A. The ventricles store blood while auricle do not
B. The auricles originated from the ventricles
C. The auricles give rise to the valves, e.g. the tricuspid and bicuspid valves
D. The ventricles are responsible for blood circulation round the blood
E. None of the above

71. What is the common name for all blood vessels which carry blood into the heart?
A. the arteries
B. the pulmonary veins
C. the veins
D. the vena cava
E. None of the above

72. Why is the blood vessels from the right ventricle carrying impure blood to the lungs called the pulmonary artery?
A. It is exceptional
B. It is a main blood vessels
C. It carries blood away from the heart
D. The blood is partly pure
E. None of the above reasons

73. What is the chief characteristics of the capillaries?
A. capillaries are found in the tissues of the body
B. they possess very thin walls
C. they contain the lymph fluid
D. they are easily destroyed
E. They cannot be seen

74. What is the name of the blood vessel carrying blood to the head of the mammal?
A. The subclavian artery
B. The anterior mesenteric artery
C. The pulmonary artery
D. The caput artery E. The carotid artery

75. The coeliac artery supplies blood to:
A. the stomach   B. the instenstines   C. the rectum
D. the spleen   E. the liver and the stomach

76. What is a portal vein?
A. This is vein that sends its blood through another organ before getting to the heart
B. This is a bottle-necked system
C. This is a vein that carries food substances
D. This is a blood system found only in the rat
E. None of the above

77. Trace the movement of a blood cell from the right kidney to the left arm
A. Right kidney→Posterior vena cava→Heart→Lungs→Aorta → Left arm
B. Right kidney→Posterior vena cava→Heart→Lungs→ Heart → Aorta → Left arm
C. Right kidney→Liver →Posterior vena cava→Heart→Lungs →Heart→ Aorta→Subclavain artery→ Left arm
D. Right kidney→Posterior vena cava →Heart→Lungs→Heart →Aorta→Subclavian artery → left arm
E. None of the above

78. What is the importance of the lymph nodes?
A. the nodes store iron
B. the nodes store fats and oils
C. the nodes manufacture white blood corpuscles
D. the nodes produce vitamin D and A for the body
E. None of the above

79. The pigments of the lymph responsible for the transport of oxygen is
A. Haemoglobin   B. Lumphane   C. Lymphatics
D. Lymph corpuscles   E. None of the above

80. Fatty acids and glycerol enter the blood stream through
A. vena cava   B. aorta
C. carotid valve   D. left jugular vein
E. none of the above

81. Oxygen and digested food are transported to the legs and gonads through
A. iliac artery
B. mesenteric vein
C. renal vein
D. hepatic portal vein
E. vena cava

CHAPTER 9

RESPIRATION

Understand the following after reading this chapter:

> Outline of the process and its significance
> Respiratory organs and surfaces
> Mechanism of gaseous exchange
> Aerobic respiration
> Anaerobic respiration

RESPIRATION

Respiration is the process of releasing energy from the breakdown of glucose. Respiration takes place in every living cell, all of the time and all cells need to respire in order to produce the energy that they require.

What is the energy is used for?

The energy produced during respiration is used in many different ways, some examples of what it is used for are:

- Working of muscles
- Growth and repair of cells
- Building larger molecules from smaller ones i.e. proteins from amino acids
- Allowing chemical reactions to take place
- Absorbing molecules in active transport
- Keeping body temperature constant
- Sending messages along nerves

Types of Respiration

Respiration is divided into two phases: External (gaseous) respiration and internal (cellular) respiration.

RESPIRATORY SURFACES OR ORGANS

Respiration surfaces or organs are the media of gaseous exchange between organism and their surrounding

	Organism	Respiratory Surface
1.	Protozoa	Body Surface
2.	Earthworm	Skin
3.	Fishes, tadpoles	Gills
4.	Insects	Trachea
5.	Spiders	Lung-book
6.	Land vertebrate	Lungs (Alveoli)

CHARACTERISTICS OF RESPIRATORY SURFACES OR ORGANS

Property of surface	Reason
Thin (ideally one cell thick)	short distance to diffuse
Large surface area	many molecules can diffuse at the same time
Moist	cells die if not kept moist
Well ventilated	concentration gradients for oxygen and carbon dioxide are kept up by regular fresh supplies of air
Close to blood supply	gases can be carried to/from the cells that need/produce them

MECHANISM OF GAS EXCHANGE

In Insects

Gas exchange is accomplished mostly by simple diffusion through the cell walls. Air enters the spiracles, and moves through the tracheal system. Each tracheal tube ends in a moist tracheole, a specialized cell for exchanging gases with another cell in the body.

When air reaches the tracheole, oxygen dissolves into the tracheole liquid. Through simple diffusion, oxygen then moves to the living cell and carbon dioxide enters the tracheal tube. Carbon dioxide, a metabolic waste, exits the body through the spiracles.

In Toads/Frogs

All frogs start life as aquatic tadpoles, breathing underwater through internal gills and their skin. Then later most develop into land animals with lungs for breathing air. But in all stages breathing is controlled by pulsing the throat. Most frogs lose their gills when they metamorphose.

Frogs breath with their mouths closed. Their throat movement pulls air through the nostrils to the lungs. Then breathe out with body contractions.

Lungs can also help in water. Filling the lungs with air gives frogbetter buoyancy, making it float more easily.

Frogs can also breathe through their skin, with tiny blood vessels, capillaries, under the outer skin layers. The African 'Hairy' frog, Trichobatrachusrobustus, has small lungs and during breeding seasons the males get hair like projections on their back legs. This is because of the high oxygen needs at this time.

In Fishes

Fish obtain oxgen that is dissolved in the water. This occurs via the gills. Gills are feathery structures inundated with blood vessels which allow for a large surface area in contact with the oxygen present in the water.

Contraction of the mouth muscles lowers the floor of the mouth lowering its atmospheric pressure. Water (with dissolved oxygen) moves into the mouth and at the same time the operculum remain closed.

The mouth then open and the operculum muscles relaxes; causing them to bulge open; this increases the volume but lowers the pressure in the gill region.

Water from the cavity moves into the gill region due to the reduced pressure; and bathes the gill filaments. Oxygen diffuses into the blood capillaries due to its high concentration in the gill region than the blood capillaries; it combines with hemoglobin and is transported as oxyhaemoglobin to the respiring tissues.

Carbon dioxide which is of higher concentration in the blood than the gill filaments diffuse into the gills and is exhaled through the water that moves out when the operculum opens.

In Mammals

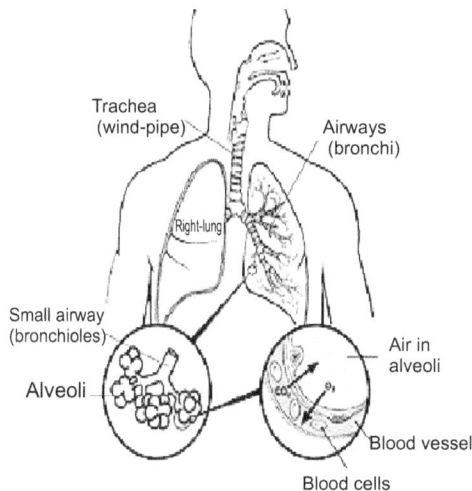

Breathing in:
1. external intercostal muscles contract – pulls rib cage upwards and outwards
2. diaphragm muscles contract – diaphragm moves upwards
3. Lung volume increases – and pressure falls (Boyle's law: pressure and volume are inversely proportional)
4. Air rushes in – to equalize pressures.

Breathing out:
1. external intercostal muscles relax – rib cage falls downwards and inwards
2. diaphragm muscles relax – returns to dome shape
3. Lung volume decreases – and pressure increases
4. Air is forced out

Inhaled (inspired)air: 21% oxygen, 0.04% carbon dioxide, 78% nitrogen and water vapour vary to climate.
Exhaled(expired) air: 18% oxygen, 3% carbon dioxide, 78% nitrogen, and saturated water vapour.

INTERNAL RESPIRATION
Cellular respiration is the oxidation and decomposition of simple carbohydrates such as glucose to release energy.
$C_6H_{12}O_6 + 6O_2$? $6H_2O + 6CO_2 +$ Energy.
It occurs in two main stages viz; Glycolysis and Krebs's cycles.

GLYCOLYSIS(ANAEROBIC RESPIRATION) involves the conversion of simple sugar, e.g. glucose molecules in the absence of oxygen to pyruvic acid.
$2C_6H_{12}O_6$ $2C_2H_3OCOOH + 2H_2$ (2ATP)

In plants, the pyruvic acid is converted to ethyl alcohol, carbon dioxide and other organic acids.
$C_2H_3OCOOH+H_2$ $C_2H_5OH + CO_2$ (0 ATP)

In animals, the pyruvic acid is converted directly into lactic acid. When alcohol or lactic acid is the end product, the process is called fermentation.
$C_2H_3OCOOH+H_2$ C_2H_5OCOOH (0 ATP)
Glycolysis occurs in the cytoplasm of the cell. It yields only 2 ATPs.

Anaerobic respirations occur in yeast, during vigorous exercise, some intestinal bacteria, germinating seeds. Fermentation in Yeast (a fungus) yields alcohol and carbon dioxide. The alcohol from the process is used to brew beer and wine and the carbon dioxide to bake bread.

KREB'S CYCLE(AEROBIC RESPIRATION) involves the use of oxygen to completely breakdown pyruvic acid in the mitochondria to release energy. Kreb's cycles yield about 36 ATPs.
$C_2H_3OCOOH+ H_2 + O_2$ $CO_2 + H_2O$ (36 ATP)
Therefore, a complete oxidation of one molecule of glucose yields a maximum of 38 ATPs. The reactions in glycolysis and Kreb's cycle are each catalyzed by specific enzymes.

RESPIRATION IN PLANTS
Plants respire by means of openings on the leaves called stomata and lenticles which lead to the intercellular spaces and mesophyll cells.

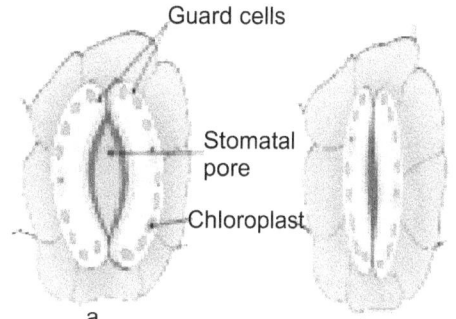

a
b
An open and a closed stomata

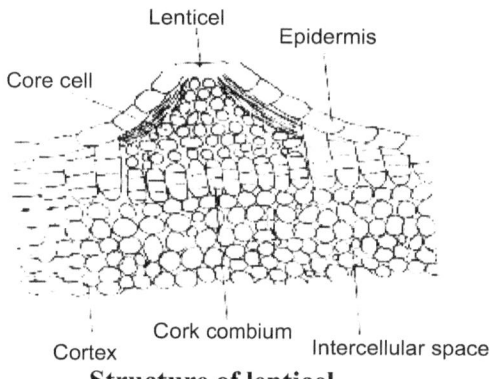

Structure of lenticel

EXPERIMENTS ON RESPIRATION

Experiment :1 To demonstrate that carbon (IV) oxide is given out by man during expiration.

Materials required: Two conical flasks, rubber stopper, rubber tubes, glass tubes, lime water.

Method: The apparatus is set up as shown fig. 40.1. Both flasks contain the same amount of lime water. Air is breathed in and out through the mouth piece. The air goes into the flasks 1 and expired air passed through flasks 2.

Observations: In flasks 1,the lime water turns slightly milky, and in flasks 2, the lime water turns milky, but the milky is more pronounced.

Conclusion: The slight milky colour in flask 1 is due to the traces of carbon (IV) oxide present in the ordinary air. The heavier milky colour in the flask 2 is due to the fact that expired air contains carbon (IV) oxide and is of a higher percentage than ordinary air. This to show that carbon (IV) oxide is give out by man during expiration.

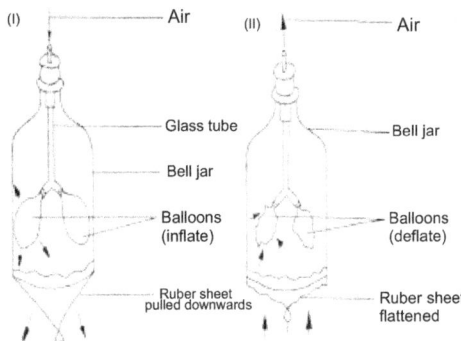

Experiment to demostrate inhalation and exhalation

Experiment: 2To demonstrate that carbon (IV) oxide gives out by a mammal during expiration.

Materials required: 1 conical flask, 2 jars, bent glass tubes, capillary tube, rubber stopper, caustic soda solution, lime water and a living small mammal such as a rat.

Method: The apparatus is set up as shown in fig. 40.2. The first jar (1) contains caustic soda solution. The second and third jars contain lime water. The conical flask contains a well fed rat. The third jar is connected to a filter pump or aspirator. A second experiment is set up to serve as control. The control does not have the small mammal. Air is sucked slowly into the flask by means of a filter pump or an aspirator. The caustic soda in the first jar is to free the air of any carbon (IV) oxide and it is confirmed by the lime water in conical flask makes use of the air (O_2) from the second jar and the air it exhales passes into the 3rd jar.

Observations: As the air passes through the jar and flask it is observed.

   (a) That the caustic soda in the first jar turned milky.
   (b) That the limewater in d second jar did not show any change in colour.
   (c) That the lime water in the third jar also turned milky.
   (d) That the lime water in the third jar (the control experiment) remained unchanged in colour.

Conclusion: Since the air inhaled by the rat was free of carbon (IV) oxide and the lime water in the third jar turned milky, it follows that the rat must have exhaled carbon (IV) oxide which passed into the third jar. This shows that carbon (IV) oxide is given out by a mammal during expiration.

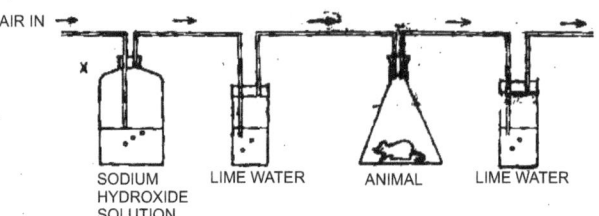

AIR IN

SODIUM HYDROXIDE SOLUTION   LIME WATER   ANIMAL   LIME WATER

Experiment:3To show that germinating seeds or grains utilise oxygen.

Materials required: A beaker, 1 conical flask, 1 test tube, a glass tube, germinating maize seed, caustic soda solution.

Method: The experiment is set up as shown in fig. 40.3. Some maize seed which has been soaked for a few days are kept in a conical flask. Caustic soda solution is poured into a test tube and it is place in the flask. The caustic soda solution removes carbon (IV) oxide which is formed by the germination seed, from the flask. The mouth of the flask carries a rubber stopper with a glass tube. One end of the glass tube is directed into a beaker which contains water. This set up is noted as experiment 1. Second experiment to serve as control, is set up but the flask contains dead grains (boiled grains). Both experiments are left for a while. Next, a glowing splint is dipped into both flasks to test for oxygen.

Observation: Experiment 1. As the germinating seed develop, a rise in the level water in the glass tube takes place. The glowing splint goes out. Control experiment. There is no rise in the tube. The glowing splint is rekindled.

Conclusion: In the first experiment where the glowing splint goes out, indicates that oxygen is absent there and it must have been utilised by the germinating seed for respiration.

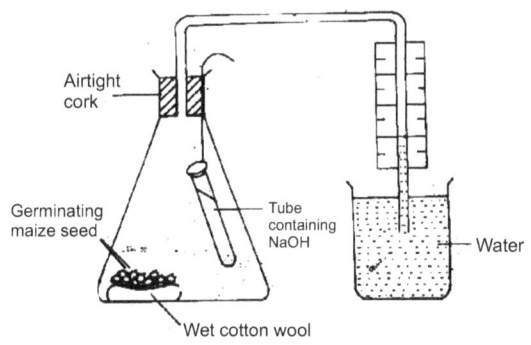

Experiment: 4 To demonstrate that germinating seeds produce heat energy.

Materials required: Two vacuum flask, some living seed, thermometers, cotton wool.

Method: Some living seed which have been soaked for some time are put in a vacuum flask. A thermometer is place among the seeds and the thermometer is supported by placing in cotton wool round it in the mouth of the flask. Second experiment, like the one above, to serve as control, is set up. This vacuum flask however, contains dead seen. These seed are disinfected to prevent bacteria and fungi from acting on them so that heat is not produced. Readings on the thermometer are taken each day for about seven days.

Observation: The thermometer in the first vacuum flask shows a rise in temperature while that of the second vacuum flask remains more or less unchanged.

Conclusion: The rise in temperature of the thermometer in the first vacuum flask shows that germinating seeds produce heat energy.

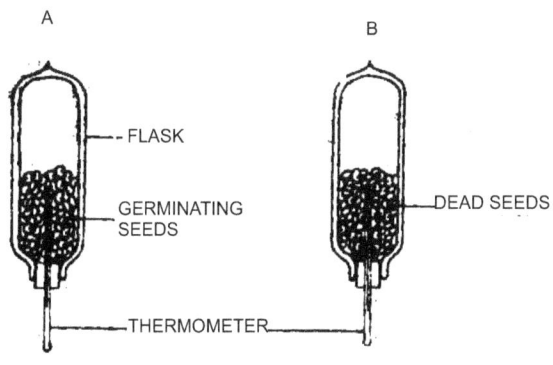

Experiment: 5(a) To demonstrate respiration in yeast cell.

Materials required: Brewer's yeast, glucose, lime water, 2 flat bottom flasks, rubber bungs, cotton wool bent glass tubes, test tubes.

Method: Some amount of brewer's yeast is mixed with 500ml of 10% glucose in a large flat bottom flask. The mouth of the glass is then fitted with a rubber bung that carries a glass tube. The glass tube dips into a test tube which contains some lime water. The mouth of the tube is covered by some cotton wool. A similar experiment (control) is set up but the glucose in the flask does not contain yeasts. Both experiments are left in a warm place.

Observation: Within a short time, the mixture of the glucose and yeasts froths. There is odour of alcohol. The lime water in the test tube turns milky.

Conclusion: The milkiness in the test tube is due to carbon (IV) oxide produced by the yeast cells during respiration. This is confirmed by the control experiment where the lime water remains unchanged. This shows that carbon (IV) oxide is liberated by the yeast cells.

Experiment: 5(b) To demonstrate respiration in yeast cell.

Materials required: Some quantity of fresh palm wine (which contains yeast cells), a round bottom flask, rubber stopper, glass tube, test tube and lime water.

Method: Some fresh palm wine is poured into a round bottom flask. A rubber stopper which carries a glass tube is inserted into the mouth of the flask. The free end of the tube is directed into a test tube which contains lime water. The experiment is then left for some time.

Observation: The lime water in the test tube goes milky.

Conclusion: The milkiness is due to carbon (IV) oxide produced by the activities of the yeast cells. This shows that carbon (IV) oxide is liberated by the yeast cells.

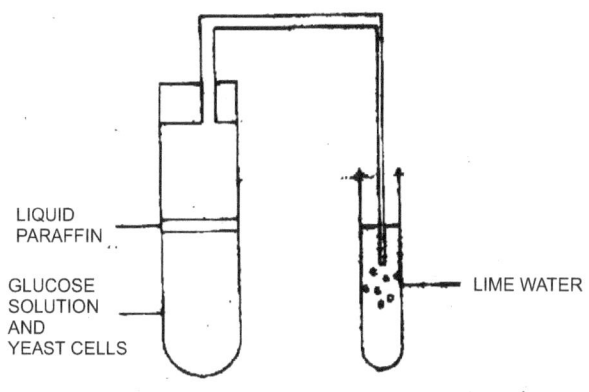

9.1. OUTLINE OF RESPIRATORY PROCESS AND SIGNIFICANCE

1. UME/92/17
In the absence of oxygen, the pyruvic acid produced during glycolysis is converted to CO₂
and A. water   B. glycerol
C. ethanol   D. citric acid

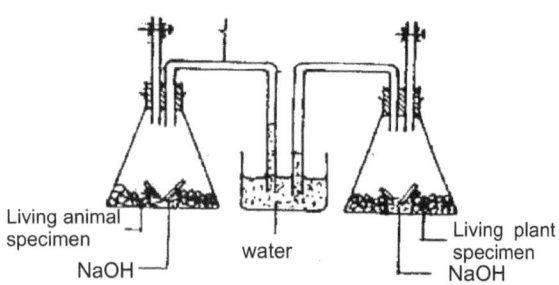

Living animal specimen
water
NaOH
Living plant specimen
NaOH

2. UME/94/19
The most appropriate title for the set up is
A. quantitative measurement of respiration in plant and animals
B. measurement of respiration rates in living organisms
C. comparism between photosynthesis and respiration
D. comparison of respiratory rates in plants and animals.

3. UME/94/20
The part labeled J is called
A. porosimeter
B. photometer
C. manometer
D. auxanometer.

4. UME/95/16
The end product of glycolysis in plants and animals is
A. pyruvic acid   B. citric acid
C. aspartic acid   D. malic acid.

5. PCE/2004/19/Type:9
The significance of respiration in organisms is the
A. oxidation of food and release of useable energy
B. breakdown of waste materials for elimination
C. breathing in of oxygen and breathing out of carbon (IV) oxide
D. build-up of useful materials in the body

9.2. RESPIRATORY ORGANS AND SURFACES

6. UME/81/22
What is the name of the respiratory organ of the crab?
A. Tubules
B. Trachea
C. Spiracles
D. Lungs
E. Gills

7. UME/82/14
A grasshopper respires by means of
A. lung – books
B. gills
C. lungs
D. antennae
E. tracheal tubes.

8. UME/82/12
The glottis is the opening which leads to the
A. oesophagus   B. larynx   C. nostrils
D. pharynx   E. mouth

9. UME/85/23
Identify which of the following are characteristics of the vertebrate respiratory surface
1. Moist 2. Vascularized 3. Semi permeable
4. Freely-permeable 5. Dry
A. 1, 2, 3   B. 1, 2, 5
C. 2, 3, 5   D. 2, 4, 5
E. 1, 3, 5

10. UME/88/22
In the mammalian respiratory system, exchange of gases occurs in the
A. lungs   B. bronchi
C. brochioles   D. alveoli

11. UME/88/13
Lung books are used for respiration in
A. spiders   B. insects
C. millipedes   D. snails

12. UME/92/16
Gaseous exchange in Aves occurs in the
A. tracheoles   B. bronchi
C. air sacs   D. trachea.

13. PCE/2003/50/Type:N
The main respiratory organ of whales is the
A. skin   B. lungs
C. spiracles   D. gills

9.3. THE MECHANISM OF GASEOUS EXCHANGE

9.3.1. INSECTS

14. PCE/2004/20/Type:9
The correct pathway for the movement of air during respiration in insects is
A. spiracle → trachea → tracheoles → body tissues
B. trachea → spiracle → tracheoles → body tissues
C. air sacs → spiracle → trachea → body tissues
D. spiracle → air sacs → trachea → body tissues.

9.3.2. FISH

15. PCE/2004/10 Type:9
Gaseous exchange in the bony fish is carried out through the
A. nostrils   B. lateral lines
C. gills   D. skin

9.3.3. TOADS

16. PCE/2001/18 Type:T
The method of respiration found in an adult toad are
A. cutaneous, buccal and olfactory
B. cutaneous and pulmonary
C. buccal and pulmonary
D. cutaneous, buccal and pulmonary

17. PCE 07/15
Gaseous exchange in tadpoles takes place in the
A. lungs
B. gills
C. mouth
D. operculum

9.3.4. MAMMALS

18. UME/82/38
Which of the following statements is TRUE of inhalation of air by man?
A. the ribs are not raised
B. diaphragm is raised
C. intercostal muscles relaxes
D. pressure of the thoracic cavity increases
E. size of thoracic cavity increase.

19. UME/86/27
Which of the following statements is NOT correct with respect to inhalation in mammals? The
A. intercoastal muscles contract
B. diaphragm is raised
C. ribs are raised
D. pressure of thoracic cavity decreases.

20. UME/2006/5
An increase in air pressure in the lungs is due to the
A. increase in the volume of the thoracic cavity
B. upward movement of the ribs
C. relaxation of the diaphragm
D. contraction of intercostals muscles

21. UME 2008/20
Exhaled air differs from inhaled air in that it
A. contains less amount of carbon (iv) oxide
B. is usually lower in temperature
C. often has more oxygen
D. usually has more water vapour

22. UME 2010/20/TYPE:
The sheet of muscle that separates the thoracic and the abdominal cavities is the
A. intercostal muscle
B. pleural membrane
C. pericardium
D. diaphragm

23. PCE/99/14
When the intercostals muscles contract during inspiration, the ribs move
A. upward and forward
B. downward and inward
C. upward and inward
D. downward and forward

24. PCE/06/14
Diffusion of oxygen in the mammalian lung is enhanced by the
A. presence of heamoglobin on the red blood cells
B. low diffusion gradient between gases in the alveoli
C. low concentration of oxygen in the blood
D. absence of haemoglobin in the red blood cells

9.3.5. PLANTS

25. UME/95/17
During respiration, air circulates round plant tissues via the
A. lenticels
B. stomata
C. guard cells
D. intercellular spaces

26. UME/78/11
The function of lenticels is
A. to remove excess water in the plant
B. to absorb water from the atmosphere
C. for gaseous exchange
D. to absorb light
E. to store food.

27. UME/99/23
In woody stems, gaseous exchange takes place through the
A. microphyles   B. stomata
C. lenticels   D. vessels

28. PCE/95/21
Investigation of respiration in green plants is carried out in the dark because
A. Respiration occurs in plants only in the dark.
B. Light is destructive to potted plants
C. Respiration is faster than photosynthesis in the presence of light.
D. Carbon dioxide released in the presence of light is used up in photosynthesis.

29. UME/2004/2/Type:4
The opening of the stoma is controlled by the
A. presence of guard cells
B. decrease in solute concentration
C. increase in solute concentration in the guard cell
D. presence of a pore.

30. UME 2006/12
The formation of water in the tissue respiration results from the
A. breakdown of water molecules
B. reduction of oxygen by hydrogen
C. reduction of carbon (iv) oxide
D. combination of water molecules

9.4. AEROBIC RESPIRATION
31. UME/87/22
The oxidative part of the respiratory process takes place in the
A. mitochondria   B. ribosomes
C. endoplasmic reticulum   D. golgi bodies.

32. UME/2001/16/Type:T
The gas produced during tissue respiration can be identified by using
A. calcium hydroxide
B. calcium carbonate
C. copper sulphate
D. sodium hydroxide

33. UME/2002/40/Type:Y
Tissue respiration is important for the
A. absorption of oxygen into the alveoli
B. release of carbon (IV) oxide into the lungs
C. release of energy for body use
D. exhalation of carbon (IV) oxide from the lungs

34. PCE/2001/19/Type:4
The life-saving respiratory technique applied to persons who are nearly drowned, electrocuted or shocked is
A, anaerobic respiration   B. artificial respiration
C. aerobic respiration   D. tissue respiration.

35. UME/2004/14/Type:4
The anaerobic fermentation of a glucose molecule yields
A. pyruvic acid and alcohol   B. 38 ATP molecules
C. water and carbon (iv) oxide
D. 2 ATP molecules and alcohol

9.4. ANAEROBIC RESPIRATION
36. UME/81/25
In a bakery, yeast is added to flour to make bread rise. This is possible because yeast produces
A. alcohol   B. oxygen
C. carbon dioxide   D. energy
E. ethanol

37. UME/82/34
When freshly tapped palm wine was kept overnight it was found to be more alcoholic. This effect must have been due to
A. the addition of saccharin
B. the addition of sugar
C. the addition of whisky or local gin
D. some complex hydrolysis reactions
E. anaerobic respiration of an organism

38. UME/84/18
Anaerobic respiration in yeast produces
A. carbon dioxide and ethanol
B. carbon dioxide and water
C. carbon dioxide and oxygen
D. carbon dioxide and glucose
E. ethanol and water.

39. UME/85/22
Which of the following events does NOT occur during anaerobic respiration of glucose?
A. Muscle cells produces lactic acid
B. Carbon dioxide is produced
C. Milk bacteria produce lactic acid
D. Energy is not produced
E. Germinating seeds produced alcohol.

40. UME/86/28
The equation that can be used to summarized the process of anaerobic breakdown of sugar is
A. $C_6H_{12}O_6 \rightarrow 2 C_2H_5OH + 2CO_2$
B. $6 CO_2 + 6H_2O \rightarrow C_6H_{12}O_6 + 6O_2$
C. $C_6H_{12} + 6O_{12} \rightarrow 6CO_2 + 6H_{12}O + Energy$
D. $C_6H_{12}O_6 \rightarrow 2C_2H_5OH + 2CO_2 + Energy$

41. UME/87/23
Fatigue of leg muscles may occur after riding many kilometers on a bicycle because of
A. insufficient glucose   B. excess carbon dioxide
C. excess protein   D. insufficient oxygen

42. UME/97/19
Anaerobic respiration differs from aerobic respiration by the production of
A. less amount of energy and water
B. greater amount of energy and alcohol
C. less amount of energy and alcohol
D. greater amount of energy and water.

43. UME 2009/20
The enzymes of the glycolytic pathway are located in the
A. mitochondria   B. gastric juice
C. plasma   D. cytoplasm

44. UME 2010/19/TYPE: D
Yeast respires anaerobically to convert simple sugar to carbon (iv) oxide and
A. acid
B. oxygen
C. water
D. alcohol

MISCELLANEOUS QUESTIONS (9)

45. The surface of an alveolus in a mammal is well supplied with tiny blood vessels known as
A. capillaries
B. arteries
C. arterioles
D. venules
(UME/2004/20.Type:4)

46. The anaerobic fermentation of a glucose molecule yields
A. pyruvic acid and alcohol
B. 38ATP molecules
C. Water and Carbon dioxide
D. 2TP molecules and alcohol (UME/2004/14/Type:4)

47. A test tube containing yeast in glucose solution was suspended in a covered conical flask containing alkaline pyrogallol. The bubbles of carbon (iv) oxide produced indicate that the yeast cells are
A. respiring in the absence of oxygen
B. liberating oxygen on their own
C. living and consuming oxygen
D. being killed by the alcohol produced
(UME/2005/20/Type: D)

48. When yeast respires anaerobically, it converts simple sugars to carbon (iv) and
A. oxygen
B. acid
C. alcohol
D. water (UME/2001/19/type: T)

49. The respiratory surface in amphibians are modified for gaseous exchange by being
A. rough and tough
B. moist and rough
C. moist and tough
D. moist and vascularised (PCE/2002/27/Type:5)

50. The tiny holes in the cuticle of insects used for gaseous exchange are known as
A. tracheoles
B. book lungs
C. spiracles
D. tracheae (PCE/96/16)

51. Which of the following uses diffusion as the principal method of gaseous exchange?
A. lizard   B. grasshopper
C. rat   D. earthworm   (UME 2007/32)

52. The main respiratory substrates of all cells are
A. oligosaccharides   B. disaccharides
C. polysaccharides   D. monosaccharides
(PCE/92/22)

53. During respiration, the incomplete oxidation of glucose to produce pyruvic acid is known as
A. Krebs cycle   B. aerobic respiration
C. glycolysis   D. autolysis (PCE/93/22)

54. The life-saving respiratory technique applied to person who is nearly drowned, electrocuted or shoked is
A. anaerobic respiration
B. artificial respiration
C. aerobic respiration
D. tissue respiration   PCE/01/19/TYPE: 4

55. The word respiration when simply used means the taking in and expulsion of air from the lungs. But in the widest sense means
A. Removal of energy
B. Loss of energy by food
C. Provision of energy
D. Oxidation of food to provide energy
E. None of the above

56. Apart from the stomata, through which other way can oxygen enter the shoot of the plan?
A. Oxygen can enter the shoot of the plant by simple diffusion
B. It can enter through any hole the stem
C. It can enter through cork splits of stems
D. It can enter through the lenticels
E. It can enter through the scars of leaves

57. Respiration is a process that makes use of oxygen releasing carbon dioxide. A plant respiring in darkness releases quite a large quantity of carbon dioxide. This means:
A. That respiration is a dark process
B. That respiration takes place only in the night
C. That respiration can be effective in the dark
D. That when plants are respiring, carbon dioxide can only be released in the night
E. None of the above

58. Roots can only take in oxygen when it is:
A. In gases state   B. Liquid state
C. In chemical combination with other elements
D. In solution with water
E. None of the above

59. What is the name of the structure through which air gets into the body of the insects?
A. Trachea
B. Tracheoles
C. Spiracle
D. Ostia
E. None of the above

60. The only section of the respitaory system of the insect that is not cuti-cularized is:
A. The trachea
B. Tracheole
C. Spiracle
D. Stigmata
E. None of the above

61. The most essential element in the respiratory structure of the fish is:
A. The gill
B. The gill rakers
C. The gill arch
D. The gill filament
E. The gill ray

62. In the bony fish, that large plate that covers the respiratory structure is called:
A. The gill cover
B. The spiracle
C. The gill slit
D. The operculum
E. None of the above

63. The passage of water over the gills is due to:
A. The closing of the mouth of the fish
B. The pressure of water on the operculum
C. The horizontal disposition of the gill filaments
D. The raising of the floor of the mouth
E. The shutting and raising of the floor of the mouth of the fish

64. The respiratory apparatus of the bird is made complex by the possession of:
A. Air bladder
B. Diaphragm
C. Air sacs
D. Hollow bones
E. None of the above

65. Of what importance are the air sacs to bird?
A. The air sacs are useful in gaseous exchange
B. They are useful in the storage of air
C. They are blood purifiers
D. They are speed accelerators
E. They sustain the sound box of birds

66. In man, the nasal cavity is separated from the buccal cavity by a bone called:
A. The turbinal bone
B. The palate complex
C. Soft palate
D. The tongue bone
E. Hyoid apparatus

67. The trachea of the respiratory system is fortified by:
A. Tiny bones
B. Cartilage bones
C. Elastic skin
D. Ligaments
E. Glottis

68. The tiny air spaces found in the lungs of mammals are called:
A. Air sacs
B. Alveoli
C. Tracheoles
D. Bronchioles
E. Pleura

69. The auditory structures (vocal cords) of mammals are located in
A. The trachea   B. The glottis
C. The wind-pipe   D. The larynx
E. The lung

70. What structures are responsible for the intake and output of air from the lungs?
A. The air sacs and trachea
B. The larynx and bronchi
C. The ribs and intercostals muscles
D. The ribs and the diaphragm
E. None of the above

71. To take in air into the lungs the diaphragm must be:
A. Dome-shaped   B. Flattened
C. Oblique   D. Normal
E. None of the above

72. What happens to air in the lungs?
A. It diffuses into lungs
B. It goes into solution at the respiratory surface
C. It goes into solution with haemoglobin
D. It is compressed into liquid
E. None of the above

73. Oxygen is taken round the body as
A. Oxyhaemoglobin   B. Haemoglobin
C. Haem   D. Blood
E. Lymph

CHAPTER 10

EXCRETION

Understand the following after reading this chapter:

> Types of excretory structures
> Excretory mechanism
> Excretory products of plants and animals

EXCRETION
Excretion is the elimination of waste products of metabolism from the cells.

Metabolism, on the other hand, is the sum total of chemical activities in the cell. It involves **anabolism** (simple molecules are built up into larger complex one e.g. photosynthesis) and **catabolism** (complex molecules are broken down into smaller, simpler ones e.g. respiration).

EXCRETORY ORGANELLES/ ORGANS IN ANIMALS

Animal	Organelle /Organ
1. Protozoa	Contractile vacuole
2. Flat worms	Flame cell
3. Roundworms	Nephridia
4. Insects	Malpighian tubules
5. Fishes	Gills
6. Land vertebrates	Kidney, skin, lung, liver and *large intestine.

*excrete iron and calcium.

THE HUMAN EXCRETORY SYSTEM
The urinary system is made-up of the kidneys, ureters, bladder, and urethra. The nephron, an evolutionary modification of the nephridium, is the kidney's functional unit. Waste is filtered from the blood and collected as urine in each kidney. Urine leaves the kidneys by ureters, and collects in the bladder. The bladder can distend to store urine that eventually leaves through the urethra.

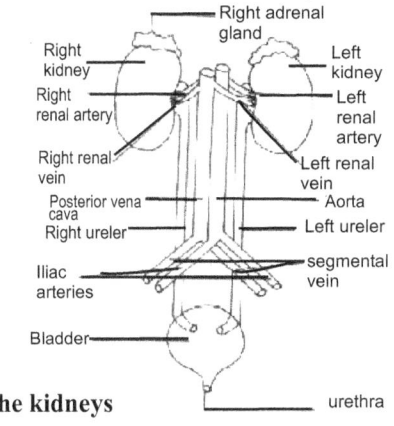

The kidneys

The Nephron
The nephron consists of a cup-shaped capsule containing capillaries and the glomerulus, and a long renal tube. Blood flows into the kidney through the renal artery, which branches into capillaries associated with the glomerulus. Arterial pressure causes water and solutes from the blood to filter into the capsule. Fluid flows through the proximal tubule, which include the loop of Henle, and then into the distal tubule. The distal tubule empties into a collecting duct. Fluids and solutes are returned to the capillaries that surround the nephron tubule.

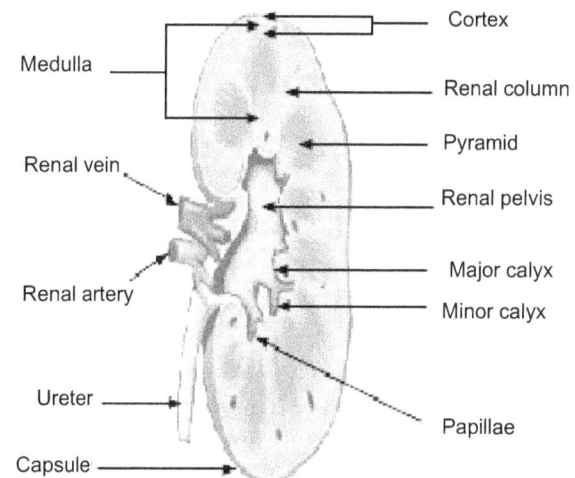

Longitudinal section of the kidney

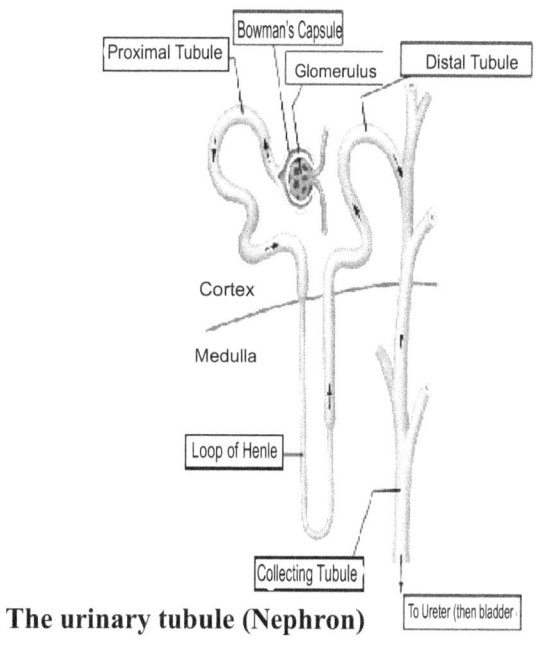

The urinary tubule (Nephron)

The nephron has three functions:
1. Glomerular filtration of water and solutes from the blood.
2. Tubular reabsorption of water and conserved molecules back into the blood.
3. Tubular secretion of ions and other waste products from surrounding capillaries into the distal tubule.

Nephrons filter 125 ml of body fluid per minute; filtering the entire body fluid component 16 times each day. In a 24 hour period nephrons produce 180 liters of filtrate, of which 178.5 liters are reabsorbed. The remaining 1.5 liters forms urine.

Urine Production
1. Filtration in the glomerulus and nephron capsule.
2. Reabsorption in the proximal tubule.
3. Tubular secretion in the Loop of Henle.

Components of the Nephron
- Glomerulus: mechanically filters blood
- Bowman's Capsule: mechanically filters blood
- Proximal Convoluted Tubule: Reabsorbs 75% of the water, salts, glucose, and amino acids
- Loop of Henle: Countercurrent exchange, which maintains the concentration gradient
- Distal Convoluted Tubule: Tubular secretion of H ions, potassium, and certain drugs.

Kidney Function
Kidneys perform a number of homeostatic functions:
1. Maintain volume of extracellular fluid
2. Maintain ionic balance in extracellular fluid
3. Maintain pH and osmotic concentration of the extracellular fluid.
4. Excrete toxic metabolic by-products such as urea, ammonia, and uric acid.

EXCRETORY SYSTEM IN INSECTS
Body fluids are drawn into the Malphighian tubules by osmosis due to large concentrations of potassium inside the tubule. Body fluids pass back into the body, nitrogenous wastes empty into the insect's gut. Water is reabsorbed and waste is expelled from the insect.

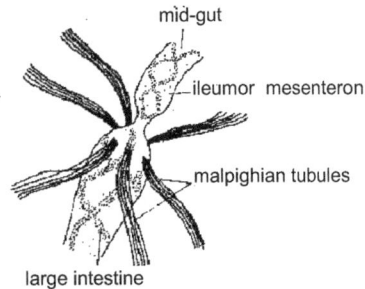

Malphigian tubule

EXCRETORY SYSTEM IN FLAT WORMS
The simplest tubular excretory system is the flame-cell system of flatworms (Phylum Platyhelminthes). These animals have neither circulatory systems nor coeloms, and so the flame-cell system must regulate the contents of the interstitial fluid directly.

The apparatus consists of a branched system of tubules ramifying throughout the body. Each of the smallest tubules at the tips of this excretory tree is capped by a bulbous cell called a flame cell. Interstitial fluid bathing the tissues of the animal passes through the flame cell and enters the tubular system.

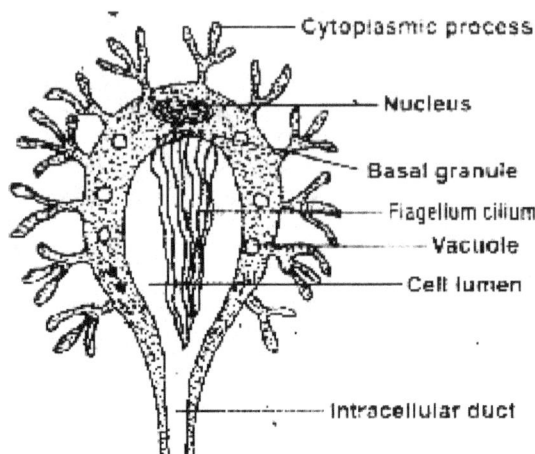

A flame cell

EXCRETORY SYSTEM IN EARTHWORMS
In earthworm excretion occurs through nephridium. The excretory system in earth worm is closely associated with its closed circulatory system. Each segment of earthworm consists of a pair of nephridia or in some species cluster of nephridium. The coelemic fluid or body fluid enters the nepridium through a membrane which is cilliated called nephrdiostome. The body fluid after reabsorption from the nephridia opens into a large bladder which opens to exterior through nephridiopore. Ammonia and water are waste products excreted by nephridia while CO_2 is excreted through general body surface.

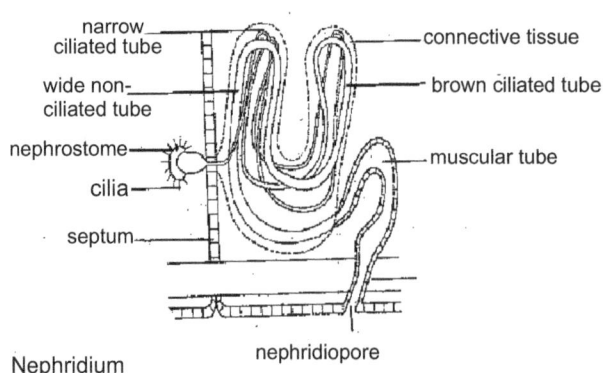

Nephridium

EXCRETION IN PLANT

Plants excrete water, oxygen and carbon dioxide by diffusion through the stomata and lenticels.

Solid wastes such as tannins, mucilage, gum, alkaloids etc. are stored at the bark of the stem or the leaves which are eliminated through peeling and leaf fall respectively.

Revision Questions

0.1. TYPES OF ECXRETORY STRUCTURES

1. UME79/16
Which of the following is NOT an excretory organ?
A. Lungs   B. Kidney   C. Leaf
D. Large Intestine   E. Skin

2. UME92/18
The excretory organ in insects is
A. Kidney   B. Malpighian tubule
C. Flame cell   D. Nephridium

3. UME93/8
Flame cells are the
A. Excretory system of worms
B. Excretory and respiratory systems of flatworms
C. Secretory system of flatworms
D. Excretory system of flatworms

4. UME98/20
In which of the following groups of animal is the malpighian tubule found?
A. Lizards, snakes and frogs
B. Crickets, houseflies and grasshoppers
C. Millipedes, centipedes and scorpions
D. Earthworms, roundworms and flatworms

5. PCE/00/18/TYPE: A
Nephridium is used as an excretory structure in
A. crustaceans
B. annelids
C. platyhelminthes
D. coelenterates

10.2. EXCRETORY MECHANISM
10.2.1. KIDNEYS

6. UME/80/22
Which part of the following parts of the mammaliam body is most closely associated with the production of urine?
A. Malpighian capsule
B. Urinary bladder
C. Ureter
D. *vas deferens*
E. Urethra

7. UME/80/29
During excretion of urea there is also a corresponding re-absorption of water into the blood stream. This reabsorption takes place in
A. Uriniferous tubules   B. Bowman's capsule
C. Glomerulus   D. Malpighian capsule
E.Renal artery

8. UME/87/24
The function of the loop Henle is to
A. increase the flow of urine
B. concentrate amino acids in the kidney tissue
C. concentrate sodium chloride in the medulla of the kidney
D. increase the volume of urine

9. UME/88/24
In the kidney, the malpighian corpuscle is located in the
A. medulla   B. helium
C. cortex   D. pelvis

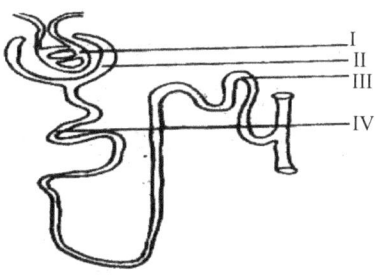

10. UME/2000/10/Type M
The parts labeled I and II make up the
A. glomerulus   B. convoluted tubules
C. malpighian body   D. Bowman's capsule

11. UME/2000/11/Type M
In mammals, reabsorption of salt takes place in
A. IV   B. III   C. II   D. I

12. UME 2006/15
Mammals are capable of producing hypertonic urine mainly because of reabsorption in the
A. Bowman's capsule   B. urethra
C. Ureter   D. loop of henle

13. UME 2007/28
The correct sequence of the movement of urea during urine formation is
A. glomerulus → Bowman's capsule → convoluted tubule → henle's loop → collecting tubule
B. glomerulus → Bowman's capsule → convoluted tubule → henle's loop → convoluted tubule → collecting tubule
C. convoluted tubule → glomerulus → henle's loop → Bowman's capsule → collecting tubule
D. convoluted tubule → Bowman's capsule → henle's loop → glomerulus → collecting tubule

14. UME 2008/21
In insects, the structure that performs the same function as the kidney in man is the
A. nephridium
B. flame cell
C. malphigian tubule
D. trachea

15. UME 2010/22 TYPE: D
The outer layer of the kidney where the bowman's capsules are found is the
A. pelvis   B. medulla
C. pyramid   D. cortex

16. PCE/01/21/TYPE: A
The structure that connects the kidney with the urinary bladder is the
A. urethra   B. ureter   C. pelvis   D. pyramid

10.2.2. SKIN
17. UME/88/25
The skin, through the sweat glands, functions as
A. an excretory organ
B. a respiratory organ
C. a sensory organ
D. a protective organ

18. UME/2004/15/Type 4
The function of the part labeled III is to
A. produce oil for the skin
B. carry blood and nitrogenous waste
C. contract to pull the hair erect
D. conduct nervous impulses

19. UME/2004/16/Type 4
The sweat gland is the structure labeled
A. IV   B. III   C. II   D. I

20. UME/86/31
In the mammalian skin, melanin and keratin are contained in the
A. sebaceous gland
B. sweat gland
C. subcutaneous gland
D. malpighian layer

21. UME 2010/22/TYPE: D
The oily substance that lubricates the mammalian hair to keep it flexible and water repellent is secreted by the
A. sebaceous glands
B. fatty cells
C. granular layer
D. sweat glands

10.3. EXCRETORY PRODUCTS OF PLANTS AND ANIMALS

10.3.1. PLANTS
22. UME/84/30
Which of the following is NOT a waste product of plants?
A Tannins
B. Oxygen
C. Carbondioxide
D. Sap
E. Alkaloids

23. UME/88/23
The main waste products formed in plants are
A. Alkaloids, tannins and resins
B. Water, alkaloids and carbondixoide
C. Oxygen, carbondioxide and alkaloids
D. Water, carbondioxide and oxygen

24. UME/90/24
Excretory products responsible for the red, purple and blue colours of flowers are called
A. alkaloids   B. tannins   C. anthocyanins   D. resins

25. UME/94/22
Which of the follow waste products in plants is excreted through the stomata and lenticels?
A. Carbondioxide   B. Alkaloids
C. Tannins   D. Anthocyanins

26. PCE/2000/19/Type A
The excretory products of higher plants which is also used for another metabolic process is
A. nitrogen   B. oxygen
C. carbon (iv) oxide   D. water vapour

27. UME 2007/31
The waste product of plants used in the conversion of hide to leather is
A. gum   B. alkaloid
C. tannin   D. resin

28. UME/82/43
Which of the following is NOT associated with excretion in mammals
A. Glomerulus   B. Urea
C. Faeces   D. Carbondioxide
E. Bowman's Capsule

29. PCE/2003/40/Type N
The major excretory product of insect is
A. sodium chloride
B. ammonia
C. Urea
D. Uric acid

MISCELLANEOUS QUESTIONS (10)
30. Reabsorption of chemical solutes in the kidney takes place in
A. Bowman's capsule   B. Tubules
C. Glomerulus   D. Nephron (PCE/95/22).

31. The urinary tubules of the kidney function through
A. Osmosis and diffusion
B. active transport and osmosis
C. ultra filtration and selective reabsorption
D. active transport and cytoplasmic streaming
(UME/2005/21/ype D)

32. The kidney is connected to the bladder by the
A. Cortex   B. Urethra
C. Urinary bladder   D. Ureter
(PCE/2005/19/Type P)

33. The excretory products in plants include
A. gum, uric acid and mucilage
B. tannin, anthocyanin and resin
C. anthocyanin, sweat and tannin
D. resin, mucilage and urea (PCE/2005/12/type P)

34. In the mammalian kidney, the Bowman's capsules are located in the
A. Ureter   B. Medulla
C. Pelvis   D. Cortex (PCE/94/21)

35. In a young human embryo, the excretory function is performed by the
A. chorion   B. allantois
C. amnion   D. embryonic membrane
(PCE92/21)

36. Excretion is defined as:
A. The voiding out from the body of obnoxious substances such as faeces.
B. The elimination of only urea from the body
C. The elimination of water and salts from the skin as sweat.
D.The elimination from the body of poisonous by products of metabolism
E. None of the above

37. The excretory structures in flatworms are the
A. contractile vacuoles   B. nephridia
C. flame cells   D. kidneys (PCE/93/23)

38. One a cold day, more urine is produced because
A. little sweat is produced
B. metabolic rate is reduced
C. water intake is more
D. water absorption is less (PCE/96/17)

39. Substances from the glomerulus pass to the Bowman's capsule by
A. selective reabsorption   B. diffusion
C. absorption   D. ultrafilteration
(PCE/03/32/TYPE: N)

40. The correct sequence in the formation of urine in the mammalian kidney is
A. secretion reabsorption ultrafilteration
B. reabsorption ultrafilteration secretion
C. ultrafilteration reabsorption secretion
D.ultrafilteration secretion reabsorption
(PCE/04/22/TYPE: 9)

41. Which of these is not an excretory products?
A. Urea
B. Ammonia
C. Carbon dioxide
D. Water
E. None of the above

42. How is urea formed in the body?
A. The ammonia obtained from the deamination of amino-acids reacts with carbon dioxide.
B. Urea is a product of uric acid decomposition
C. Urine reacts with water to produce urea
D. Proteins are broken down into urea
E. None of the above

43. Which of these is not an excretory organ?
A. The kidney   B. The skin
C. The liver   D. The lungs
E. None of the above

44. What is the name given to that part of the kidney in which the pyramids are found?
A. The medulla
B. The Cortex
C. The pyramidal zone
D. The pelvis
E. None of the above

45. The outer zone of the kidney is:
A. The medulla
B. The periphery
C. The cortex
D. The capsular zone
E. The cortical membrane

46. Which of these structures is in no way connected with the kidney?
A. The malpighian capsule
B. The glomerulus
C. The pyramid
D. The renal vein
E. None of the above

47. What is the function of the pyramids of the kidney?
A. The pyramids store fat
B. The pyramids bear the collecting tubes of the kidney
C. The pyramids strengthen the openings of the collecting canals
D. None of the above reasons

48. In what way is the lung as excretory organ?
A. It is the responsible for the elimination of carbon dioxide
B. It is responsible for the loss of the nitrogen of the air from the tissues
C. It eliminate water from the tissues of body
D. It regulates salt metabolism
E. None of the above

49. On a cool and wet day it is observed that a large quantity of urine is passes out by a large of students. This is because
A. The kidney is over-active on wet days
B. Not much water is lost from the body by way of sweat
C. The uriniferous tubules can no longer absorb water
D. The convoluted tubules and the loop of Henle are usually inactive on wet days
E. None of the above

50. In a very young embryo, the excretory materials are stored in:
A. The bladder
B. The yolk
C. The allantois
D. The placenta
E. The embryonic membranes

51. In what way is the liver involved in excretion?
A. It eliminate iron from the body
B. It stores urea
C. It converts urea into urine
D. It manufactures urea
E. It converts uric acid into urea

52. In the insect the organ of excretion is:
A. The hindgut
B. The hepatic caeca
C. The crop
D. The salivary gland
E. The malpighian tubules

53. The function of the malpighian layer of the skin is:
A. To protect the skin
B. To separate the epidermis from the dermis
C. To support the skin
D. To produce new epidermal cells
E. None of the above reasons

54. The excretory product of insects is:
A. Urine   B. Urea
C. Ammonia   D. Urates   E. Uric acid

55. In insects, a large proportion of nitrogenous waste is stored in:
A. The reproductive system   B. The fat bodies
C. The blood   D. The gut
E. The crop

56. In which of these processes occurring in insects is nitrogenous waste NOT lost?
A. Ecdysis
B. Pupation
C. Moulting
D. Metamorphosis
E. Shedding of pupal case

57. How is an excretory material lost in the amoeba?
A. Through the contractile vacuole
B. Through the general body surface
C. Through the food vacuoles
D. Through the temporary anus
E. None of the above

58. In the fish, apart from the kidney another excretory structure is:
A. The skin
B. The gills
C. The lining of the buccal cavity
D. The lateral lines
E. None of the above

59. The sebaceous gland of the skin produces
A. Urea
B. Water
C. A gelatinous substance
D. A salt
E. An oily substance

CHAPTER 11

SUPPORT AND MOVEMENT

Understand the following after reading this chapter:

> Supporting tissues in plants
> Movement in plants
> Supporting tissues in animals
> Types and functions of skeleton

SUPPORTING TISSUES IN PLANTS
Specialized cells and tissues are used to provide support in plants. They include parenchyma, collenchyma, sclerenchyma and xylem tissues

Parenchyma tissue: consists of loosely-packed, thin-walled rounded cells with air spaces between them. It is the principal tissue of the cortex and pith.
1. It makes up the ground tissue of many organs and therefore provides support.
2. It stores starch and other substances.
3. Support the plant by turgor.
4. Conducts gases and water across the plant.

Sclerenchyma tissue: the walls of the sclerenchyma cells are evenly thickened with lignin.
1. It strengthens and supports the plant.

Collenchyma tissue: the walls of the collenchyma cells are thickened at the corners.
1. It provides support for young sterms and leaves to the lignin it contains.
2. It provides mechanical support to the plant.

Xylem tissue or wood: the conducting cells in the xylem are the dead enlonged lignified tubes called vessels and tracheids.
1. It strengthens the tissues
2. It conducts water and other solutes

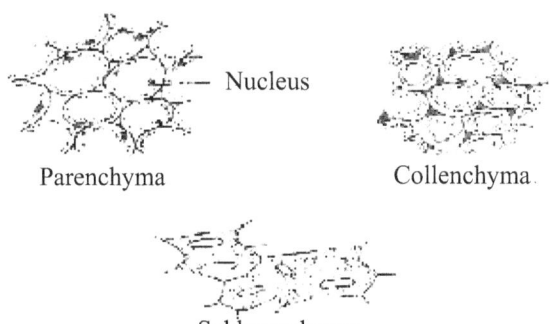

Parenchyma Collenchyma

Schlerenchyma

Supporting tissues in plants

MOVEMENT IN PLANTS
Movement in plants nearly always involves growth.

Tropism: This is the response of parts of plants towards or away from a unilateral stimulus. Tropic movements are directional and positive if organism moves towards the stimulus and negative if it moves away from it. The response to light is phototropism, to water, hydrotropism, to gravity, geotropism and to touch, thigmotropism. Tropism is easily demonstrated in actively growing structures. Shoots of plants are positively phototropic, roots of plants are positively hydrotropic and positively geotropic.

Nastism: This is the non-directional response of parts of plants of plants to a diffuse (general) stimulus. It may not involve growth. Examples: (1) Petals of sun-flower which open in light and close in the dark; (2) Leaves of plants (e.g. Oxalis) which spread out in light but fold up in the dark (3) Collapse of *Mimosa pudica*plant to touch this is also referred to as **sleep movement**.

Taxism: This is the response of a whole organism toward or away from stimulus. Tactic movements are directional and positive if organism moves towards the stimulus and negative if it moves away from it. Examples are phototaxism, chemotaxism and thermotaxism.

SKELETON AND SUPPORTING TISSUES IN ANIMALS
Skeleton is the bony framework which provides support for organisms.

Types of Skeleton
1. **Exoskeleton**: This is the outer supporting tissues of the body. It is found mostly in arthropods.
2. **Endoskeleton**: This is the inner supporting tissue of the body. It is found mostly in vertebrates.
3. **Hydrostatic skeleton**: It is made up of fluid that keeps the body turgid. Example is found in earth worm.

Functions of Skeleton
1. It provides support for the body
2. It provides rigidly and shape for the body
3. It provides a base for the attachment of muscles
4. Bones store salts, notably, calcium and phosphorus salts
5. Red blood cells are produced in the marrow of the long bones
6. It protects important organs of the body
7. It enhances movement of the body

In mammals, the endoskeleton is divided into two viz: the axial skeleton and the appendicular skeleton.

Axial Skeleton

The axial skeleton comprises the skull, vertebral column, ribs and sternum. The skull is made up of the cranium which houses the brain, the facial skeleton which anchors the nose, eyes and the muscles of the cheek and the jaws.

Each bone of the vertebral column is called a vertebra. The bones of the vertebra column is classified into five, based on their characteristic features.

Thus:

1. Cervical vertebrae   –   Neck region
2. Thoracic vertebrae   –   Chest region
3. Lumber vertebrae   –   Upper trunk region
4. Sacral vertebrae   –   Lower trunk region
5. Caudal vertbebrae   -   Tail region

(Memory Tip: Come To London Study Commerce)

Cervical vertebrae

The atlas and axis are the first and second cervical vertebrae respectively. They posses special modification

Atlas

1. It has a highly reduce centrum
2. It has a highly reduced spine
3. It has a large neural canal
4. It has two large depressions on its anterior surface which fit into two projections at the base of the skull, so, enabling the head to nod
5. It has reduced transverse process

Axis

1. It has a prominent spine
2. It has reduced transverse process
3. It has a broad and flat centrum
4. It has a large neural canal
5. It has an elongated odontoid process which fits into the neural canal of the atlas, thus, allowing rotational movement of the head

Typical/Ordinary Cervical Vertebra

1. It has a centrum
2. It has a short neural spine
3. It has branched transverse processes
4. It has a large neural canal
5. it has vertebraterial canals

Thoracic Vertebra

1. It has a long neural spine projected backwards
2. It has reduced transverse processes that articulate with the ribs.
3. It has articular facets on the centrum and transverse processes which articulate with the head and tubercle of ribs respectively

Lumber vertebra

1. It has a short and flat neural spine
2. it has well developed transverse processes
3. it has a thick centrum
4. it has a well developed pre-and-post-zygapophyses

Sacral vertebrae

1. The sacral vertebrae consist of fused bones
2. The transverse process of the first and second vertebrae articulate with the pelvic girdle
3. The neural canal diminishes gradually from the first vertebra until it disappears in the last.

Caudal vertebra

1. Usually reduced to form small centrum bones
2. It lacks neural canal

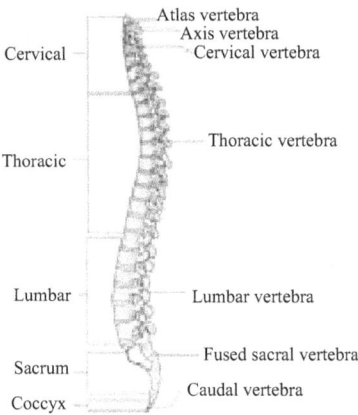

Vertebral column of man

Ribs and Sternum

The ribs are long and curved bones consisting of two main parts; a bony part known as the vertebral parts and a sternal part, made up of cartilage part which is attached to the sternum.

The ribs, sternum and the thoracic vertebrae from the rib cage which protect the heart and lungs.

Appendicular Skeleton

The appendicular skeleton consist of the pectoral girdle, the pelvic girdle, the forelimbs and the hind limbs

The Pectoral Girdle is made up of two halves with each consisting of a triangular bone called scapular or shoulder blade and the clavicles or collar bone

The Pelvic girdle consists of two halves known as innominated bone. Each consists of three fused bones namely, ilium, ischium and pubis.

The Forelimb is made up of a single bone nearest the body called humerus. The humerus is followed by the bones of the fore-arm called radius and ulna. The wrist comprises about 7-8 rod-like bones called carpals. The bones of the palm are called metacarpals, while the jointed one forming the fingers and thumb are called phalanges.

The Hind Limb consists of a single upper bone forming the thigh called femur and two bones of the lower leg called tibia and fibula. The small bones of the ankle are called tarsals. The rod-like bones of the sole of the foot are called metatarsals while the jointed bones of the toes are called phalanges.

Joint

A joint is a point where two or more bones meet. The bones are held together by a tissue called ligament. A joint is lubricated by the synovial fluid.

There are two types of joints, namely:

1. **Fixed or immoveable Joint**: This occurs between the flattened bones of the skull e.g. suture

2. **Moveable Joint**: This permits certain degree of movement. There are different types of moveable joints e.g.

i. **Hinge joint** e.g. elbow and knee joints

ii. **Ball and socket joint** e.g. hip and shoulder joints

iii. **Sliding or gliding joint** e.g. wrist and ankle joint

iv. **Pivot joint** e.g. neck where atlas rotate on odontoid process of axis.

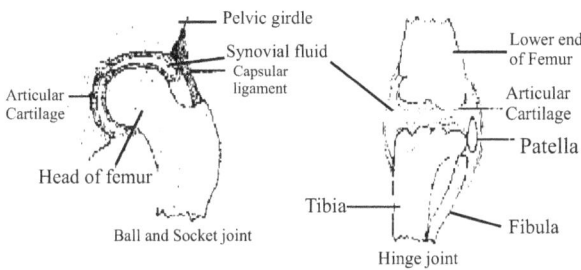

Two types of joints

EXPERIMENT ON TROPISM

Experiment: 1 To demonstrate hydrotropism.

Materials required. A trough, top soil, a porous pot, seedlings.

Method. A trough is taken and it is two third filled with moist top soil. Some amount of soil is removed from the centre and an empty porous pot is placed where the soil has been removed. A few seedlings, say 2 or 3 are planted on the sides of the pot. The seedlings are planted about 6cm away from the pot. Without watering, the seedlings are then left for some days. Water is poured into the porous pot when the plants start to wither. A similar experiment to serve as control is also set up but the porous pot does not contain water. Both experiments are then left for many days. After this period, the young plants are dug out, exercising care so as not to damage the roots. The soil around the roots is carefully washed out to expose the roots.

Observations. The main root curves towards the direction

of the porous pot. There is no curving of the root in the case of the control experiment. Here, the root is straight.

Conclusion. The curved root is in response to water which found its way out of the porous pot. This indicates that the root shows positive hydrotropism.

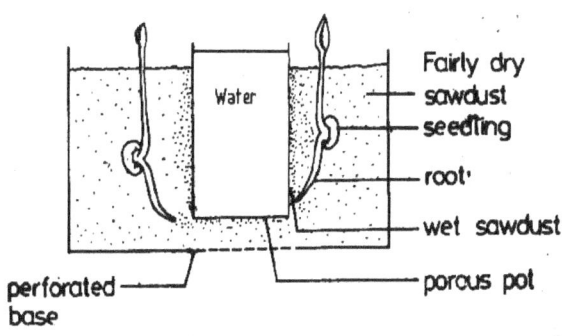

Experiment: 2 To demonstrate geotropism.
(a) Negative geotropism (shown by most plant shoots)

Materials required. Bean seeds, container, soil, cupboard.

Method. A container is collected and top soil is packed inside it. Two or three bean seeds are sowed in the soil and the seeds are left to germinate. When the plants are at the seedling stage, the container is kept in a dark cupboard where the seedlings cannot receive light. The container is placed on its side and it is left like that overnight. The seedlings are then checked the following day.

Observations. The young shoot turns upwards.

Conclusion. The upward turn shown by the shoots is as a result of negative response to the force of gravity.

(b) Positive geotropism (shown by most roots)

The above experiment is repeated. However, the container is left on its sides for about 3 days. After this period the soil around the main root is removed. It is noticed that the roots bend downwards in response to the force of gravity. The response by the root is described as **positive geotropism**.

Experiment: 3 To demonstrate phototropism.

Materials required and method. The material required for this demonstration is a pot of seedlings. The pot is placed beside a window, where light can get to the seedlings. They are left for a few days. After this period, the shoots are observed to bend towards the source of light. The side that is directed towards the light is the made to point towards another side, i.e. be directed away from the light source. The shoots again turn and face the light.

Conclusion. The turning towards the light by the shoots shows positive hydrotropism.

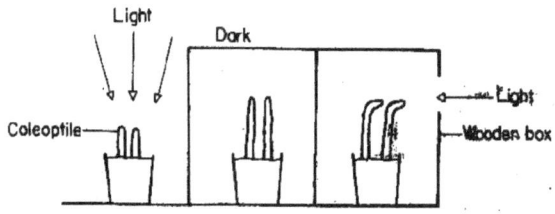

Fig. 65. Experiment to show phototropism

Experiment: 4 To demonstrate geotropic response of the shoot and root using a klinostat.

Materials required. A klinostat, a bean seedling.

Method. The cork of a klinostat is adjusted to a vertical plane, so that the bean seedling is horizontal. The apparatus is then rotated for a period of time and then stopped and left for sometime.

Observations. There was neither a bending of the shoot (plumule) upward, nor the root (radicle), downward, i.e. the shoot and root remained straight. But when the rotation stopped, the shoot and the root curved upwards and downwards respectively, after a time.

Conclusion. Both shoot and root did not respond to geotropic effect because the effects of the stimulus were diffused by the rotation of the cork disc. This shows that the force of gravity affects the growth of shoot and root. Secondly, the experiment proves that growth is due to unequal distribution of auxin.

NOTE: A klinostat is an instrument which supports the fact that phototropism and geotropism are due to unequal distribution of auxin in the shoot and root respectively. The instrument removes or neutralises the effects of directional stimuli of light and force of gravity on the plant organ.

The parts of the klinostat include a clockwork mechanism. This bears a rod which carries a metal disc and to the metal disc is fitted a cork disc. During an experiment, seedlings are pinned to the cork disc. Both discs can be adjusted to vertical or horizontal position. When the motor is on, the disc rotates slowly, thus turning the seedlings. Usually, there is one turn in every 15 minutes.

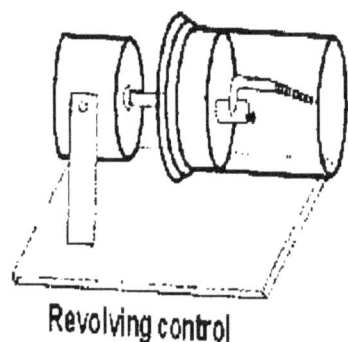

Revolving control

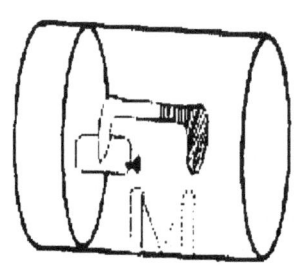

Stationary (Positive geotropism)

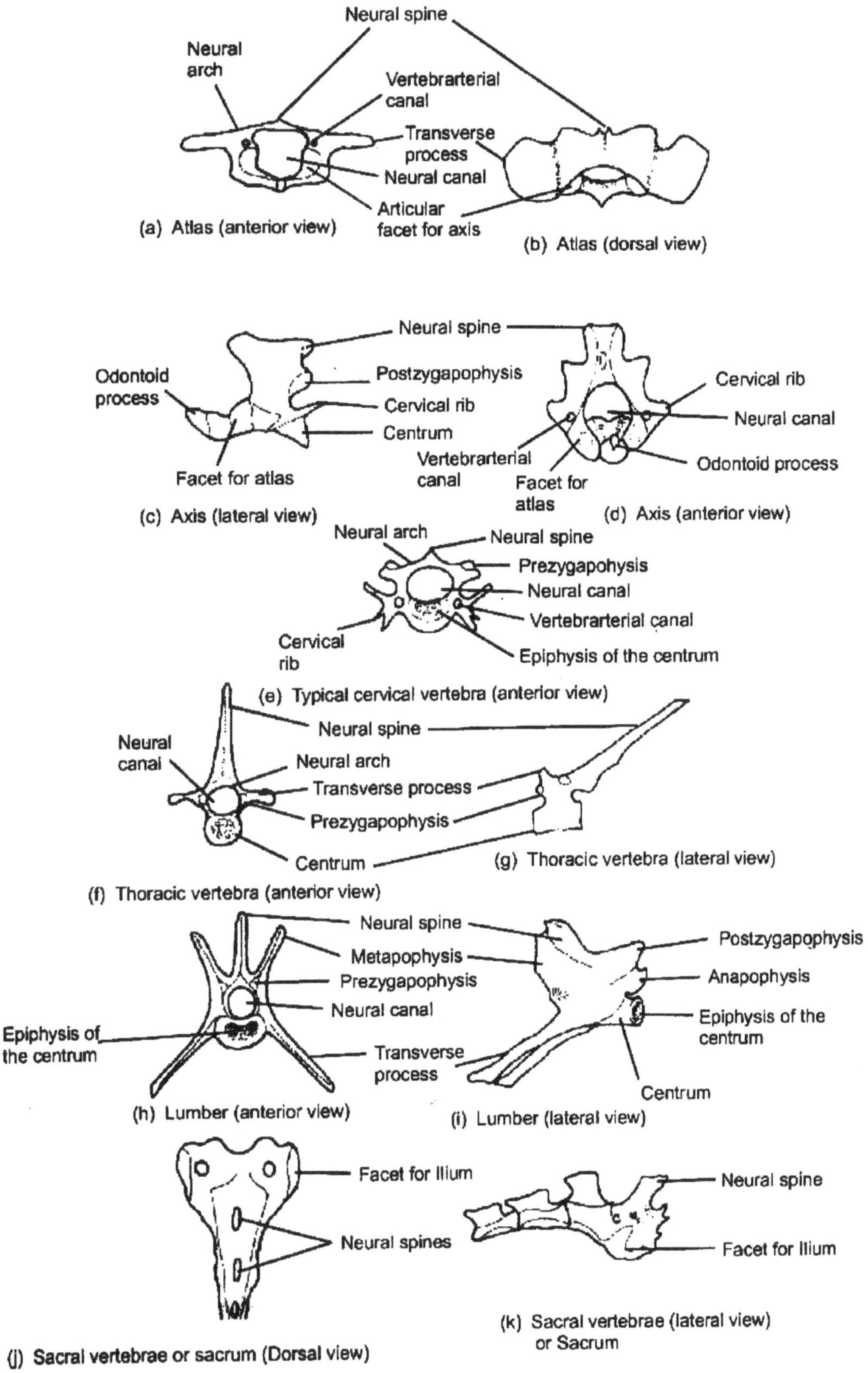

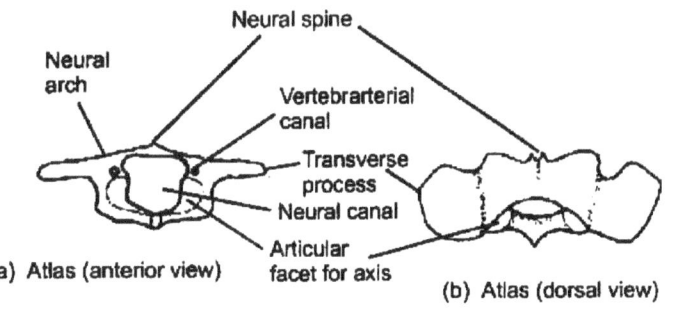

(a) Atlas (anterior view)

(b) Atlas (dorsal view)

(c) Axis (lateral view)

(d) Axis (anterior view)

(e) Typical cervical vertebra (anterior view)

(f) Thoracic vertebra (anterior view)

(g) Thoracic vertebra (lateral view)

(h) Lumber (anterior view)

(i) Lumber (lateral view)

(j) Sacral vertebrae or sacrum (Dorsal view)

(k) Sacral vertebrae (lateral view) or Sacrum

Parts of the axial skeleton (a–k)

101

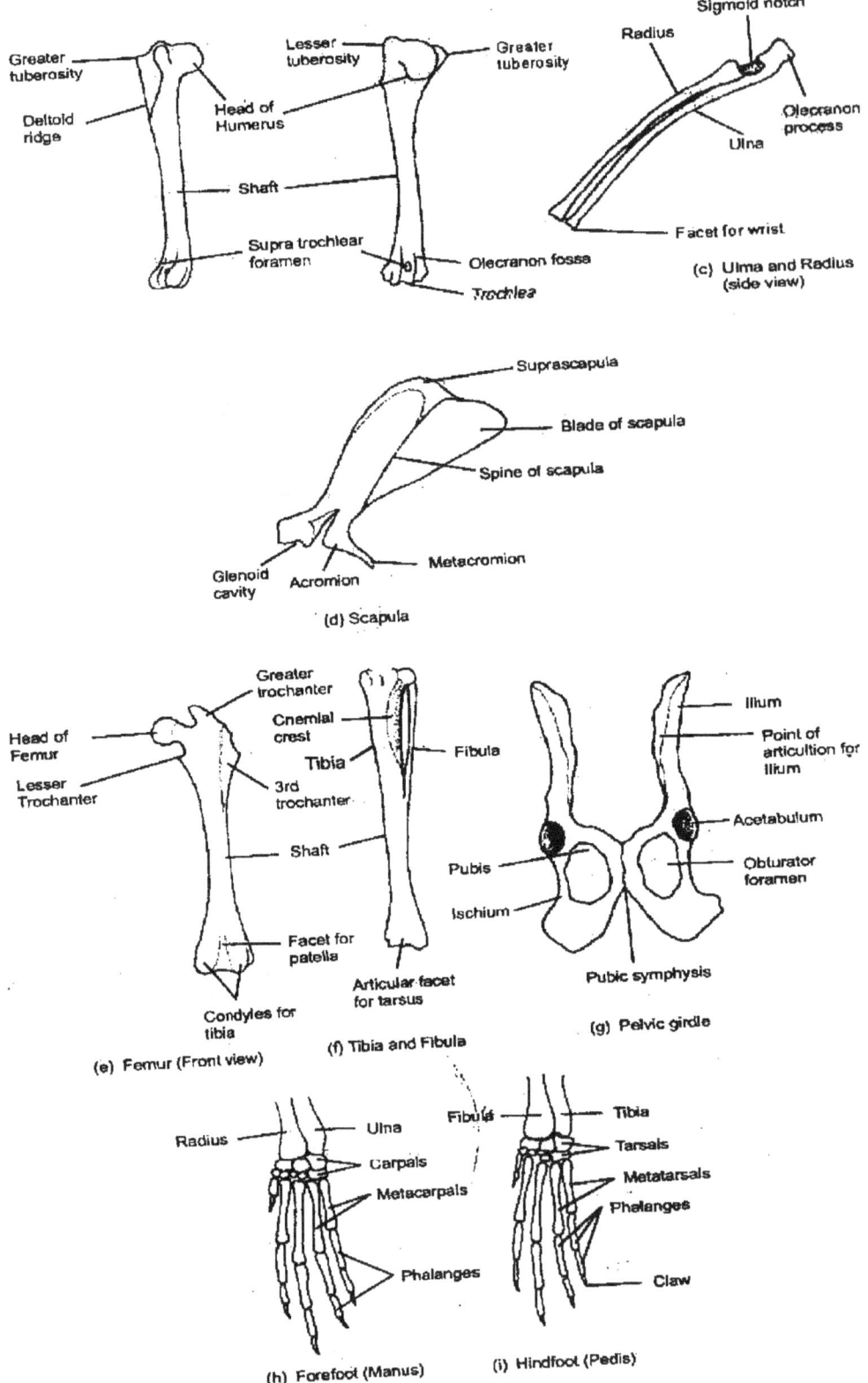

Parts of Appendicular Skeleton (a-I)

11.1. SUPPORTING TISSUES IN PLANTS

1. UME/79/3
Which of these tissues serves the function of support and water conduction?
A. Parenchyma   B. Collenchyma
C. Sclerenchyma   D. Xylem
E. Phloem

2. UME/87/27
Sclerenchyma cells are lignified to
A. strengthen and support the plant
B. transport synthesized food
C. conduct water and salt
D. protect the plant from injury

3. UME/88/26
Parechyma cells serve as supporting tissues when they
A. contain chloroplast   B. have crystals
C. become flaccid   D. become turgid

4. UME/94/17
The phloem parenchyma is sometimes used for
A. food storage
B. supporting the stem
C. production of the sieve tube
D. transporting water

5. PCE/2000/20/Type A
The plant tissue that is functionally similar to animal bones is
A. endodermal tissue
B. sclerenchymatous tissue
C. epidermal tissue
D. parenchymatous tissue

6. PCE/96/19
Which of the following contributes to the rigidity and strength of the stem?
A. Parenchyma   B. Cambium
C. Sclerenchyma   D. Xylem

7. UME 2008/22
A plant parenchyma cell also acts as a supporting tissue when it
A. becomes flaccid   B. contains crystals
C. becomes turgid   D. is pigmented

11.2. MOVEMENT IN PLANTS

8. UME 81/42
Growing yam tendrils climb for support. The growth response is
A. haptotropism   B. geotropism
C. phototropism   D. hydrotropism
E. chemotropism

9. UME/82/16
Roots of plants are normally
A. positively phototropic
B. negatively geotropic
C. negatively hydrotropic
D. positively hydrotropic
E. negatively chemotropic

10. UME/83/37
In a positive phototropic response of a coleoptile, the region of greatest curvature is brought about by the
A. Movement of auxins away from the region of curvature
B. Even distribution of auxins in all parts of the coleoptile
C. Inhabitation of growth by auxins in the region of smaller curvature
D. Concentration of auxins in the region of curvature
E. Absence of auxins in the coleoptile

11. UME/88/27
Taxism differs from tropism because
A. the whole organism is affected
B. it is a directional movement
C. it is a response to multi-directional stimuli
D. part of the organism is affected

12. UME/91/30
The movement of the whole organism to an external stimulus is termed
A. tropism   B. a taxis
C. a nastic movement   D. a phototropic movement

13. UME/93/23
The response shown by the tips of the root and shoot of a plant to the stimulus of gravity is
A. haptotropism   B. phototropism
C. hydrotropism   D. geotropism

14. UME/95/20
When a healthy shoot of a flowering plant is illuminated from one side, auxins accumulate on the
A. non-illuminated side of he shoot
B. illuminated side of the shoot
C. upper side of the shoot
D. lower side of the shoot

15. UME/2003/25/Type A
The response of plants to external stimuli in a non-directional manner is known as
A. nastic movement   B. phototropism
C. tactic movement   D. geotropism

16. UME 2010/23/TYPE: D
Which of the following stimuli is likely to elicit a nastic response in an organism?
A. light intensity   B. chemical substances
C. gravity   D. touch

17. Tropism is defined as:
A. Movement to a source of light
B. Growth away from or to source of gravity
C. Response to water
D. Growth curvature to diffused light
E. Growth towards or from a unilateral stimulus

18. Nastic movement e.g. is seen in the opening of flowers and buds, is a response to the external stimulus of light. This stimulus, however, does not determine:
A. The opening of the flowers or buds
B. The direction of the movement of flowers and buds
C. The period of opening of the flowers and buds
D. The duration of opening of the flowers and buds
E. None of the above

19. This is a good example of negative geotropism:
A. The downward growth of roots of the palm tree
B. The prop roots of the screw pine
C. The roots of epiphytic plants
D. The breathing roots of mangrove trees
E. The clasping roots of pepper

20. Rhizomes and some stolons growing horizontally are
A. Positively geotropic
B. Negatively geotropic
C. Nearly negatively geotropic
D. Neutral
E. None of the above

21. The klinostat is used in geotropic experiments to illustrate:
A. Positive geotropism
B. How geotropic responses could be prevented
C. Negative geotropism
D. Horizontal geotropic curvature
E. None of the above

22. If young seedlings are illuminated from the top, the seedlings will:
A. Bend towards the top
B. will bend sideways
C. Remain straight
D. Grow taller
E. None of the above

23. When the sensitive plant *Mimosa pudica* is touched its leaves drop. This is due to:
A. The inherent nature of the plant
B. Contact stimulus
C. The activity of the pulvinus of leaf bases
D. Differential changes in turgor in the lower and upper sides of the pulvinus
E. None of the above

24. After a storm, the maize plants in a field are found blown down. A day after this, all the plants straightened up at some nodes of the stems. This is an advantage of:
A. Negative geotropism
B. Negative phototropism
C. Wind activity
D. Hydrotropism
E. Positive geotropism

25. If a young seedling is stimulated by unilateral light, the seedling bends towards the source of the light. The accumulated hormones on the shaded side of the seedling caused
A. Contraction of the cells on the shaded side of seedling
B. The elongation of the cells on the illuminated side
C. A greater elongation of the cells on the shaded side
D. The seedling to bend away from light
E. None of the above

26. The opening and closing of the stomata of leaves are controlled by a number of factors such as light, humidity of the surrounding air and water absorbed from the soil. Light is responsible for:
A. The evaporation of water from the cells
B. The formation of carbohydrates in the stomata
C. The formation of carbohydrates in the guard cells
D. The formation of the chloroplasts in the guard cells
E. None of the above

11.3. SUPPORTING TISSUES IN ANIMALS
27. Which of these is not true of the skeleton?
A. The skeleton supports the body and its associated structures
B. The skeleton, especially the long bones, gives rise to the red blood cells
C. The skeleton serves as structure for the insertion of muscles
D. The skeleton, especially the skull and the ribs, protects the delicate structure within them
E. None of the above

28. Which of these is not an external skeleton?
A. the scales   B. the carapace
C. the carpals   D. the hoofs
E. None of the above

29. Which of these make up the internal skeleton of an animal?
A. Bone and cartilage   B. Ligament
C. Water   D. Elastin
E. None of the above

30. What type of skeleton has the earthworm?
A. it is a bony skeleton
B. it is an internal skeleton
C. it is a hydrostatic skeleton
D. it is an external skeleton E. None of the above

31. PCE/98/21/TYPE C
The difference between tendons and ligaments is that
A. Tendons join bones to muscle while ligaments join bone to bone
B. Ligaments join bone to muscle while tendons join bone to bone
C. Both are strong muscle fibres but tendons are stronger
D. Ligaments are found in the hip bone while tendons are found in the shoulder bone

32. UME/84/15
Exoskeleton is NOT found in the
A. maggot   B. mosquito larva
C. earthworm   D. caterpillar   E. termite

33. UME/84/39
The axial skeleton of a mammal does not include the bone of the
A. skull   B. tail
C. limbs   D. back
E. neck

34. UME/85/29
The appendicular skeleton is made up of the
A. limbs
B. skull and limbs
C. phalanges
D. ulna and radius
E. girdles and limbs

35. UME/93/24
Which of the following is the correct order of the vertebrae along the spinal column?
A. Axis →Atlas →Thoracic → Lumbar→ Cervical →Sacral
B. Atlas →Cervical →Axis → Thoracic → Lumber → Sacral
C. Atlas→ Axis→ Cervical →Thoracis → Lumber → Sacral
D. Axis → Atlas → Cervical →Thoracic→ Sacral → Lumbar

36. UME/82/19
Which vertebra has a projection called odontoid process?
A. Atlas   B. Axis
C. Thoracic   D. Lumbar
E. Caudal

37. UME/2000/16/TYPE M
The scapula and ischium are part of the
A. pectoral girdle
B. pelvic girdle
C. appendicular skeleton
D. hind limb

38. PCE/97/20
The function of zygapophysis is for the
A. free movement of the bones
B. support of the internal organs
C. protection of the spinal cord
D. articulation of successive vertebrae

39. PCE/97/19
The part of the typical vertebra through which spinal cord passes is the
A. centrum   B. neural arch
C. neural canal   D. vertebraterial canal

40. PCE/99/24/Type E
The ball and socket joint allows movement
A. only in one direction
B. up and down with slight rotation
C. in two directions   D. in several directions

41. PCE/93/24
The type of joints found between the bones of the ankle is
A. sliding or gliding   B. ball and socket
C. hinge   D. rotatory

42. Which of the follow is NOT a function of the mammalian skeleton?
A. Protection   B. Respiration
C. Transportation   D. Support.

43. UME 2009/22
The axial skeleton is found in the
A. skull, ribs, vertebral column and breastbone
B. skull, humerus, vertebral column and ribs
C. breastbone, clavicle, ribs and vertebral column
D. femur, sternum, ulna and skull

44. PCE/96/18
The major difference between the atlas and the axis of the cervical vertebrae is that only the axis possesses
A. facets for attachment to the skull
B. an odontoid process
C. a long neural spine
D. a large transverse process

45. Which of these structures is not common to both atlas and axis vertebrate?
A. the vertebraterial canals   B. the neural arch
C. the neural spine   D. the postzygapophyses
E. the centrum

46. What is the importance of the neural canal?
A. it carries the neural fluid
B. it carries the spinal cord
C. it carries the neuron
D. it carries the odontoid process of the axis bone
E. none of the above

47. The atlas and axis bones differ from other cervical bones in this respect
A. they have no neural spines
B. they have no vertebraterial canals
C. they have no centra
D. they have no transverse processes
E. None of the above.

48. The thoracic bones are distinguished from other bones by the possession of
A. long neural spines
B. capitulum and tuberculum
C. vertebraterial canals
D. demi – facets
E. articular surfaces

49. What is the interveterbral disc?
A. it is a pad in the centrum of a bone
B. it is a pad between two centra
C. it is cartilage bone in the body
D. it is a soft pad of bone between the thoracic bones
E. None of the above

50. The lumbar bones are large and stout. This is because
A. they are the most important bones of the skeleton
B. they carry the weight of the animal
C. they are at the back
D. they support the bones of the limbs
E. They support the large muscles of the back

51. What is the characteristic of the sacrum?
A. The bones are flattened out
B. The transverse processes are small and slender
C. The bones are fused together
D. The neural canal is often very large
E. None of the above

52. What is the name of the longest bone of the lower arm?
A. It is the radius   B. It is the ulna
C. It is the tibia   D. It is the intermedium
E. It is the ulnare

53. The long bone of the thigh is:
A. the thigh bone   B. the humerus
C. the femur   D. the tibia
E. the fibula

MISCELLANEOUS QUESTIONS (11)
54. Turgidity of cell in herbaceous plants is important as it
A. provides mechanical support
B. enables absorption of more water
C. enables the intake of mineral salts
D. prevents plasmolysis of the cells
(PCE/2001/22/type:4)

55. The liquid that serves as a lubricant to the mammalian joints is the
A. amniotic fluid
B. spinal fluid
C. seminal fluid
D. synovial fluid
(PCE/2005/10/Type:P)

56. When bacterial swim from cold to warm region, this is known as
A. negative chemotaxis
B. positive thermotaxis
C. positive phototaxis
D. negative phototaxis
(UME/2005/23/Type: D)

57. Hydrostatic skeleton is the type of supporting system found in
A. mammals
B. reptiles
C. oligochaetes
D. arthropods (UME/2005/23/Type D)

58. Tropic response due to the stimulus of touch is known as
A. geotropism
B. phototropism
C. thigmotropism
D. hydrotropism (PCE/2001/23/Type: 4)

59. Taxism differs from tropism because
A. only a part of the organism is affected
B. the whole organism is affected
C. it is a response to multi-directional stimulus
D. it is a directional movement (PCE/2000/21/Type:A)

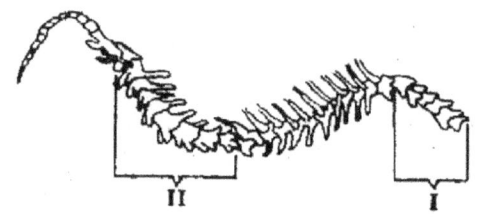

60. The type of joint between adjacent bones in the part labeled II is the
A. hinge joint   B. suture joint
C. sliding joint
D. ball- and- socket joint (UME/2004/8/Type:4)

61. The bones labeled II are called
A. thoracic vertebrae
B. lumbar vertebrae
C. cervical vertebrae
D. sacral vertebrae (UME/2004/9/Type:4)

62. The chitin in the exoskeleton of many arthropods is strengthened by
A. calcium compounds
B. organic salts
C. lipids
D. proteins (UME/99/26/Type:D)

63. The presence of endoskeleton is characteristics of
A. invertebrata
B. vertebrata
C. insecta
D. coelenterata (UME/2000/30/Type:Y)

64. When the forearm is raised, which muscle contracts n the process?
A. cardiac muscle
B. biceps
C. triceps
D. involuntary muscle (PCE/92/24)

65. The type of joints found between the bones of the ankle is
A. sliding or gliding
B. ball and socket
C. hinge
D. rotator (PCE/93/24)

66. The closing up of the lid of a pitcher plant on the entry of an insect is referred to as
A. tactic movement
B. tropic movement
C. nastic movement
D. heliotropic movement (PCE/96/22)

67. Which of the following is made up of dead cells
A. cambium
B. cork
C. palisade
D. meristem (PCE/98/20/TYPE: C)

68. Muscles are usually attached to bones by the
A. cartilage
B. ligaments
C. membranes
D. tendons (PCE/03/44/TYPE: N)

69. Stomata closes in the night because
A. The air is dry
B. No water is absorbed by the roots
C. Transpiration is at a standstill
D. There is no more formation of carbohydrates
E. The chloroplasts of guard cells cease to function

CHAPTER 12

REPRODUCTION

Understand the following after reading this chapter:

REPRODUCTION
Reproduction is the process through which living organisms give rise to young ones of their kind. There are two types of reproduction viz, asexual and sexual reproduction.

ASEXUAL REPRODUCTION
Asexual reproduction is the formation of new individuals from the cell(s) of a single parent.
It is very common in plants; less so in animals.

Budding
Some cells split via budding (for example baker's yeast), resulting in a 'mother' and 'daughter' cell.

Spore formation
Many multicellular organisms form spores during their biological life cycle in a process called *sporogenesis*. Plants and many algae on the other hand undergo *sporic meiosis* where meiosis leads to the formation of haploid spores rather than gametes. These spores grow into multicellular individuals (called gametophytes in the case of plants) without a fertilization event. These haploid individuals give rise to gametes through mitosis.

Fragmentation
Fragmentation is a form of asexual reproduction where a new organism grows from a fragment of the parent. Each fragment develops into a mature, fully grown individual. Fragmentation is seen in many organisms such as animals (some annelid worms, turbellarians and sea stars), fungi, and plants.

Natural Vegetative Propagation
All plant organs have been used for asexual reproduction, but stems are the most common.

Stems
In some species, stems arch over and take root at their tips, forming new plants.
The horizontal above-ground stems (called **stolons**) of the strawberry, for example, produce new daughter plants at alternate nodes.
Underground stems
- rhizomes
- bulbs
- corms and
- tubers

They are used for asexual reproduction as well as for food storage.
Irises and day lilies, for example, spread rapidly by the growth of their rhizomes.

Leaves
The typical example is the leaves of the common ornamental plant *Bryophyllum* (also called Kalanchoë). Mitosis at meristems along the leaf margins produce tiny plantlets that fall off and can take up an independent existence.

Roots
Some plants use their roots for asexual reproduction. The dandelion is a common example. Trees, such as the poplar or aspen, send up new stems from their roots. In time, an entire grove of trees may form — all part of a **clone** of the original tree.

Artificial Vegetative Propagation
Commercially-important plants are often deliberately propagated by asexual means in order to keep particularly desirable traits (e.g., flower color, flavor, resistance to disease).

Grafting: In grafting 2 plants are used to develop a new plant with combined traits from the 2 parent plants. In grafting the **scion** is the above ground part of one plant. The **scion** is attached to the **stock** which is the rooted part of the second plant.

Cuttings: Cuttings are part of the plant that is cut off of the parent plant. Shoots with leaves attached are usually used. New roots and leaves will grow from the cutting. The shoot is cut at an angle. A growth promoter may be used to help with the growth of the roots

Layering: In layering a shoot of a parent plant is bent until it can be covered by soil. The tip of the shoot remains above ground. New roots and eventually a new plant will grow. These plants can then be separated.

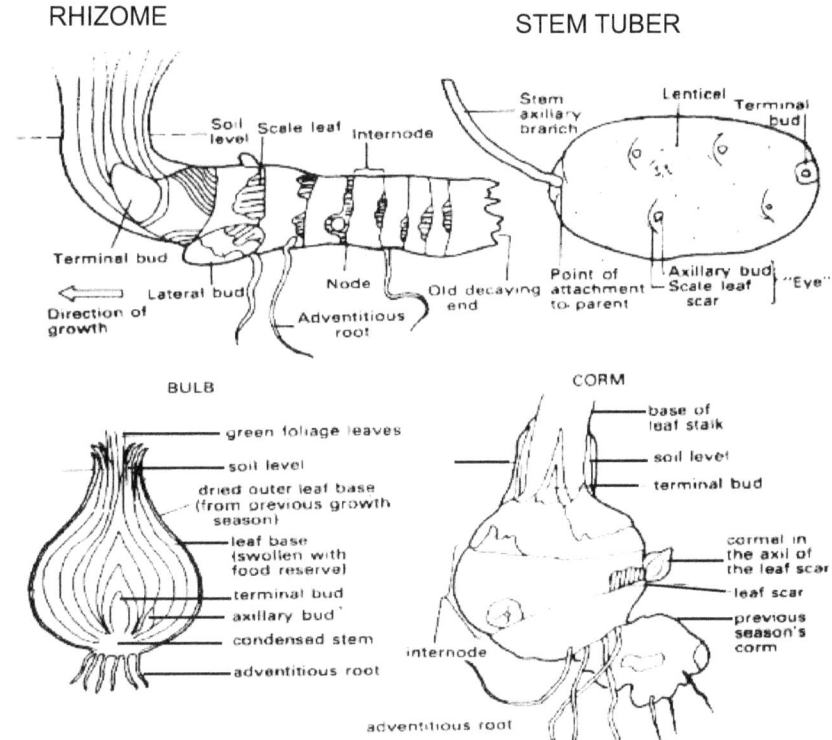

Vegetative propagation by means of rhizome, stem tuber, bulb and corm

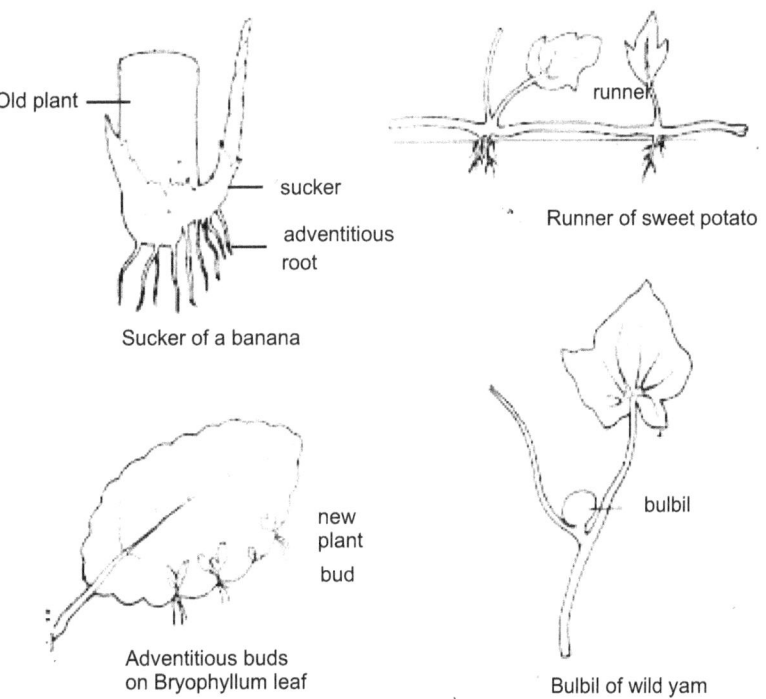

Vegetative propagation by means of sucker, runner, brophyllum and bulbil

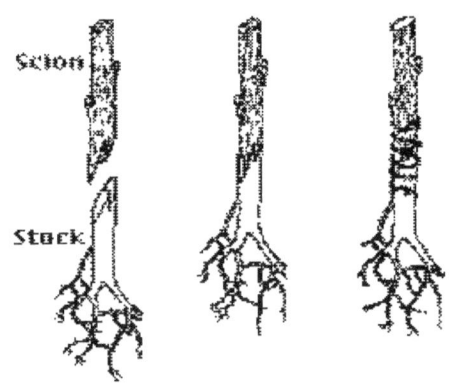

Scion

Stock

ASEXUAL REPRODUCTION IN ANIMALS
Fission
An important form of fission is binary fisson but there is some other type of fission that called mutifission. In binary fission, the parent organism is replaced by two daughter organisms, because it literally divides in two.

In multiple fission, the <u>nucleus</u> of the parent cell divides by <u>mitosis</u> several times, producing several nuclei. The cytoplasm then separates, creating multiple daughter cells.

Budding
Here, offspring develop as a growth on the body of the parent. In some species, e.g., hydra, <u>jellyfishes</u> and many <u>echinoderms</u>, the buds break away and take up an independent existence. In others, e.g., <u>corals</u>, the buds remain attached to the parent and the process results in **colonies** of animals. Budding is also common among parasitic animals, e.g., tapeworms.

Fragmentation
As certain tiny worms grow to full size, they spontaneously break up into 8 or 9 pieces. Each of these fragments develops into a mature worm, and the process is repeated.

Parthenogenesis
In parthenogenesis ("virgin birth"), the females produce eggs, but these develop into young without ever being fertilized. Parthenogenesis occurs in some fishes, several kinds of insects, and a few species of frogs and lizards. It does not normally occur in mammals because of their <u>imprinted genes</u>. However, using special manipulations to circumvent imprinting, laboratory mice have been produced by parthenogenesis.

Cell Division by Mitosis
Mitosis is a type of cell division that produces genetically identical cells. During mitosis DNA replicates in the parent cell, which divides into two new cells, each containing an exact copy of the DNA in the parent cell. The only source of genetic variation in the cells is via mutations.

Interphase		· This is when the cell is <u>not</u> dividing, but is carrying out its normal cellular functions.
		· chromatin not visible
		· DNA, histones and centrioles all replicated
the		· Replication of cell organelles e.g. mitochondria, occurs in cytoplasm.
Prophase		· chromosomes condense and become visible – this prevents tangling with other chromosomes.
		· Due to DNA replication during interphase, each consists of two identical sister <u>chromatids</u> connected at the centromere
chromosome		· centrioles move to opposite poles of cell
		· nucleolus disappears
		· phase ends with the breakdown of the nuclear membrane
Metaphase		· <u>spindle fibres</u> (microtubules) connect centrioles to chromosomes
		· chromosomes align along equator of cell and attaches to a spindle fibre by its centromere.

Anaphase		· centromeres split, allowing chromatids to separate · chromatids move towards poles, centromeres first, pulled by kinesin (motor) proteins walking along microtubules (the track) · Numerous mitochondria around the spindle provide energy for movement
Telophase		· spindle fibres disperse · nuclear membranes from around each set of chromatids · nucleoli form
Cytokinesis		· In animal cells a ring of actin filaments forms round the equator of the cell, and then tightens to form a <u>cleavage furrow,</u> which splits the cell in two.
		· In plant cells vesicles move to the equator, line up and fuse to form two membranes called the <u>cell plate</u>. A new cell wall is laid down between the membranes, which fuses with the existing cell wall.

Mitosis and Asexual Reproduction

Asexual reproduction is the production of offspring from a single parent <u>using mitosis</u>. The offspring are therefore genetically identical to each other and to their "parent"- in other words they are <u>clones</u>. Asexual reproduction is very common in nature, and in addition we humans have developed some new, artificial methods. The Latin terms *in vivo* ("in life", i.e. in a living organism) and *in vitro* ("in glass", i.e. in a test tube) are often used to describe natural and artificial techniques.

Meiosis

Meiosis is the special form of cell division used to produce gametes. It has two important functions:

• To form haploid cells with half the normal chromosome number
• To re-arrange the chromosomes with a novel combination of genes (genetic recombination)

Meiosis comprises two successive divisions, without DNA replication in between. The second division is a bit like mitosis, but the first division is different in many important respects. The details are shown in this diagram for a hypothetical cell with 2 pairs of homologous chromosomes (n=2):

111

First Division	Second Division
Interphase I · chromatin not visible · DNA & proteins replicated	**Interphase II** · Short · no DNA replication · chromosomes remain visible.
Prophase I · chromosomes visible · homologous chromosomes join together to form a <u>bivalent</u> · chromatids <u>cross over</u> (chiasmata- see below)	**Prophase II** · centrioles replicate and move to new poles.
Metaphase I · bivalents line up on equator	**Metaphase II** · chromosomes line up on equator
Anaphase I · chromosomes separate (not chromatids-centromere doesn't split)	**Anaphase II** · centromeres split · chromatids separate.
Telophase I · nuclei form · cell divides · cells have 2 chromosomes, not 4 chromatids.	**Telophase II** · 4 haploid cells, each with 2 chromatids · cells often stay together to form a <u>tetrad</u>.

Meiosis and Sexual Reproduction

Sexual reproduction is the production of offspring from two parent using <u>gametes</u>. The cells of the offspring have two sets of chromosomes (one from each parent), so are diploid. Sexual reproduction involves two stages:

- <u>Meiosis</u>- the special cell division that makes haploid gametes
- <u>Fertilisation</u>- the fusion of two gametes to form a diploid zygote

There are differences between meiosis II and mitosis:

- DNA will not replicate before meiosis II

- The sister chromatids in meiosis II will likely not be identical due to crossing over

- The number of chromosomes at the center of the cell in meiosis II is half the number of chromosomes in mitosis.

Meiosis results in 4 genetically different haploid nuclei

SEXUAL REPRODUCTION

Sexual reproduction: the process involving the fusion of haploid nuclei (23 chromosomes) to form a diploid zygote (46 chromosomes) and the production of genetically dissimilar offspring.

Advantages: produces genetically different offspring so

they don't all die from change in the environment. Disadvantage: it takes lots of time and energy, good characteristics can be lost.

Sexual reproduction in humans

Testes: have many coiled tubes which produce**sperm**, and the cells between tubes produce**testosterone**.

Scrotum: holds testicles

Spermduct: carries sperm from testicles to urethra.

Prostate gland: makes **seminal fluid** (semen =99.5% seminal fluid and 0.5% sperm) (*actually it makes prostate fluid and the seminal vesicles make seminal fluid)

Urethra: carries semen from sperm duct to tip of penis

Penis: the male sex organ, used to transfer semen tothe female. In most mammals, it is also used to expel urine from the body.

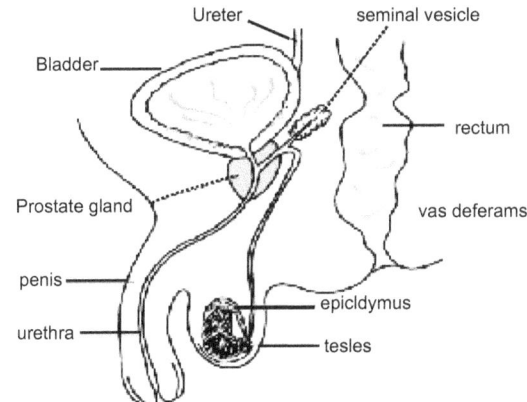

Male reproductive system

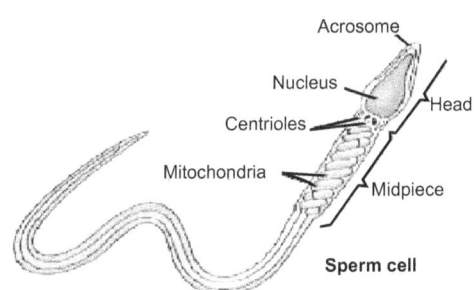

Sperm cell

Structure of sperm cell

Ovary: contains follicles which develop into the ova andproduces **progesterone** and **oestrogen** (hormones).

Oviduct: aka fallopian tube, carries the ovum to the uterus.Fertilisation occurs in the first 1/3.

Uterus: aka womb, where the fetus develops.

Cervix: neck of uterus: a strong rigid muscle, moist by mucuswith a small opening

Vagina: aka birth canal, receives the penis during intercourse,and is the way out for the baby at birth. Moist tube of muscle about 8cm long, it is very flexible and secretes slippery mucus when aroused.

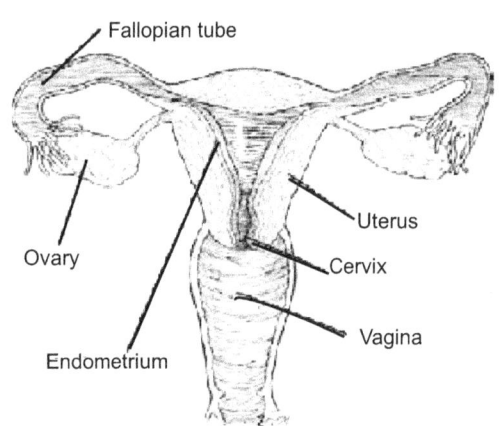

Female reproductive system

Sexual intercourse/sex/coitus/copulation: Penis fills with blood and becomes erect - hard enough to enter vagina.

Vagina walls secrete a lubricant. Rubbing of the **glans** (end of penis) against the vagina wall sets of a reflex action, causes sperm to be released from the **testes**, and is transported by peristalsis along sperm ducts and urethra, where

seminal fluid is added to make **semen**. The exit of semen from the penis is called **ejaculation**. Sperm then swim through the cervix and oviducts to the first third of the oviduct (1st third from the ovary) where one combines with the egg.

Fertilisation: the joining together (**fusion**) of an ovum and a sperm (**gametes**) to form a **zygote.**

Development of zygote:

1. One sperm penetrates
2. The ovum membrane alters to form a barrier against sperm.
3. Head of sperm (male nucleus) approaches and then fuses with the nucleus of the ovum.
4. Zygote divides over and over, to make a ball of cells called an **embryo** (6 days after fertilisation).
5. It implants itself in the wall of the nucleus (**implantation**) which is followed by **conception** (development into an individual)

Development of fetus: zygote is changed through **growth** (increase in number of cells by mitosis) and **development** (organisation of cells into tissues and organs)

Umbilical cord: contains umbilical artery which carries deoxygenated blood and waste products e.g. urea from the fetus to placenta and umbilical vein which carries oxygenated blood and soluble food such as iron, glucose and amino acids rom placenta to fetus. It contains the blood of the fetus

Placenta: organ for exchange of soluble materials such as foods, wastes and oxygen between mother and fetus, it is the physical attachment between the uterus and the fetus.

It has the mother's blood in it. (*It also serves as a form of protection for the fetus from mother's immune system and blood pressure difference, and it secretes hormones to maintain the uterus since the corpus luteum degenerates on the third month)

Amniotic sac: membrane which encloses amniotic fluid, broken at birth.

Amniotic fluid: protects fetus against mechanical shock, drying out and temperature fluctuations.

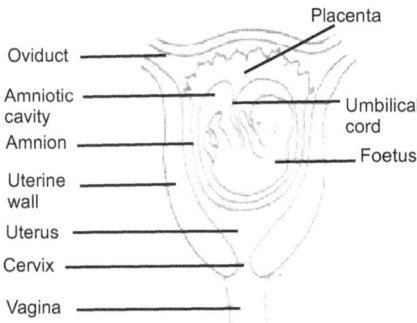

Development of foetus in the uterus

Sexual Reproduction in Flowering Plants
Flower structure

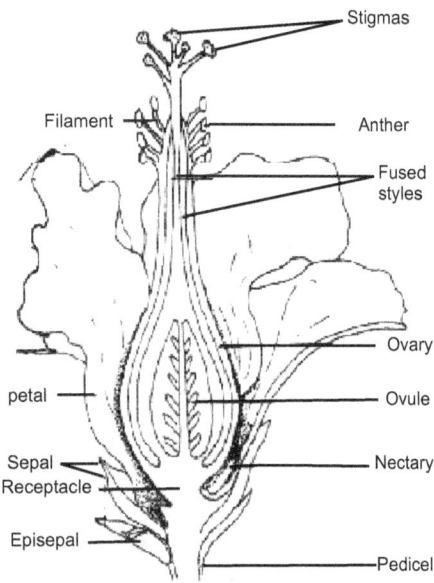

Structure of an hibiscus flower

Sexual reproduction in flowering plants centres around the flower. Within a flower, there are usually structures that produce both male gametes and female gametes.

Sepal: protect the flower bud.

Petal: brightly coloured and scented and may have nectaries which are all used to attract insects, petals in wind pollinated flowers are tiny, and used for pushing the bracts (leaf-like structures) apart to expose stamens and stigma

Anther: has pollen sacs with pollen grains which contain the male nucleus (male gamete).

Stigma: platform on which pollen grains land

Ovary: hollow chamber, ovules grow from the walls.

Pollen tube: pollen grain lands on stigma and creates a tunnel down the style, through the micropyle, to the ovules. Structure of non-endospermic seed:

Development of the ovule and female gamete

Inside the ovary there may develop one or more **ovules**. Each ovule begins life as a small projection into the cavity of the ovary. As it grows and develops it begins to bend but remains attached to the ovary wall by a placenta.

At the start, the ovule is a group of similar cells called the **nucellus**. As it develops, the mass of cells differentiates to form an inner and an outer **integument**, surrounding and protecting the nucellus within, but leaving a small opening called the **micropyle**.

At the centre of the ovule is an **embryo sac** containing the haploid egg cell (the female gamete).

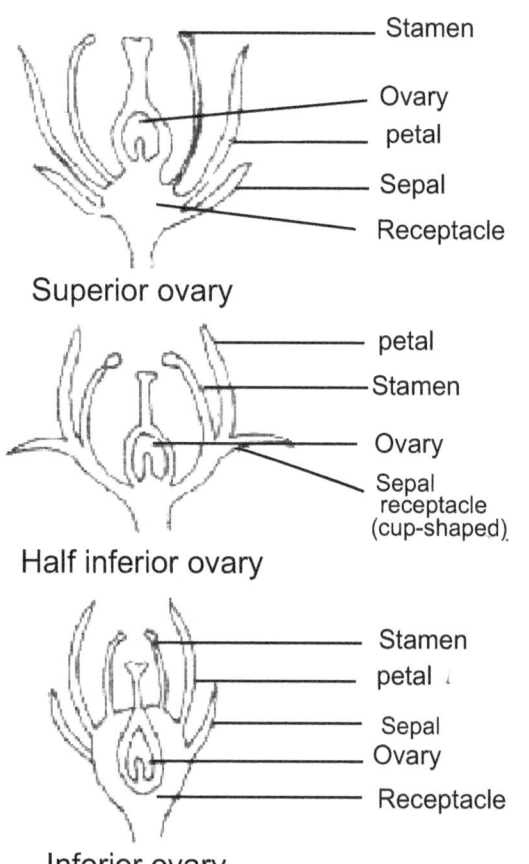

Types of ovary

Development of the male gamete

Each anther contains 4 pollen sacs. Many pollen grains develop inside each pollen sac. It begins with a mass of large **pollen mother cells** in each pollen sac. All are diploid.

114

In each pollen grain the wall thickens and forms an inner layer (the **intine**) and an often highly sculptured outer layer (the **exine**). The surface pattern is different on pollen grains from different species. When the pollen grains are mature, the anther dries out and splits open (a process called **dehiscence**) and the pollen is released.

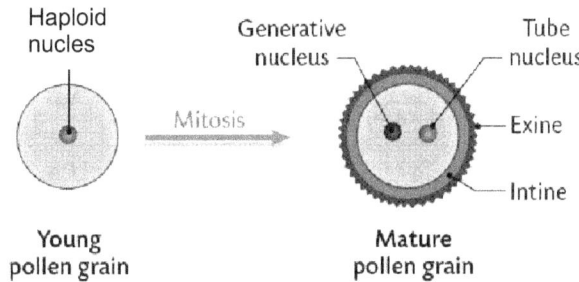

Haploid nucles — Mitosis — Generative nucleus — Tube nucleus — Exine — Intine

Young pollen grain   Mature pollen grain

Pollination

Many plants favour **cross-pollination**, so pollen must be transferred to the stigma of another plant if sexual reproduction is to take place. Some flowers rely of the wind to carry pollen grains others rely on insects.

Self-pollination is where the pollen is transferred to the stigmas of the same flower or the stigma of another flower on the same plant. Self-pollination is obviously more reliable, particularly if the nearest plant is not very close.

A potential drawback is that both gametes come from the same parent. If the plant is well adapted to a stable environment, the production of uniform offspring may be advantageous. However, inbreeding will result and if there are disadvantageous recessive characteristics in the parent, they are much more likely to be exposed than if the plant cross-pollinates.

Cross-pollination is less reliable and more wasteful than self-pollination, but it is genetically favourable because genes are transferred and variation increases.

Strategies to favour cross pollination:

- **Dioecious plants**: Some plants have flowers that are only male - they have only **stamen**. Other plants of the same species have flowers that are only female - they have only **carpels** e.g. Pawpaw
- **Monoecious plants**: Some flowers on a plant are only male; other flowers on the same plant are only female. So, self pollination is avoided by a difference in the timing of their development e.g. Maize
- **Protandry**: Anthers on some plants mature first. Pollination of immature stigma on the same plant is therefore not possible.
- **Protogyny**: The stigmas mature first.
- **Self-incompatibility**: Pollination can occur but the pollen tube does not grow well, if at all, so no fertilisation takes place.

For those plants that cross-pollinate, some are wind pollinated, others are insect pollinated. Here are some of the differences:

Feature:	Wind pollinated flowers:	Insect pollinated flowers:
Petals:	Small inconspicous, sometimes absent. If present, not brightly coloured.	Large, brightly coloured, conspicous and attractive to insects.
Scent:	None.	Often scented.
Nectary:	Absent.	Present.
Pollen:	Produced in large quantities, light, smooth pollen grains.	Less produced, pollen grains larger, sculptured walls to ald attachment to insects and to stigma.
Anthers:	Move freely, so pollen is easily dispersed.	Fixed to filaments and positioned to come into contact with visiting insects.
Stigma:	Large often branched and feathery, hanging outside the flower to trap pollen.	Small inclosed within the flower, positioned to come into contact with visiting insects.

Fertilisation

If the pollen grain lands on a compatible stigma, a **pollen tube** will grow so that eventually the egg cell, hidden away in the embryo sac, can be fertilised. A tube emerges from the grain, its growth being controlled by the tube nucleus at the tip of the tube. It may grow downwards in response to chemicals made by the ovary (a response known as chemotropism).

During the growth and extension of the tube, the **generative nucleus**, behind the tube nucleus, divides by mitosis to produce **2 male haploid gametes**. The pollen tube enters the ovule through the micropyle and penetrates the embryo sac wall. The tip of the tube bursts open, the tube nucleus dies and what follows is called **double fertilisation**.

1 male gamete fuses with the egg cell to produce a **diploid zygote**.

1 male gamete fuses with both the polar nuclei to produce the **triploid primary endosperm nucleus**.

Immediately after fertilisation, the ovule is known as the **seed**.

The following happens:

1. The zygote divides many times by mitosis to produce an **embryo**. It differentiates to become a **plumule** (young shoot), **radicle** (young root) and either 1 or 2 **cotyledons** (seed leaves). It is attached to the wall of the embryo sac by a suspensor.

2. The primary endosperm nucleus divides many times by mitosis to produce **endosperm tissue**. In some seeds this endosperm is a food store for later use by the seed. In others it may gradually disappear as the cotyledons develop.

3. To accommodate all this growth the embryo sac expands and the nucellus is crushed out of existence, giving its nutrients to the embryo and endosperm.

4. The integuments surrounding the embryo sac become the tough and protective **testa** (seed coat). The micropyle remains though so that oxygen and water can be taken in at germination.

5. The water content of the seed decreases drastically so the seed is prepared for dormancy.

6. The ovary wall becomes the **pericarp** - the fruit wall, the whole ovary now being the fruit. The function of the fruit is to protect the seeds and to aid in their dispersal, e.g. by an animal. That is why they can be brightly coloured and sweet; animals will eat them and scatter the seeds either at the time of eating or when they are passed out of the gut in defecation, unharmed.

The arrangement of ovules within the ovary is called **Placentation**. The main types of placentation are:

1. Marginal e.g. Crotalaria
2. Parietal e.g. Pawpaw
3. Axile e.g. Tomato
4. Free-central e.g. Waterleaf

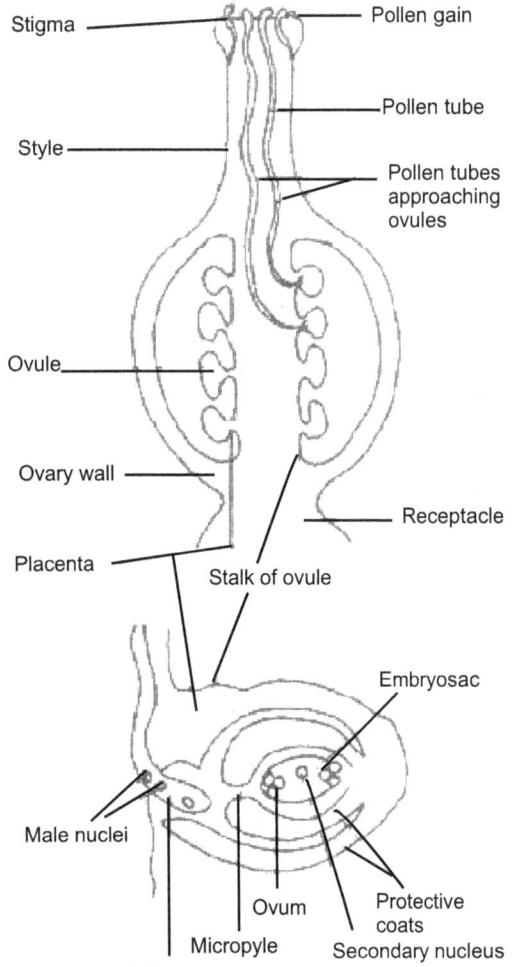

Fertilization in flower and structure of ovule

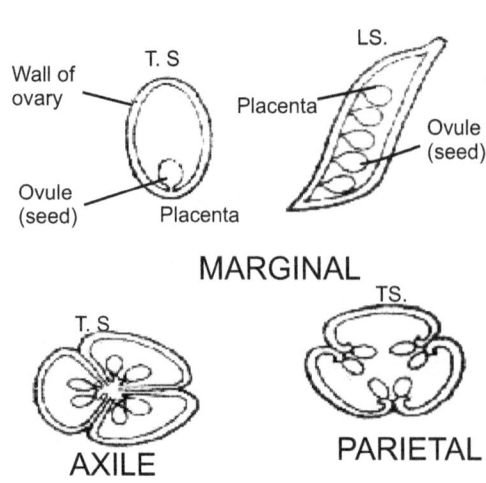

Types of placentation

116

TYPES OF FRUITS

There are two types of fruits viz:

1. True fruits: These are fruit developed from the ovary alone e.g mango, orange etc

2. False fruit: These are developed from the ovary and other floral parts such as the calyx, sepal or receptacle e.g. apple, pear etc

CLASSIFICATION OF FRUITS

Fruits are broadly classified into two as: **Dry fruits** and **succulent fruits.** Dry fruits are further divided into **dry dehiscent** and **dry indehiscent**

Types of dry dehiscent fruits

1. Follicle e.gcretis
2. Legumes e.g beans, flambouyantetc
3. Capsules e.g poppy, castor oil, Okra etc
4. Schizocarps e.gdesmodium, cassia etc

Types of dry indehiscent fruits

1. Achene e.g sunflower
2. Nut e.g cashew nut
3. Caryopsis e.g maize
4. Cypsela e.gTridax

Types of succulent fruits

1. Berry e.g tomatoes, orange etc
2. Drupe e.g mango, oil palm, coconut etc

DISPERSAL OF FRUITS AND SEEDS

Wind dispersal (Anemochory) - plants like members of Family Asteraceaehave structures that made their seeds easily blown by wind. That structure is called pappus. That is the bristle wing like structure of the seed-fruit e.g. Tridax.

Water dispersal (Hydrochory) - this is for plants that can float on water e.g. coconut. Most of the magrove species are dispersed by this means.

Animal dispersal (Zoochory) - this can be epizoochory or endozoochory. In epizoochory the seed is carried by animals accidentally on their body because the seed/fruit of the plants has structures that made them adhesive on animal skin/fur e.g. Triumfetta, *Desmodium*. In endozoochory the plant has very attractive fruit that can be eaten by the animal. The seed will be also consumed by the animal but not digestable e.g. guava, tomato.

Mechanical dispersal - dispersal caused by dehiscence of fruit. The fruit is usually pressurized and when ripe will explode. A good example is the fruit of okro – *Hibiscus esculentus*.

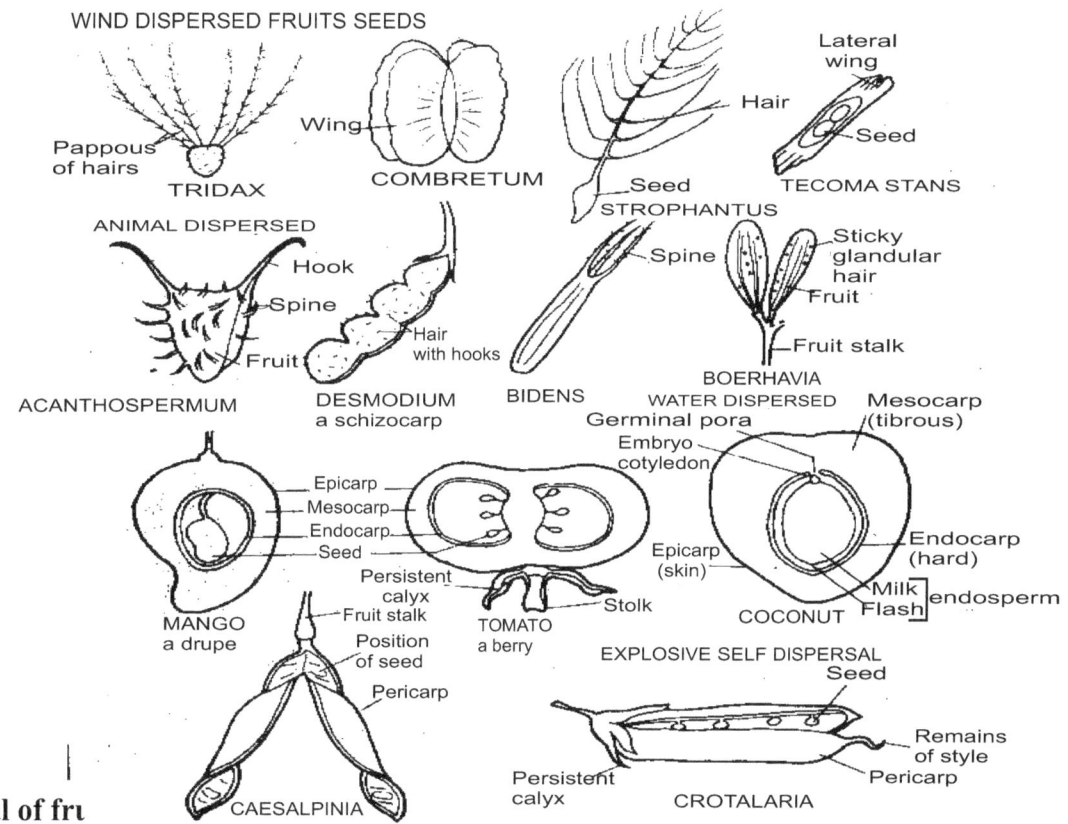

Dispersal of fru

117

12.1. ASEXUAL REPRODUCTION
12.1.1. FISSION AND BUDDING
1. UME/81/16
Spirogyra reproduces vegetatively by
A. spore formation   B. fragmentation
C. multiple fission   D. budding
E. binary fission

2. UME/89/31
Exponential increase in the population of an organism is a characteristic feature of
A. binary fission   B. sexual reproduction
C. budding   D. vegetative propagation

3. UME/97/42
During binary fission in lower organism, the nucleus is known to undergo
A. mitosis   B. meiosis
C. fragmentation   D. mutation

4. UME/2004/17 Type: 4
The type of reproduction that is common to both hydra and yeast is
A. grafting   B. budding
C. conjugation   D. binary fission

5. UME/84/34
Asexual reproduction does NOT occur in
A. Mucor, Spirogyra and Paramecium
B. Penicillium, Paramecium and Amoeba
C. Mucor, Rhizopus and Penicillium
D. Amoeba, Spirogyra and Mucor
E. Rhizopus, Ascaris and Amoeba.

6. UME/88/31
Which of these plant groups are normally propagated by asexual means
A. Banana, yam, pineapple and cassava
B. Yam, cassava, rubber and banana
C. Yam, cassava, orange and banana
D. Banana, cashew, coffee and pineapple

7. UME 87/32
The plantain reproduces asexually by
A. suckers   B. buds   C. fragments   D. spores their nodes
E. they lie horizontally on the soil

15. Which of these is propagated by means of the sucker?
A. the canna lily
B. the cocoyam
C. the onion
D. the Indian bamboo tree
E. the plantain

16. How is the cassava plant propagated?
A. by means of its seeds
B. by means of roots
C. by means of leaves
D. by means of cuttings
E. None of the above

17. Which of these plants can be propagated by the leaves?
A. Spinach   B. Onion
C. *Bignonia*   D.Liliac
E. *Bryophyllum pinnatum*

18. Which of these is NOT the advantage of vegetative reproduction?
A. The new plants are not dependent on pollination and other non-certain devices
B. The plants grow bigger than they would otherwise
C. Development is usually rapid
D. The new plants cannot survive in suitable conditions
E. Large areas can be covered in a relatively short period.

12.1.3. ARTIFICIAL VEGETATIVE PROPAGATION

19. UME/88/30
In vegetative propagation, which of the following requires part of another plant to develop
A. Scion   B. Bulb   C. Rhizome   D. Sucker

20. In grafting, the desired tree is grafted on to an old-established plant. This old-established plant is known as:
A. the host   B. the stoke
C. the scion   D. the stick
E. the stock

21. What do you understand by scion?
A. It is a plant part
B. It is a term used in propagation
C. It is the needed tree that is grafted on to the stoke
D. It is the desired plant
E. None of the above

12.2. SEXUAL REPRODUCTION

22. UME 93/25
Which of the following is TRUE of the process of conjugation in Paramecium?
A. Micronucleus disintegrate
B. Each ex-conjugant divides only once
C. Macronucleus undergoes division
D. Each micronucleus divides twice

23. PCE/97/21
The stage of mitosis associated with the formation of spindle fibre is the
A. prophase   B. metaphase
C. anaphase   D. telophase

24. PCE/97/22
The development of young ones from unfertilized eggs is known as
A. metamorphosis   B. oviparity
C. viviparity   D. parthenogenesis

25. PCE/92/25
Which of the following processes results in gamete formation?
A. mitosis   B. meiosis
C. cell enlongation   D. cell differentiation

26. PCE/92/26
During sexual reproduction in spirogyra the protoplast that moves during conjugation is referred to as the
A. gametophyte   B. zygospore
C. male gamete   D. female gamete

27. PCE/96/20
The stage of meiosis at which chiasmata occurs is
A. prophase I   B. metaphase I
C. perigyny   D. protogyny

28. PCE/06/2/TYPE: A
In meiosis, the separation of daughter chromatids occur at
A. anaphase II   B. metaphase I
C. telophase II   D. anaphase I

29. PCE 2000/23 Type A
The process by which a zygote divides repeatedly to produce many cells is
A. mitosis   B. fission
C. gametogenesis   D. cleavage

30. UME 2003/26 Type B
Homologous pair of chromosome separate during
A. Cytolysis   B. Cleavage
C. Mitosis   D. Meiosis

31. At what phase of mitosis are the chromosomes drawn to the middle?
A. Interphase   B. Prophase
C. Telophase   D. Anaphase
E. Metaphase

12.3. SEXUAL REPRODUCTION IN FLOWERING PLANTS
12.3.1. PARTS OF A FLOWER AND THEIR FUNCTIONS

32. UME 80/21
For pollination and fruit formation, the essential part(s) of the flower should be the
A. corolla   B. ovary
C. pistil (gynoecium)   D. ovules
E. receptacle

33. UME 93/27
The main function of the petal of a flower is to
A. attract pollination agents
B. protect the flower while still in the bud
C. serve as landing stage for insects
D. protect the inner part from dessication

34. PCE 2006/8 Type A
The calyx in a flower serves for
A. fertilization   B. attraction
C. pollination   D. Protection

35. PCE/95/23
A flower with all its carpels or ovaries fused together is described as
A. syncarpous   B. apocarpous
C. monocarpous   D. polycarpous

36. PCE/97/23
The two whorls in the flower that are directly related to the production of gamete are
A. calyx and androecium
B. androecium and gynoecium
C. calyx and gynoecium
D. calyx and corolla

37. What is an inflorescence?
A. It is a group of flowers
B. It is a cluster of flowers
C. It is the arrangement of flowers on a stem
D. It is the branching system of flower heads
E. None of the above

38. In some flowers the petals are exactly alike and are equally spaced. Such flowers are said to be:
A. Zygomorphic   B. Gamopetalous
C. Polypetalous   D. Actinomorphic
E. Petaloid

39. PCE/96/3
The condition in which male and female parts of a flower mature a different times is
A. homogamy   B. autogamy
C. dichogamy   D. cleistogamy

40. UME 87/30
A flower showing radial symmetry is said to be
A. pentamerous   B. protandrous
C. protogynous   D. actinomorphic

41. PCE 94/25
An example of a dioecious plant is
A. pawpaw   B. maize
C. oil palm   D. castor oil

12.3.2. POLLINATION

42. There are many ways in which plants prevents self-pollination. One such way is that the pistil of flowers ripen first. This is known as :
A. Protogyny
B. Heterostyly
C. Dichogamy
D. Protandry
E. None of the above

43. UME 85/31
f a flower is protandrous then it
A. must be unisexual
B. has an undeveloped androecium
C. has no androecium
D. must be insect-pollinated
E. can prevent self-pollination

44. UME 89/28
Wind-pollinated flowers usually have
A. rough pollen grains
B. sticky stigmas
C. small and short stigmas
D. long styles

45. Which of these is NOT true of wind pollinated flowers?
A. The flowers are inconspicuous
B. Vast quantities of pollen are released
C. The stigma of the flowers are sticky
D. The anthers are pendulously attached in the flower head
E. The pollen grains are smooth

46. PCE 2004/25 Type 9
The offspring resulting from cross pollination usually exhibit
A. variable characteristics
B. identical characteristics
C. high resistance
D. low resistance

47. UME 99/25
Insects visit flowers in order to
A. feed on the nectar
B. deposit pollen on the stigma
C. pollinate the flower
D. transfer pollens from the anthers

48. UME 85/32
For pollen to be released in Crotalaria the insect must depress the
A. wing
B. keel
C. standard
D. antepetalous stamen
E. antesepalous stamen

12.3.3. FERTILIZATION

49. What is the first noticeable change in the flower after fertilization?
A. The calx is shed
B. The petals fall
C. The style imbibes water
D. The stigma becomes unreceptive to pollen grains
E. The ovules ripen

50. If a fruit bears above it withered sepals, stamens and the stigma, the ovary is:
A. Epigynous
B. Hypogynous
C. Perigynous
D. Semi-perigynous
E. None of the above

51. UME 92/25
Double fertilization in higher plants is significant because it ensures
A. formation of a fertile embryo
B. formation of a fertile embryo and endosperm
C. development of the seed
D. development of the fruit

52. UME 90/31
Fruits which develop without fertilization are described as
A. simple
B. parthenocarpic
C. aggregate
D. compound

53. UME 84/29
How many nuclei are found in the pollen tube during fertilization?
A. 2   B. 3   C. 5   D. 6   E. 7

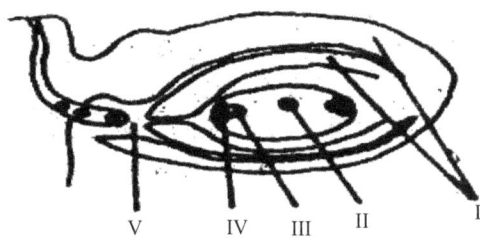

V   IV   III   II   I

54. UME 97/26
The function of the part labeled V is for the passage of
A. pollen tube and pollen nucleus
B. air, water and pollen nucleus
C. air, antipodal cells and ovum
D. synergids and egg cell

55. UME 97/27
The female gamete is represented by
A. I
B. II
C. III
D. IV

12.3.4. PRODUCTS OF SEXUAL REPRODUCTION
56. What are simple fruits?
A. These are fruits formed from the gynaecium having only a single ovary
B. These are fruits whose gynaecium has free carpels
C. These are fruits formed from separate carpels
D. These are fruits formed from a gynaecium with many ovules
E. None of the above

57. What are indehiscent fruits?
A. These are fruits which have soft and juicy mesocarp
B. These are fruits with dry coverings
C. These are fruits with stony mesocarp
D. These are fruits which do not open when ripe
E. These are fruits with stony endocarps

58. Which of these has a type of fruit known as the caryopsis?
A. The maize
B. The bean
C. The mango
D. The tomato
E. None of the above

59. What is the name of the dry fruits which split open along both ventral and dorsal sutures?
A. It is the follicle
B. It is the achene
C. It is the legume
D. It is the capsule
E. It is the nut

60. What is a drupe?
A. It is fruit with soft epicarp
B. It is fruit with soft mesocarp
C. It is fruit with fibrous mesocarp
D. It is fruit with juicy cotyledon

61. Which of these is a multiple or compound fruits?
A. The plum
B. The bean
C. A bunch of oranges
D. The pineapple
E. None of the above

62. What is understood by placentation in plants?
A. It is the method through which plants feed
B. It is the arrangement of the ovules or seeds in the ovary
C. It is the method the ovary feeds
D. It is the way in which ovules take up their food
E. It is the arrangement of fruits on the branch of a plant

63. In the parietal placentation, the seeds or ovules are:
A. Arranged on the axis of the ovary
B. Arranged on one suture of the fruit
C. Arranged on the free central axis
D. Arranged on the edges of the carpels of the ovary
E. None of the above

64. UME 95/21
The type of placentation shown is
A. axial
B. marginal
C. parietal
D. central

65. UME 95/22
An example of a plant having the placentation shown is
A. allamander
B. hibiscus
C. water lily
D. pride of Barbados

66. UME 79/38
The maize grain is regarded as a fruit and not a seed because
A. it is covered by a sheath of leaves
B. the testa and fruit wall fuse after fertilization
C. it has both endosperm and cotyledon
D. it has coleorhiza and coleoptiles
E. the pericarp and seed coat are separate.

67. UME 81/35
A fruit formed from a single flower having several free carpels is
A. a multiple fruit
B. a dry dehiscent fruit
C. a dry indehiscent fruit
D. a simple fruit
E. an aggregate fruit

68. UME 82/30
Any fruit which can break into several parts each containing one seed is a
A. capsule
B. aggregate fruit
C. legume
D. schizocarp
E. follicle

69. UME 83/19
Groundnut is not really a 'nut' in the biological sense because
A. it is harvested from inside the ground
B. its pericarp is not hard and tough
C. its fruit is not succulent
D. it is an achene

70. UME 83/20
What type of fruit is formed from a single flower having several free carpels
A. Multiple fruit
B. Simple fruit
C. Aggregate fruit
D. Dehiscent fruit
E. Indehiscent fruit

71. UME 85/30
The maize grain is a fruit and not a seed because it
A. has a large endosperm
B. is formed from an ovary
C. is a monocotyledon
D. has no plumule and radicle
E. has a hypogeal germination

72. UME 87/28
The pineapple fruit is best described as
A. aggregate, succulent and indehiscent
B. aggregate, succulent and dehiscent
C. multiple, succulent and indehiscent
D. multiple, succulent and dehiscent

121

73. UME 87/29
The flower shown above is
A. complete, regular, hermaphroditic with superior ovary
B. incomplete, regular, staminate with superior ovary C. complete, regular, hermaphroditic with inferior ovary
D. incomplete, irregular, pistillate with superior ovary

74. UME 87/31
A samara differs from a cypsela by having
A. an extended pericarp
B. a hard pericarp
C. a pericarp fused with the seed coat
D. some hairy out growth on the pericarp

75. UME 88/28
A dry dehiscent fruit which breaks up into one-seeded parts is a
A. schizocarp   B. capsule
C. follicle   D. legume

76. UME 89/29
The components of castor oil seed and the maize grain are similar EXCEPT for the
A. number of cotyledons B. location of the embryo
C. number of radicles   D. numbers of plumules

77. UME 91/33
A collection of achenes formed from several carpels of a flower is
A. a multiple fruit   B. an aggregate fruit
C. a schizocarp   D. a simple fruit

78. UME 95/23
The term caryopsis is used to describe a fruit in which the
A. testa and pericarp are separate
B. seed and endocarp are fused
C. testa and pericarp are fused
D. seed coat and fruit wall are impermeable

79. UME 97/29
Coconut and oil palm fruits can be grouped as
A. berry   B. legume
C. capsule   D. drupe

80. UME 98/25
A dry fruit formed from two or more carpels containing several seeds is a
A. follicle   B. legume
C. capsule   D. schizocarp

81. Cotton seeds are dispersed by wind. What attribute or structure assists in its dispersal?
A. Possession of flat seeds
B. Possession of "wings"
C. Possession of a floss
D. Lightness of the seeds
E. Hairiness of the seeds

82. Which of these seeds are dispersed by the explosive mechanism?
A. The para-rubber
B. The mango
C. The oil palm
D. *Tecoma stans*
E. *Triumfetta*

12.4. REPRODUCTION IN MAMMALS

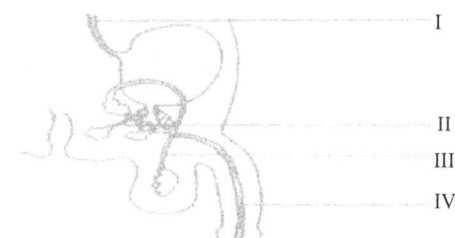

83. UME 2005/25 Type D
The structure labeled I originates from the
A. liver   B. kidney
C. small intestine   D. stomach

84. UME 2005/26 Type D
Birth control by vasectomy is achieved by severing the structure labeled
A. I   B. II   C. III   D. IV

85. UME 2009/25
The reproductive system of a male mammal is made up of
A. claspers, prostate gland, sperm duct and vas deferens
B. testis, prostate gland, sperm duct and vas deferens
C. oviduct, urethra, testis and sperm duct
D. testis, uterus, prostate gland and sperm duct

86. UME 2010/24/TYPE: D
In the male reproductive system of a mammal, sperm is stored in the
A. urethra   B. epididymis
C. semniferous tubules   D. *vas deferens*

87. What is the function of the vesicular seminalis?
A.   It is for the storage of urine
B.   It is for the storage of excretory materials
C.   It is for the storage of spermatozoa
D.   It is for the storage of effete blood cells
E.   None of the above

88. What is the function of the *vas deferens*?
A. It transports both urine and spermatozoa
B. It transport only the spermatozoa
C. It transports only urine
D. It transports excretory materials from the kidney
E. None of the above

89. The duct through which urine and spermatozoa pass out in the male mammal is:
A. the ureter   B. the penis
C. the urethra   D. the *vas deferens*
E. None of the above

90. What is the function of the bladder in mammals?
A. It is for storage of urea
B. It is for the storage of urine and spermatozoa
C. It stores toxic substances
D. It is for the storage of water for the body
E. None of the above

91. UME/2003/23/Type: A
In mammalian males, the excretory and reproductive systems share the
A. ureter   B. testes
C. *vas deferens*   D. urethra

92. PCE 2001/25 Type: 4
The spermatozoa produced in the testes is often stored in a coiled tube called
A. spermatic cord   B. *vas deferens*
C. scrotum   D. epididymis

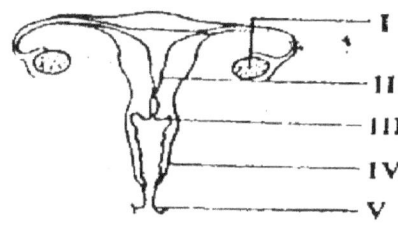

93. PCE 98/22
The structure labeled III is the
A. ovary   B. cervix
C. vagina   D. vulva

94. PCE 98/23
The fertilized ovum is implanted in the part labeled
A. I   B. II
C. III   D. IV

95. UME 2008/25
During ovulation, an egg is released from the
A. corpus luteum   B. ovarian funnel
C. Graafian follicle   D. fallopian tube

96. UME 2010/30/TYPE: D
The barrier between maternal and foetal blood is the
A. liver   B. umbilical cord
C. uterine wall   D. Placenta

97. UME 99/24
Fertilization in humans usually takes place in the
A. lower part of the uterus
B. upper part of the uterus
C. lower part of the oviduct
D. upper part of the oviduct

98. UME 95/24
The sex of a foetus is determined during
A. meiosis   B. copulation
C. fertilization   D. placentation

99. PCE/98/25/TYPE: C
Identical twins are formed when one ovum
A. is fertilized by one sperm
B. is fertilized by two sperms
C. divides after fertilization
D. divides before fertilization

100. PCE/06/16/TYPE: A
Ectopic pregnancy may result from
A. the fertilization of two ova
B. a foetus developing in the uterus
C. the non-fertilization of ovum
D. a zygote developing in the fallopian tube

101. When the embryo has remained in the womb for a period, it becomes a miniature adult and it is finally expelled from the womb. This process is known as
A. The gestation   B. The after-birth
C. Parturition   D. Foetalization
E. None of the above

102. PCE/2005/13/Type: P
Which of the following provides the mammalian embryo with nutrients before the formation of the placenta?
A. Amnion   B. Ovary
C. Uterus   D. Oviduct

103. PCE/2002/43/Type: 5
In a developing mammalian embryo, external shock is absorbed by the
A. yolksac   B. placenta
C. umbilical cord   D. amniotic fluid

104. PCE/2002/43/Type: 5
In a mammalian development, the function of the yolk is to
A. supply nutrient to the embryo
B. act as shock absorber to the foetus
C. supply air to the embryo
D. facilitate the process of excretion in the foetus

105. PCE/93/27
The hormones contained in birth control pills used by females are
A. progesterone and testosterone
B. oestrogen and progesterone
C. oestrogen and testosterone
D. progesterone & follicle stimulating hormone

106. PCE 2000/24 Type A
Within the mammalian female reproductive system, a ring of muscles which closes the lower end of the uterus is called
A. cervix
B. vulva
C. vagina
D. oviduct

107. UME 81/33
Which of the following statements is NOT true of menstruation?
A. It occurs monthly
B. The discharge contains amniotic fluid
C. It signifies no conception
D. It involves the lining of the uterus
E. It may cause pains

108. UME 86/37
Gestation in mammals is the period
A. required for growth after birth
B. between the formation of foetus and birth
C. of development from zygote to birth
D. before the formation of the zygote

109. UME 88/32
In a mammal, the placenta performs functions similar to those of the
A. lungs, kidneys and digestive system
B. lungs, heart and nervous system
C. liver, intestines and reproductive system
D. intestines, heart and digestive system

110. UME 90/29
During the mammalian embryo development, large amounts of oestrogen and progesterone are produced in the
A. umbilical cord
B. amnion
C. placenta
D. amniotic fluid

111. UME 98/26
The outermost embryonic membrane in the mammal is the
A. amnion
B. chorion
C. allantois
D. yolk sac

MISCELLANEOUS QUESTIONS (12)
112. An example of a product of sexual reproduction is
A. yam tuber
B. cassava tuber
C. ginger
D. groundnut (PCE/97/21/Type: E)

113. When a peacock displays its colourful feathers, it is
A. ready for a fight
B. protecting itself from predators
C. protecting its mate from predators
D. courting a female (UME/2005/47/Type: D)

114. In which of the following groups of plants is the pericarp inseparable from the seed coat
A. Nut
B. Follicle
C. Cypsela
D. Caryopsis (UME/2002/38/Type: Y)

115. Tomatoes, guava and pepper belong to a group of fruits called
A. berry
B. drupe
C. schizocarp
D. caryopsis (PCE/92/27)

116. Placentation in fruit describes the
A. arrangement of fruits on the branches
B. development of the ovary wall to form the pericarp
C. arrangement of ovules in the ovary
D. classification of fruits into different types (PCE/93/26)

117. The crown of leaves on a pineapple fruit is known as
A. offset   B. suckers
C. runners   D. bulbils (PCE/95/24)

118. Maturation of the stigma of a flower before the anther is described as
A. protandry
B. hypogyny
C. perigyny
D. protogyny (PCE/98/24/TYPE: C)

119. What makes yam an underground stem?
A. The fact that it store starch
B. The fact that it has no roots
C. The fact that it bears dormant buds known as "eyes"
D. The fact that it is swollen
E. None of the above

120. That part of the leaf that secures it to the stem is:
A. The peduncle
B. The pedicel
C. The petiole
D. The patula
E. The perianth

121. The leaves of the onion are modified into a storage organ known as:
A. The rhizome
B. The corm
C. The rubber
D. The stem
E. None of the above

122. The food stored by the onion is:
A. Starch
B. Sugar
C. Protein
D. Oils
E. Fats

123. The plant *Bryophyllum pinnatum* can be propagated easily because
A. The leaf stores food
B. It does not die
C. The stem is stout
D. It is immortal
E. None of the above

124. What are monoecious plants?
A. These are plants having only the male reproductive parts present in one individual
B. These are plants with both the male and female reproductive parts present in a single plant
C. These are plants with only pistillate flowers
D. These are monocarpellary fruits
E. These are plants with many flowers

125. What is binary fusion?
A. The splitting up of an individual into two identical parts
B. The splitting up of an animal into minute forms
C. The loss of a part of an animal due to injury
D. The fusion of the gametes
E. None of the above

126. The structure from which gametes arise is:
A. The scrotum
B. The ovary
C. The ovule
D. The gonad
E. None of the above

127. Which of these animals do not show parental care over their young?
A. the mammal
B. the fowl
C. the lizard
D. the earwigs
E. the boa constrictor

128. Which of these is true of seeds dispersed by water?
A. The plants bearing them grow near water
B. They are light
C. They are buoyant
D. They contain air space in the internal cavity of the seed
E. They are hollowed

CHAPTER 13

GROWTH

Understand the following after reading this chapter:

> Meaning of growth
> Growth in plants and animals
> Germination of seeds

GROWTH

Growth is the permanent increase in volume and weight of the cells protoplasm. It occurs through food assimilation, cell expansion, cell division and cell differentiation.

Factors Affecting Growth are: nutrition/food, environment, hormones, heredity and diseases

Growth Curve

The curve can be shown appearing slowly along the line and stabilizing.

During the initial stage, i.e., during the lag phase, the rate of plant growth is slow. Rate of growth then increases rapidly during the exponential phase. After some time the growth rate slowly decreases due to limitation of nutrients. This phase constitutes the stationary phase.

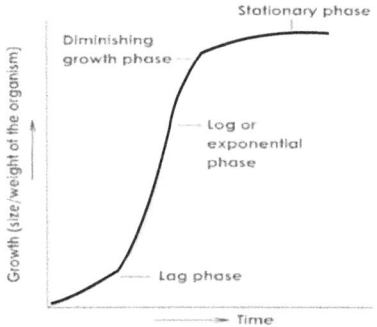

The curve obtained by plotting growth and time is called a growth curve. It is a typical sigmoid or S- shaped curve.

Patterns of Growth

Growth patterns include:

Isometric growth: when parts of an organism grow at the same rate as the whole organism e.g. plant leaves.

Allometric growth: When parts or organs of an organism grow at different rates from the whole organisms e.g. the head and brain of man during the early stage of life.

Intermittent growth: This is a step-like growth; it occurs in arthropods due to moulting.

Germination

Germination is the development and emergence of the embryo from the seed coat into a seedling.

When conditions are right, the seed will take up water through the micropyle by imbibition. This triggers the beginning of the growth of the seed. The cell swells and the testa splits. With the addition of water, large molecules of carbohydrate, protein and fat can be hydrolysed (broken down) to produce substances for respiration. The water activates such enzymes as a-amylase to catalyse this digestion. The growing embryo releases a hormone called **gibberellin** and some enzymes are produced and released in response to this. The soluble products of digestion are delivered to the cotyledons, root and shoot. They respire aerobically and grow in size. By the time the food store has been used up, the shoot has grown enough to push the first leaves into the sunlight. Photosynthesis can then start.

Conditions necessary for germination: Oxygen, warmth, moisture and viable seeds

Types of germination

Epigeal germination: The hypocotyls rapidly enlongates thereby pulling the cotyledons and the plumule above the ground e.g. cowpea and most dicot.

Hypogeal germination: The epicotyls rapidly pushing up the plumule above the ground and leaving the cotyledon underground e.g. Sorghum and most monocot.

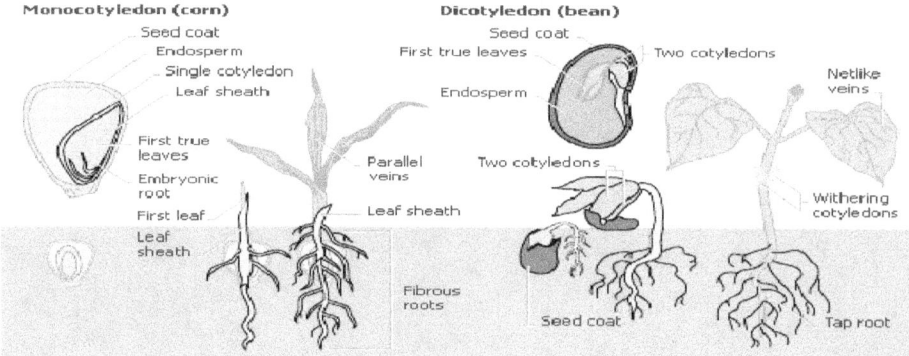

Germination in monocot and dicot seed

EXPERIMENTS ON GROWTH

Experiment:1 To demonstrate the conditions necessary for germination.
The conditions necessary for germination are warmth, oxygen, water, and viable seeds.

Materials required. Four test tubes, labels, dry cotton wool and dry seeds.

Method. Four clean test tubes are collected and dry cotton wool is placed at the bottom of each. Some dry seeds are put in the tubes and the tubes are labelled 1, 2, 3, and 4. They are treated as follows:

Tube 1: This tube is placed where there is air and warmth. Water is not added so that the seeds receive warmth and air only.
Tube 2: Water that is previously boiled and cooled is poured into the tube. The boiled water lacks air. A few drops of oil are added to the water surface to prevent air from dissolving in it. The tube is then placed where there is warmth. In this tube, warmth and water are present but no air.
Tube 3: Some water is added to moisten the cotton wool. The tube is kept where there is enough air and warmth. This tube is regularly checked and water is added to keep the cotton moist. In this tube, air, water, and warmth are present for the seeds. No condition is missing.
Tube 4: Here, the seeds are treated as in tube 3. However, this is kept in a cold place such as a refrigerator. In this tube, air and water are present for the seeds but no warmth. The four tubes are then left for a few days.

Observations. Only seeds in tube 3 show normal germination.

Conclusion. Seeds in tube 3 which show normal germination were supplied with air, warmth and water. Seeds in other tubes lacked one of the conditions and hence normal germination could not take place. This shows that air, warmth and water are necessary for seed germination.

Experiment:2 To determine the area of fastest growth in a root.

Materials required. Bean seedling, blotting paper, a gas jar, water proof Indian ink.

Method. A bean seedling whose root (or radicle) is straight is selected. The selected root is just an inch long. From the tip upwards, the root is marked across with lines. The lines are marked with a water proof Indian ink. The lines are equidistant from each other, 1mm apart. Blotting papers are moistened and are used to line a gas jar. The seedling is then placed between the glass and the moistened paper. The setup is then kept in a dark place for two days. After this period, the plant root was carefully inspected.

Observations. The root increased in length. The lines at the upper portion (i.e. lines far from the root tip, the lines moved apart widely from each other.
Conclusion. The sharp increase of the spaces between lines 2, 3, 4 and 5 shows that the region of fastest growth lies behind the root tip, a few centimetre behind.

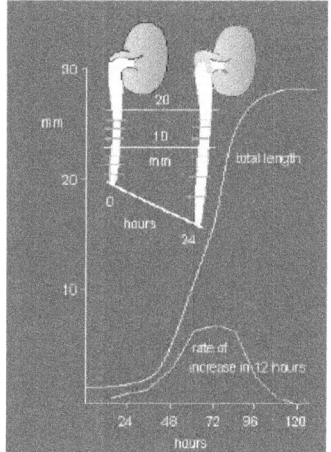

Growth rate in germinating seed

Experiment:3. To demonstrate root pressure using a manometer.

Materials required. A young vigorously growing plant, a sharp pen knife, a manometer.

Method. A young vigorously growing plant is cut at the base, say 12cm from the ground level, with a sharp pen knife. Next, a manometer (an instrument for measuring root pressure) is carefully tied to the cut surface of the stem. The manometer is held in a place by means of rubber tubing. The setup is then observed for some time.

Observations. The mercury in the manometer is seen to move up gradually.

Conclusion. The upward movement of the mercury is due to the pressure generated by the flow of sap.

Revision Questions

13.1. MEANING OF GROWTH

1. UME/89/32
The irreversible life process by which new protoplasm is added to increase the size and weight of an organism can be termed.
A. anabolism   B. catabolism
C. growth   D. development

2. UME/99/25
Biological growth refers strictly to an increase in the
A. protoplasm of an organism   B. size of an organism
C. number of organisms   D. development of form

3. UME/98/24
The most reliable estimate of growth is by measuring changes in
A. length   B. volume
C. surface area   D. dry weight

4. UME/2000/20/TYPE: M
For growth to occur in organisms, the rate of
A. food storage must be low
B. anabolism must exceed that of catabolism
C. catabolism must exceed that of anabolism
D. food storage must be high

5. PCE/2000/25/TYPE: A
A unicellular organism is said to have attained his maximum growth when it
A. has reproduced itself several times
B. has used up all the nutrients in the medium
C. can no longer reproduce itself
D. can reproduce itself

6. PCE/07/1
Growth in living things is brought about by the
A. deposition of new material from the surroundings
B. synthesis of new material from within the body
C. absorption of water and minerals
D. elasticity of the cell

13.2. GROWTH IN PLANTS AND ANIMALS

7. UME/88/33
The radicle of a bean seedling grows most rapidly in the region
A. of the root tip   B. just above the root tip
C. just around the root tip   D. just below the root tip

8. UME/2001/17/TYPE: T
A seedling grown in the dark is likely to be
A. etiolated   B. dormant
C. sturdy   D. stunted

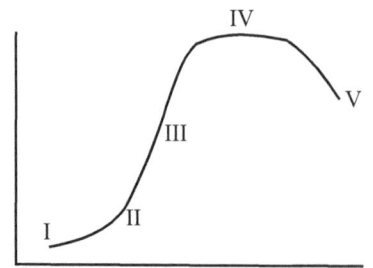

9. PCE/95/36
The area with the highest growth rate is labeled
A. V   B. IV   C. III   D. I

10. PCE/95/37
Environmental resistant is greatest at
A. II   B. III   C. IV   D. V

11. PCE/99/23
When the tip of coleoptiles is illuminated from one side, it comes towards that side because the growth hormone is
A. concentrated on that side at the tip of the coleoptiles
B. concentrated on the dark side at the tip of the coleoptiles
C. destroyed by the light rays at the tip of the coleoptiles
D. concentrated below the region of elongation

12. PCE/2002/6/TYPE: 5
Growth in animals differs from that in plants because it occurs
A. in the reproductive cells only
B. equally in all parts of the body
C. in restricted parts of the body
D. at the apex only

13. PCE/07/22
Which of the following is the correct sequence of growth in multicellular organism?
A. cell enlargement→cell division→cell differentiation
B. cell differentiation→cell division→cell enlargement
C. cell division→cell differentiation→cell enlargement
D. cell division→cell enlargement→cell differentiation

13.3. GERMINATION OF SEEDS

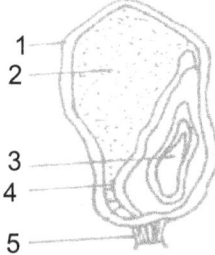

14. UME/86/34
Which of the labeled parts will develop into a new maize plant?
A. 2   B. 3   C. 4   D. 5

15. UME/86/35
The structure labeled 1 is the
A. plumule   B. radicle
C. cellmembrane   D. seed coat

16. UME/86/36
The main function of the structure labelled 2 is to
A. protect the inner parts of the seed
B. keep the inner parts moist
C. nourish the embryo and the growing parts
D. maintain the shape of the seed

17. PCE/96/23
Dormancy in seeds may be caused by
A. the rapid diffusion of oxygen into the seed
B. delayed development of the embryo
C. the rapid absorption of water
D. insufficient stored food for the embryo

13.3.1. TYPES OF GERMINATION
18. UME/86/47
In an epigeal germination, it is the
A. epicotyls that elongates fast
B. hypocotyls that elongates fast
C. hypogeal that elongates fast
D. plumule that elongates fast
E. roots that elongates fast

19. UME/85/37
Germination which results in the cotyledons being brought above ground is called
A. hypocotyls
B. epicotyls
C. epigeal
D. hypogeal
E. plumule

20. UME/92/27
Hypogeal germination is characterized by the
A. emergence of the plumule out of the groundB
elongation of the hypocotyls
C. provision of nourishment by the endosperm   D .
elongation of the epicotyls

21.   UME/2004/18/TYPE: 4
Epigeal germination of a seed is characterized by
A. lack of growth of the hypocotyls
B. more rapid elongation of the hypocotyls
C. more rapid elongation of the epicotyls than the hypocotyl
D. equal growth rate of both the hypocotyls and epicotyls

22. UME 2009/26
In a bean seed, absorption of water at the beginning of germination is through the
A. hilum   B. micropyle
C. testa   D. plumule

13.3.2. CONDITIONS NECESSARY FOR GERMINATION
23. UME 2006/13
Water is necessary for a germinating seed because it
A. promotes aerobic respiration
B. activates the enzymes
C. wets the soil for proper germination
D. protects the seeds from desiccation

24. PCE/2006/13/TYPE: A
The two most critical factors required for the growth of plant are the
A. nutrients and water content of the soil
B. light intensity and topography
C. types of plant hormones and enzymes
D. temperature and pH of the surroundings

25. PCE/95/28
The most important conditions necessary for seed germination are
A. water, temperature and oxygen
B. water, temperature and carbondioxide
C. oxygen, water and carbondioxide
D. oxygen, temperature and carbondioxide

26. UME/78/46
A germinating seed requires oxygen which is essential for
A. converting carbohydrates into glucose
B. transporting energy from one part of the plant to another
C. the production of energy by oxidizing essential carbohydrates
D. hydrolysis of proteins
E. the formation of water molecules within the germinating seed

27. UME/80/12
Water is required for seed germination to take place because it
A.   activates the enzymes
B.   liberates energy for growth
C.   softens the testa
D.   permits radicle growth
E.   allows oxygen to diffuse into the seed

28. UME/79/23
Alkaline pyrogallol was used in an experiment.
The experiment must have been connected with
A. excretion
B. germination
C. transportation
D. digestion
E. photosynthesis

29. Growth can best be determined in a population of spirogyra by measuring the
A. total lengths of the filaments
B. total widths of the filaments
C. rate of photosynthesis in the population
D. dry weight of the organism   (UME/87/33)

30. Water is required for seed germination to take place because it
A. activates the enzymes
B. liberates energy for growth
C. softens the testa
D. permits radical growth
E. allows oxygen to diffuse into the seed (UME/80/12)

31. Seed whose cotyledons are swollen as a result of stored food are described as
A. non-cotyledonous
B. endosperms
C. non-endospermous
D. albuminous (PCE/97/24)

32. Epigeal germination can be found in
A. sorghum
B. maize
C. millet
D. groundnut (UME/91/35)

33. Which of the following chemicals is a plant growth substance?
A. oxaloacetic acid
B. pyruvic acid
C. indoleacetic acid
D. oxalosuccinic acid (PCE/92/28)

34. The storage tissue in a maize grain is the
A. radicle
B. plumble
C. endosperm
D. embryo   (PCE/03/37/TYPE: N)

35. The best method of measuring the growth rate of an animal is by taking it
A. wet weight
B. dry weight
C. total volume
D. surface area   (PCE/04/27/TYPE: 9)

36. Some germinating seeds are put into a thermos flask with a cork through which passes a thermometer. This is a noticeable increase in the reading of the thermometer. The increase in temperature is due to:
A. The external temperature of the room in which the flask is
B. The heat given off by the decomposing seeds
C. The dissipated heat from the respiring seeds
D. The heat of action of the seeds
E. The rise of temperature of the air in the flask

37. In a germinating seed, the first embryonic structure to emerge is:
A. The plumule
B. The tap root
C. The coleoptile
D. The coleorhiza
E. The radicle

38. In the groundnuts, the seed leaves are usually brought above the soil level. This is:
A. Hypogeal germination
B. Hypocotyl germination
C. Epigeal germination
D. Epicotyl germination
E. None of the above

39. In this type of germination the coryledons are left in the soil, e.g. the maize:
A. Hypogeal germianiton
B. Epigeal germination
C. Hypocotyl germination
D. Epicotyl germination
E. None of the above

40. In the germinating bean, the zone between the soil and the cotyledons (seed-leaves) is:
A. The epicotyl region
B. The internode
C. The hypocotyls region
D. The plumular region
E. None of the above

41. The germinating bean plant before establishing itself, depends for its food on:
A. its roots
B. its first foliage leaves
C. The hypocotyls
D. Its radicle
E. Its seed leaves

CHAPTER 14

COORDINATION AND CONTROL

Understand the following after reading this chapter:

> Nervous coordination
> Sense organs
> Hormones
> Homeostasis

THE NERVOUS SYSTEM
The nervous system is made up of the
Central nervous system (CNS), which consist of the brain and spinal cord, and the
Peripheral nervous system (PNS), which comprises the somatic and autonomic system

THE BRAIN AND ITS FUNCTION
The brain is composed of numerous brain cells or neurons. It is confined and protected by the cranium of the skull. It is divided into three main sections, namely: the forebrain, midbrain and hind brain.

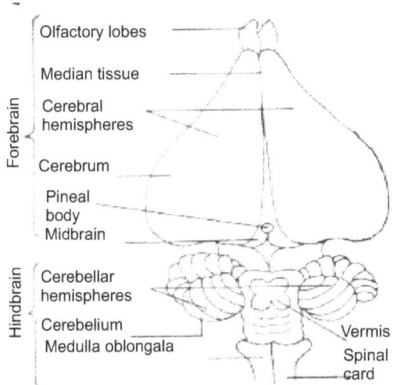

The brain of the rabbit (dorsal view)

Parts of mammalian brain	Function(s)
1. Cerebral hemisphere (Cerebrum)	Intelligence, memory, voluntary actions, sensations.
2. Hypothalamus	Regulation of body temp. and osmotic pressure in blood; appetite and emotions.
3. Pituitary gland	Secretes a number of hormones.
4. Optic lobes	Concerned with sight and movement of eyeball.
5. Cerebellum	Muscular coordination and bodily balance.
6. Medulla oblongata	Involuntary actions, e.g. heartbeat, respiratory movements, peristalsis.

THE SPINAL CORD AND ITS FUNCTION
The spinal cord is also composed of neurons. Spinal nerves emerge at intervals from the vertebraterial canals and extend to all parts of the body. It is responsible for some local reflex actions

THE PERIPHERAL NERVOUS SYSTEM
The peripheral nervous system consists of all the nerves that link the CNS to all parts of the body. It is divided into two parts.
Sensory somatic system, which receives information from sensory receptor and conduct same to the CNS.
Autonomic nervous system, consists of the **sympathetic** and **parasympathetic system**.
Neurons are specialized cells which receive information (stimuli) and transmit them in form of electrical impulse to muscles and glands. A neuron consists of the cell body which contains the nucleus and dendrites, and a single long process called axon. There are three types of neurons:
Afferent or sensory Neurons: These conduct impulses from receptors or sensory organs to the CNS.
Efferent or motor Neurons: They relay impulses from the CNS to effector organs i.e muscles and glands
Intermediate or Association Neurons: They transmit impulses within the CNS. They connect the afferent to efferent neurons.

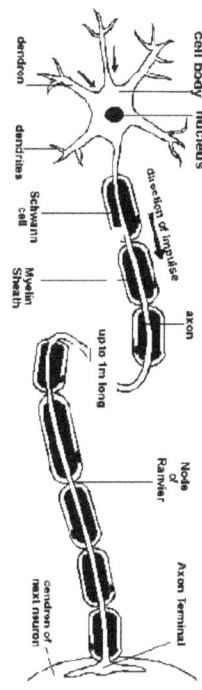

TRANSMISSION OF NEURAL IMPULSES

The transmission of neural impulses is electrochemical in nature. At resting potential, there is polarization (potential difference) on the nerve fiber (axon) membrane. The concentration of K+ inside is greater than on the membrane and the concentration of Na+ is greater on the membrane than inside. When the membrane is stimulated, it becomes highly permeable to Na+ which then rush in momentarily (depolarization) and reverse the polarity of the membrane. The flow of the Na+ is sustained down the axon producing the neural impulse or action potential.

Reflex Actions

A reflex or involuntary action is a quick and automatic response by an organ or system of organs of the body to a stimulus, e.g sneezing, blinking of the eye, laughing when tickled, knee jerk e.tc

Reflex Action	Voluntary Action
Response is stereotyped	Response is not stereotyped it varies with circumstances
It is done unconsciously i.e. it does not involve the higher centre of the brain	It is a conscious action i.e. it involves the higher centre of the brain
It is rapid/fast	It is less rapid/slow
It involves a small number of neurons	It involves a large number of neurons

The Reflex Arc

A reflex arc is the nervous pathway taken by a reflex action. It consists of receptors which receive stimulus and transmit it through the afferent neuron to the spinal cord.

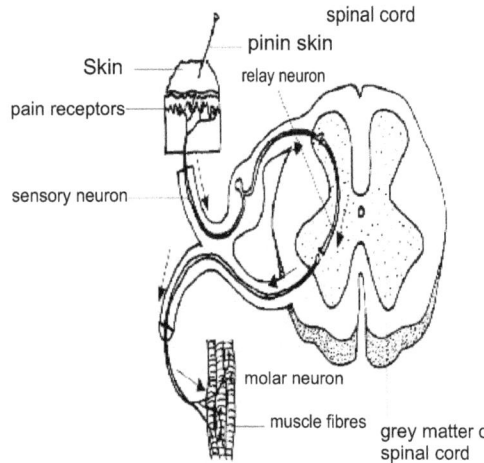

A reflex arc

Voluntary Actions

A voluntary action is an action consciously taken in response to a stimulus.

SENSE ORGAN

There are five principal sense organviz: eye, ear, nose tongue and skin.

THE EYE

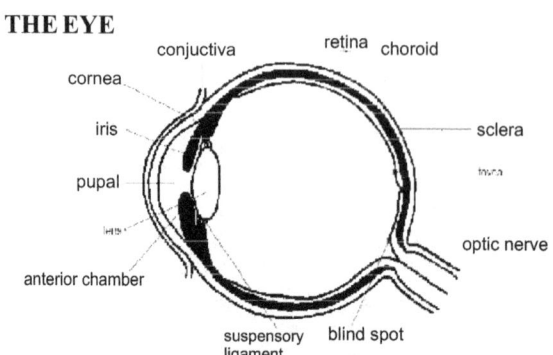

Longitudinal section of the mammalian eye

Parts of the eye	Function(s)
1). Aqueous humour	Refracts light; keeps eyeball firm.
2). Vitreous humour	Refracts light; keeps eyeball firm.
3). Cornea	Refracts light rays into pupil.
4). Suspensory ligament	Attaches lens to ciliary body.
5). Iris	Controls amount of light entering the eye.
6). Pupil	Allows light to enter the eye.
7). Lens	Focuses light rays into retina.
8). Ciliary muscle	Controls curvature or thickness of the lens.
9). Rectus muscle	Movement of eyeball.
10). Sclera	Protection against mechanical injury.
11). Choroid	Pigmented black to prevent internal reflection of light.
12). Yellow spot (fovea)	Region of acute vision
13). Optic nerve	Transmits impulses from eye to brain.
14). Blind spot	No photoreceptor cells; no vision when image falls on it.
15). Retina	Light sensitive layer, contains:

(i). **Cones** concerned with colour vision in bright light.
(ii). **Rods** concerned with vision in dim light.

Image Formation

The eye sees by reflection of light from objects. Light rays entering the eye are refracted by the cornea, aqueous humour, lens and vitreous humour and brought to focus on the retina to form an image. The image formed by the eye is real, inverted and smaller than the object.

Accommodation: This is the ability of the eye to focus both far and near objects due to the adjustment done on the

132

lens by the ciliary muscle.

Eye Defects
Short sightedness or myopia: This is the ability of the eye to see only near objects clearly but not distant objects. Caused by large or elongated eye balls. Image of distant object is formed before the retina. It can be corrected by wearing concave lenses.

Long sightedness of Hypermetropia: This is the ability of the eye to see distant objects clearly but not near objects. Caused by small or short eyeball. Image of near object is formed behind the retina. It can be corrected by the use of convex lenses

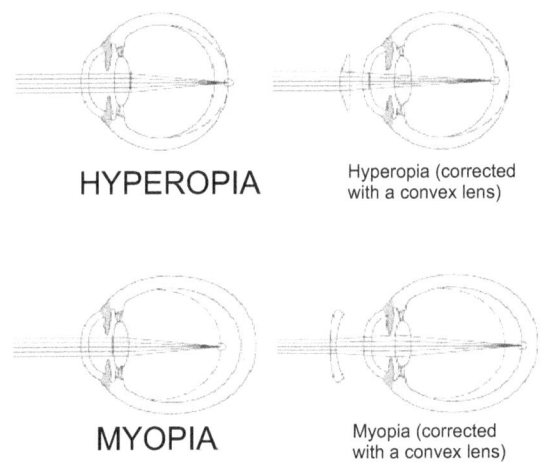

Correction of eye defects

Astigmatism: This is caused by unequal curvature of the eyeball. It can be corrected by the use of cylindrical lenses.
Cataract: This is the opacity of the lens which may be caused by bacterial infection
Conjunctivitis: This is the inflammation of the conjunctiva

Presbyopia: This is the recession of the near point with age, beyond 40cm, due to increased plasticity of the lens. It can be corrected by the use of a bi-focal lens.

THE EAR
The ear is the organ responsible for hearing and balancing. It is made up of three parts, namely, the outer ear, middle ear and inner ear.

The outer ear consists of the pinna and the external auditory meatus. The eardrum demarcates the external ear from the middle ear.

The middle ear is oval and filled with air. It contains the malleus, incus and stapes collectively known as ossicles.

The inner ear is filled with a fluid called perilymph. It contains the semicircular canals utriculus and saculus. the cochlea is made up of spirally, coiled tube containing a fluid called endolymph and the organ of corti connected to the brain by the auditory nerves.

Mechanism of Hearing
Sound waves from the air enter the ear through the auditory meatus and hit the eardrum which is set into vibration. The vibration is picked up by the bone ossicles and amplified before entering the inner ear. The oscillation of the stapes set the endolymph into vibration which in turn stimulates the auditory nerve fibres to transmit impulses to the brain. The brain then interprets the pitch, quality and loudness of the sound and relays back and the sound is heard.

Balance
The semi circular canals are sensitive to the turning movement of the head. The utriculus and sacculus contain receptors sensitive to any movement of the body. These sensations are transmitted as impulses to the brain which interpretes the position of the animal.

THE NOSE
The nose is the organ of smell. It has ciliated tips of olfactory cells on the surface of the nasal cavities which

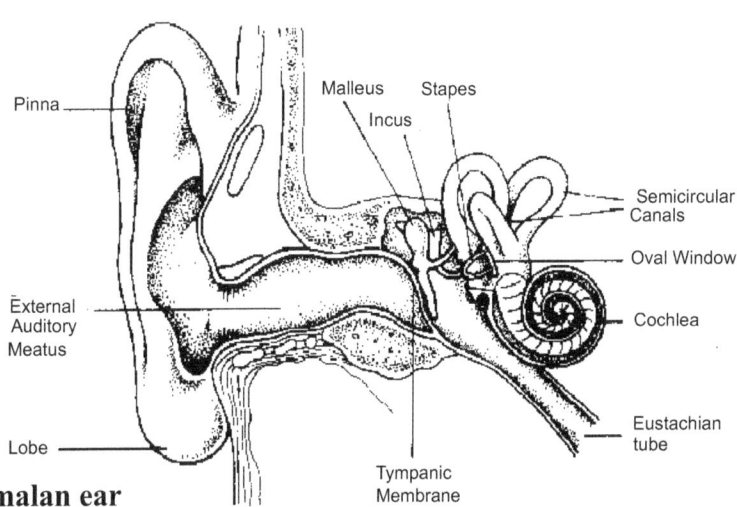

Section of the mammalan ear

respond to various chemicals in the air.

THE TONGUE

The tongue is the organ of taste. It has buds which are sensitive to four primary tastes. The back of the tongue is sensitive to bitter stimuli, the sides to sour stimuli and the tip to sweet and salty stimuli.

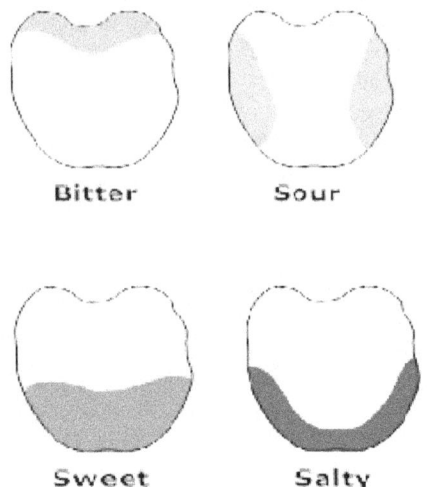

Bitter   Sour

Sweet   Salty

Different parts of sensitivity on the tongue for the four tastes

THE SKIN

The skin is a waterproof structure; One of the organs of the sense of touch. The skin has three parts:
1. Epidermis:
  i. Cornified layer;
ii. Granular layer;
iii. Malphigian layer.
2. Dermis.

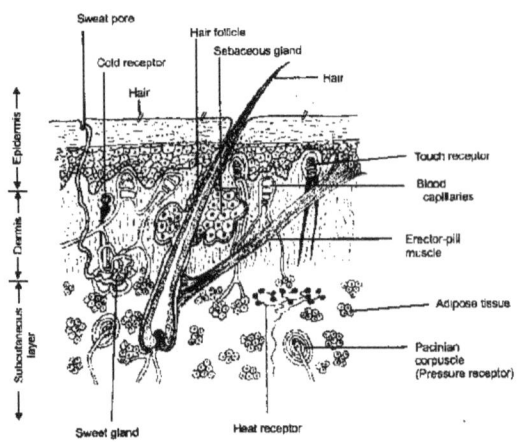

Function of the Skin
1. It protects the underlying tissues from damage and mechanical injury.
2. It protects the body against bacteria and other infective microbes.
3. It protects the body against ultraviolet rays
4. It excretes sweat
5. The nerves cells respond to change in the environment.
6. The sebaceous glands secrete sebum which prevents the skin from cracking or drying out
7. It synthesizes vitamin D.
8. It regulates body temperature through heat gain and heat loss.

Heat gain –Vaso constriction, shivering, contraction of hair erector muscles etc
Heat loss – Vasodilation and sweating

THE ENDOCRINE SYSTEM
The endocrine system is made up of the ductless glands situated in various parts of the body. Secretions of the ductless glands are called hormones. The endocrine system consists of the pituitary, thyroid, pancreatic,

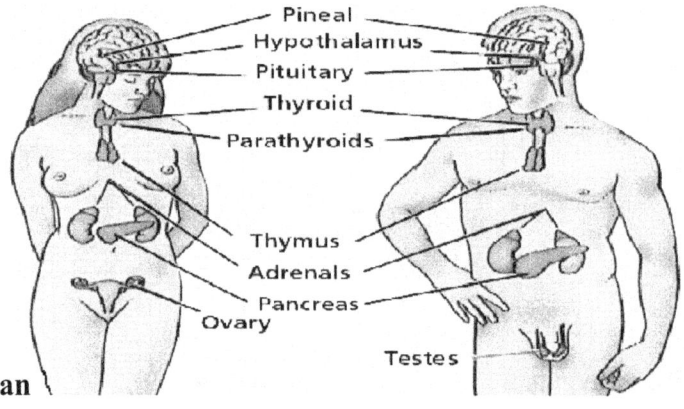

Endocrine system of man

Nervous System	Endocrine System
Transmission is through nerve fibre	Transmission is through the circulatory system
it is rapid	it is slow
Response is immediate, short-lived and very precise	Response is slow, long-lasting and diffuse
Message passed as electrical impulses along nerve	Message or information passed as chemical substances in the blood stream
Effect are usually localized and temporary	Effect may be widespread and permanent
it occurs in animal only	it occurs in plants and animals

1. **Pituitary gland**: The pituitary gland has two divisions, namely: anterior pituitary gland and posterior pituitary gland.
Anterior pituitary gland secrete the following hormones: Thyroid stimulating hormone (TSH), Follicle stimulating hormone (FSH), Luteinizing hormone (LH), Growth hormone (GH), Adenocorticotropic hormone (ACTH) and Prolactin.
Posterior pituitary gland releases two hormones: Antidiuretic hormone (ADH) and Oxytocin.
Over secretion of GH causes gigantism in infants and acromegaly in adult. Under secretion results in dwarfism in children.

2. **Thyroid gland** is located in the neck region. It secretethyroxine, which regulates rate of body metabolism. It stimulates growth in young mammals and metamorphosis in tadpole. Over secretion cause underweight, goiter and mental instability. Under secretion result in cretinism and myxoedema

3. **Parathyroid glands** are four bodies behind the thyroid gland. They secrete parathormone which regulate (increase) blood calcium.

4. **Pancreatic gland** is located between the stomach and small intestine. It helps the production of insulin and glucagon which regulates blood glucose. It also secretes digestive juice. Over secretion can cause insulin shock. Under secretion results to *Diabetes mellitus* (excess blood sugar).

5. **Adrenal glands** are located on top of each kidney. It helps the production of adrenaline or emergency hormones which prepares the body for emergencies. It also control certain behavioural patterns such as bravery and fear.

6. **Gonadal or reproductive glands**: These are the ovaries in females and testes in males.
Ovaries are located in the lower abdomen. They help

theproduction of sex hormones such as oestrogen (which maintain female secondary sexual characters), progesterone, prolactin, relaxin. Also produces eggs or ova
Testes arelocated in the groin enveloped by the scrotal sac.Help theproduction of sex hormone called testosterone or androgen, which maintain secondary sexual characters and helps theproduction of sperm.

PLANT HORMONES
Common plant hormones include Auxins, gibberellins, kinins and florigens
1. **Auxins**: They are produced at the shoot and root apices.
i. Facilitate cell expansion in the zone of elongation
ii. Promote cell division and stem elongation
iii. Regulate growth in plant's apical region.
iv. Control tropism
v. Stimulation of adventitious roots, hence used in stem cuting
vi. Retard abscission.
vii. Promote formation of pistillate (female) flowers.
Note: A synthetic form of Auxin is **Indole Acetic Acid**
2. **Gibberellins**: They are found in embryos of germinating seed and in roots
i. Breaks dormancy in buds and seeds and promote seed germination.
ii. Enhances elongation of stems at internodes in the presence of auxins.
iii. Induces synthesis of plant enzymes e.g. amylase
v. Induce flowering, fruit and seed development
vi. Retard leaf abscission.
vii. Promote formation of staminate (male) flowers.

3. **Kinins/Cytokinins**: They are found in the root apex, developing fruits and germinating seeds
i. Induce rapid cell division
ii. Promote growth of lateral buds and fruits
iii. Break seed dormancy
iv. Regulate shoot growth

4. **Abscissic Acid**: Functions are opposite to those of gibberellins
i. Induces dormancy in buds
ii. Inhibits seed germination
iii. Retards plants growth especially stemenlongation
iv. Accelerates abscission of leaves, fruits and flowers.

5. **Ethylene**:
i. Induces ripening of fruits
ii. Promotes abscission of leaves

14.1. NERVOUS COORDINATION
14.1.1 THE COMPONENTS, STRUCTURE AND FUNCTIONS OF THE C.N.S.

1. Which of the following represent the essential structures of the central nervous system?
A. The axon and proprioceptors
B. The brain and axon
C. The spinal cord and the axon
D. The spinal cord and peripheral nerves
E. The spinal cord and the brain

2. What is the function of the central nervous system?
A. It dictates the working condition of the body
B. It directs the functions of certain parts of the body
C. It controls and co-ordinates the activities of individual animals
D. It relays impulses from one part of the body to the other
E. None of the above

3. PCE/99/20
The central nervous system is connected with the receptor and effectors organs by
A. motor and sensory neurons
B. motor neurons and cranial nerves
C. cranial nerves and spinal nerves
D. spinal nerves and sensory neurons

4. UME/84/31
If an animal is very active and has good muscular control, it is likely to have well-developed
A. olfactory lobes   B. cerebral hemispheres
C. optic lobes   D. cerebellum
E. spiral cord

5. UME/87/38
The centre for controlling body temperature in the brain is the
A. cerebrum   B. cerebellum
C. medulla   D. hypothalamus

6. UME/88/35
What part of the central nervous system is concerned with answering an examination question?
A. cerebrum   B. cerebellum
C. medulla oblongata   D. spinal cord

7. UME/97/31
The part of brain that regulates most biological cycles in humans is
A. olfactory lobe   B. optic lobe
C. medulla oblongata   D. pineal body

8. PCE/93/30
The part of the brain that coordinates reproductive events is the
A. olfactory lobe   B. medulla oblongata
C. cerebellum   D. cerebrum

9. PCE/94/28
If an animal is very active and has good muscular control, it is likely to have well-developed
A. olfactory lobes   B. spinal cord
C. cerebellum   D. cerebrum

10.   PCE/94/29
A neurons is made up of a
A. cell body, dendrites and an axon
B. cell body, muscle fibre and dendrites
C. sheath, an axon and nucleus
D. synaptic knobs, cell body and dendron

11. PCE/98/30
The part of the brain which controls intellectual functions is called
A. medulla oblongata   B. cerebrum
C. cerebellum   D. olfactory lobe

12. PCE/98/27
Changes in the concentration of the blood is detected by
A. medulla oblongata   B. cerebellum
C. hypothalamus   D. cerebrum

13. The sensory cells situated on the surface of the body and which receive external stimuli are:
A. The proprioceptors   B. The external receptors
C. The exteroceptors   D. The enteroceptors
E. The externoceptors

14. Which of these functions and activities is not undertaken by the cerebral hemispheres of man?
A. Argument   B. Reasoning
C. Recitation   D. Blinking of the eyes
E. None of the above

15. Samuel was riding down a hill when he had an accident. After treatment it was discovered that he was partially paralyzed and further that he lacked a sense of feeling. The part of Samuel's body much affected by the accident was:
A. The brain   B. The cerebrum
C. The cerebellum   D. The spinal cord
E. The peripheral nerves

16. A tube of stored ammonia is passed round an audience of market women. It appeared that some of the women were indifferent to the smell of the ammonia. Which part of their brain or body was defective
A. The bones of the nasal cavity

B. The olfactory lobes   C. The olfactory organs
D. The olfactory nerves   E. None of the above

17. Kalu is a child who always has a high body temperature. Hospital investigations showed that his illness was due to a defective portion of his brain. Which part of his brain was severely hit?
A. The cerebral hemisphere   B. The ventricles
C. The cerebellum   D. The corpora striata
E. None of the above

18. The spinal cord arises from:
A. The spinal stump   B. The cerebral body
C. The medulla oblongata   D. The cerebellum
E. The brain

19. One contestant in a boxing bout after a knockout cannot lift up his right arm. Which part of his brain was severely hit?
A. The right cerebellum
B. The medulla oblongata
C. The left cerebral hemisphere
D. The right cerebral hemisphere
E. The central lobes of the cerebellum
20. The thin skin that immediately covers the spinal cord is:
A. The dura mater   B. The arachnoid skin
C. The pia mater   D. The fibrous capsule
E. None of the above

21. What is the characteristic of the ventral fissure of the spinal cord?
A. It is elongated
B. It is usually wide
C. It contains blood vessels and nerves
D. It gives rise to the ventral nerve roots
E. None of the above

22. Where in the spinal cord are the cell bodies found?
A. In the ventral and dorsal roots of the white matter
B. In the dorsal root of the grey matter
C. In the grey matter
D. In the white matter
E. In the dorsal root ganglia

23. The white matter is noted for:
A. Having nerve fibres and myelin sheaths
B. The relay of nerve impulses
C. Possessing only the fatty sheaths
D. The storage of nerve forming cells
E. Possessing the intermediate or relaying neurons

14.1.2 THE COMPONENTS AND STRUCTURE OF THE P.N.S

24. What is the function of the peripheral nervous system?
A. It control the activities of the body
B. It is a relay system
C. It co-ordinates the receptive functions of the skin
D. It serves as a link between the organs and the tissues of the body and the central nervous system
E. None of the above reasons

25. UME/93/28
In the mammal, the autonomic nervous system consists of
A. sympathetic and parasympathetic systems
B. brain and cranial nerves
C. brain and spinal nerves
D. spinal cord and spinal nerves

26. UME/2003/26/TYPE:A

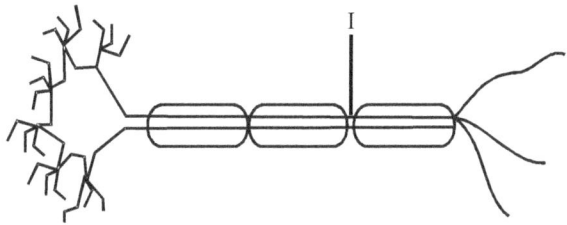

The structure can be found in the
A. peripheral and central nervous system
B. peripheral nervous system only
C. sympathetic and parasympathetic nervous system
D. central nervous system only

27. UME/2003/27/TYPE: A
The point marked I is referred to as
A. myelin sheath   B. dendrites
C. node of ranvier   D. axon

14.1.3. MECHANISM OF TRANSMISSION OF IMPULSES

28. UME/2004/11/TYPE:4
The two key cations involved in the action potentials of nervous transmission are
A. Mg^{2+} and K^+
B. Na^+ and Fe^{2+}
C. Fe^{2+} and Mg^{2+}
D. Na^+ and K^+

29. PCE/91/32
Which of the following plays a direct role in the transmission of impulses across a synapse?
A. acetylcholine
B. cholinesterase
C. sodium ion
D. potassium ion

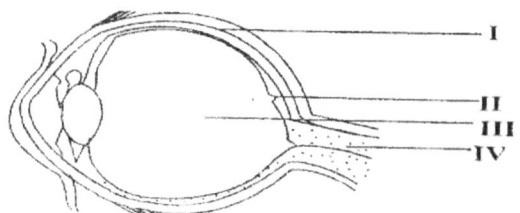

30. PCE/98/26/TYPE: C
In mammals, the correct sequence in the flow of an impulse is
A. effectors → sensory neurone→ stimulus→ motor neurone
B. stimulus → sensory neurone → relay neurone → motor neurone→effector
C. stimulus→ motor neurone→ stimulus → sensory neurone
D. stimulus →motor neurone → relay neurnone → sensory neurone, effector

31. UME/86/40
In a mammal, stimulus is transferred from the receptor muscle to the central nervous system through the
A. motor neurons   B. effector muscles
C. dendrites   D. sensory neurons
E. synapses

32. UME/98/27
The small masses of nervous tissue in which many neurons have their nuclei are called
A. dorsal roots
B. ventral roots
C. ganglia
D. synapses

14.1.4. REFLEX ACTION

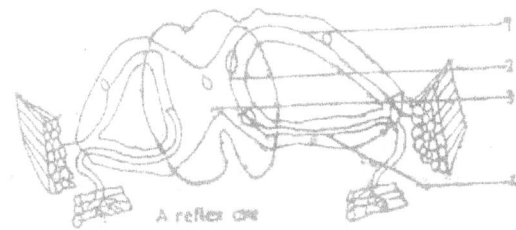

A reflex arc

33. UME/90/33
All the cell bodies in the spinal cord are found in
A. 1   B. 2   C. 3   D. 4

34. UME/90/34
In a reflex action, impulse flows from
A. 1 to 2   B. 2 to 1   C. 4 to 1   D. 4 to 2

35. UME/80/34
The following are connected with the movement of a reflex action
(1) Central nervous system (2) Muscle (3) Skin (4) Sensory nerve (5) Motor nerve.
Which of the following sequences indicates a correct path?
A. 1-2-3-4-5   B. 2-1-4-5-3
C. 4-1-5-2-3   D. 4-1-5-2-3   E. 3-4-1-5-2

14.2.   SENSORY ORGANS
14.2.1. EYE

36. PCE/2004/30/TYPE: 4
The part that provides nourishment for the eye is labelled
A. I   B. II   C. III   D. IV

37. PCE/2004/31/TYPE: 4
The part labelled II is the
A. blind spot   B. retina
C. yellow spot   D. conjunctiva

38. PCE/2003/34/TYPE: N
The three layers that form the wall of the eyeball are
A. sclera, choroid and epidermis
B. choroid, cornea and epidermis
C. retina, choroid and dermis
D. sclera, choroid and retina

39. PCE/96/24
Which of the following is found in the retina?
A. ciliated cell   B. cone-shaped cell
C. epithelial cell   D. cubical cell

40. UME/82/25
The parts of the mammalian eye that strongly bend light rays are the
A. cornea and the lens
B. cornea and aqueous humour
C. cornea and vitreous humour
D. lens and aqueous humour
E. lens and vitreous humour

41. UME/79/17
When the ciliary muscle of the eye contracts, the eye lens
A. bulges   B. contracts   C. rotates
D. flattens   E. is rounded

42. UME/97/32
The ability of the eye to focus on both near and distant objects is termed
A. image formation   B. refraction
C. hypermetropia   D. accommodation

43. UME/85/39
An old man is likely to be long-sighted because age affects the
A. optic nerves   B. retina
C. ciliary muscles   D. cornea
E. aqueous humour

44. UME/94/27
When a short-sighted person views a distant object without spectacles, the image is formed
A. on the retina   B. in front of the retina
C. behind the retina   D. on the blind spot

45. PCE/92/29
Human beings have relatively poorer vision at night than nocturnal animals because the human eyes
A. are meant for day vision
B. contain fewer rods
C. contain more rods
D. contain fewer cones

46. The function of concave lens in correcting myopia is to
A. diverge incoming rays
B. converge incoming rays
C. reflect incoming rays
D. screen incoming rays

47. PCE/94/30
A person is suffering from an eye defect in which he can see only distant objects clearly. This eye defect may be corrected by using
A. convex lens   B. biconvex lens
C. concave lens   D. biconcave lens

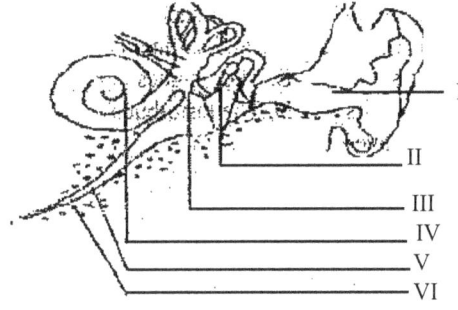

48. UME/2001/22/TYPE: 1
The parts which function together to bring about hearing are lebelled
A. IV, V and VI   B. I, II, IV and VI
C. I, II, III and IV   D. I, II and IV

49. UME/2001/23/TYPE: T
The part labelled II is the
A. fenestra ovalis
B. middle ear canal
C. internal auditory meatus
D. ear ossicles

50. UME/98/32
Which path does sound entering the human ear follow?
A. oval window, ossicles, eardrum
B. ear drum, oval window, ossicles
C. eardrum, ossicles, oval window
D. ossicles, ear drum, oval window

51. UME/2000/19/TYPE: M
The inner ear contains two main organs, namely, the
A. eardrum and eustachian tube
B. cochlea and semi-circular canals
C. oval window and ossicles

D. pinna and cochlea
52. UME/83/39
The function of the ossicles (malleus, incus and stapes) in the mammalian ear is the
A. transmission of vibrations
B. regulation of pressures
C. support of the inner ear
D. maintenance of balance during motion
E. secretion of oil

14.2.3. TONGUE
53. PCE/2001/27/TYPE: 4
In the tongue, the buds responsible for sweat taste are at the
A. sides   B. back   C. centre   D. tip

54. The structures that respond to such sensations as sweetness, bitterness, sourness and saltiness are:
A. The taste-hairs   B. The tongues
C. The gustatory hairs   D. The taste-buds
E. None of the above

14.2.4. NOSE (OLFACTORY)
55. UME/99/34/TYPE: D
The correct sequence for the perception of smell in mammals is
A. chemicalsàolfactory nerve endingsà brain
B. dissolved chemicalsà nasal sensory cellsàbrain
C. chemicals à mucous membraneàsensory cellsà brain
D. dissolved chemicals àsensory cellàolfactory nerve à brain

14.2.5. SKIN

56. UME/2004/15/TYPE: 4
The function of the part labelled III is to
A. produce oil for the skin
B. carry blood and nitrogenous waste
C. contract to pull the hair erect
D. conduct nervous impulses

57. UME/2004/16/TYPE: 4
The sweat gland is the structure labelled
A. IV   B. III   C. II   D. I

14.3. HORMONES
14.3.1. ANIMAL HORMONES
58. UME/85/38
The mammalian endocrine system is responsible for
A. transmitting impulses
B. regulatory body temperature
C. regulating osmotic pressure of blood
D. chemical co-ordination
E. the manufacture of blood

59. UME/89/34
A fundamental similarity between nervous and hormonal system is that both
A. involve chemical transmission
B. have widespread effects
C. shed chemical into the blood stream
D. evoke rapid responses

60. UME/2000/9/TYPE: M
In which part of the human body does the secretion of the growth hormone occur?
A. head region   B. waist region
C. neck region   D. gonads

61. PCE/94/32
The endocrine glands responsible for the production of antidiuretic hormone is the
A. thyroid gland   B. pancreas
C. adrenal gland   D. pituitary gland

62. PCE/99/22/TYPE: E
The hormone whose deficiency in a child would cause reduced mental and physical growth is
A. adrenalin   B. thyroxine
C. insulin   D. secretin

63. UME/95/27
The gland directly affecting metabolic rate, growth and development is known as
A. adrenal gland   B. thyroid gland
C. mammary gland   D. parathyroid gland

64. UME/80/33
Which of the following hormones is produced during fright or when agitated?
A. Insulin   B. adrenalin
C. thyroxine   D. pituitrin
E. progesterone

65. PCE/96/25
Insulin and glucagon are hormones produced by the
A. thyroids   B. adrenal gland
C. pancreas   D. pituitary gland

66. UME/86/40
The hormone which regulates the amount of glucose in the blood is called
A. adrenalin   B. auxin
C. insulin   D. thyroxine

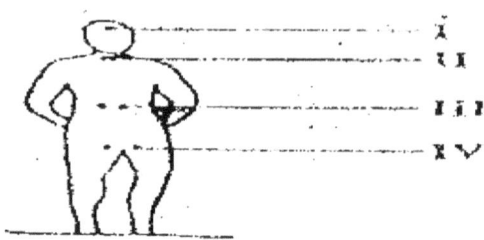

67. UME/2002/43/TYPE: Y
The gland usually found in the position labelled I is the A. adrenal   B. thyroid
C. pancreatic   D. pituitary

68. UME/2002/43/TYPE: Y
The hormone secreted at II serves to
A. facilitate the development of facial hours
B. raise the level of calcium ions in the blood
C. lower blood glucose level
D. make the body react to emergencies

69. PCE/92/30
Overproduction of somatotropic hormones in the human being could result in
A. dwarfism   B. gigantism
C. cretinism   D. obesity

70. PCE/94/31
The cells responsible for the secretion of insulin in human body are called
A. Graafian follicles   B. islets of langerhans
C. crypt of lieberkhun D. corpus luteum

71. PCE/06/5/TYPE: A
The antidiuretic hormone is secreted by the
A. posterior pituitary gland
B. anterior pituitary gland
C. Gonads
D. pancreas

72. PCE/07/23
Which of the following produces secretions that help on controlling the level of calcium in the blood?
A. gonad   B. parathyroid
C. pancreas   D. pituitary gland

14.3.2. PLANT HORMONES (PHYTOHORMONES)

73. UME/87/36
The growth of a coleoptile towards unilateral light source is due to
A. rapid rate of photosynthesis
B. unequal distribution of auxin
C. the effect of geotropism
D. the effect of photolysis

74. UME/89/33
Fruit enlargement can be induced by spraying young ovary with
A. cytokinin, abscisic acid and ethylene
B. gibberellins, ethylene and abscisic acid
C. auxin, abscisic acid and ethylene
D. auxin, cytokinin and gibberellins

75. UME/93/30
The substance that is responsible for apical dominance in plants is known as
A. gibberellins   B. tannin
C. auxin   D. kinin

76. UME/87/39
Unlike auxins, gibberellins
A. induce the formation of adventitious roots
B. do not affect leaf and fruit abscission
C. cannot stimulate stem elongation
D. are quite effectives as herbicides

77. UME/91/36
A dwarf plant can be stimulated to grow to normal height by the application of
A. Thyroxin   B. gibberellins
C. Insulin   D. Kinin

78. UME/2000/33/TYPE: M
Which of the following growth activities in plants is brought about by gibberellins?
A. Rapid cell division
B. tropic response
C. cell elongation
D. main stem elongation

79. UME/2003/21/TYPE: A
The most important hormone that induces the ripening of fruit
A. Cytokinin   B. indole acetic acid
C. Ethylene   D. gibberellins

80. PCE/06/9/TYPE: A
Weeds can be effectively controlled by the use of
A. abscisic acids   B. gibberellins
C. auxins   D. cytokinins

81. One of the following pairs of plant hormones helps in seed germination
A. Cytokinin and Auxin
B. Gibberellin and Abscissic acid
C. Auxin and Gibberellin
D. Cytokinin and Gibberellin
E. Abscissic acid and Cytokinin

82. One of the following pairs of hormones is responsible for both cell elongation and apical dominance
A. Cytokinin and Auxin
B. Gibberellin and Cytokinin
C. Florigens and Abscissic acid
D. Auxin and Gibberellin
E. Gibberellin and Abscissic acid

14.4. HOMEOSTASIS

83. UME/82/7
An homeothermic animal kept in a room where the temperature is lower than the body temperature may lose heat by four physical processes. Which of the following processes is NOT connected with body temperature regulation?
A. Radiation   B. sweat
C. Evaporation   D. conduction
E. convection

84. UME/83/17
The kidneys of all vertebrates act as osmo-regulators. This means that they
A. keept the composition of the plasma constant
B. regulate osmotic processes
C. control the volume of blood entering the kidneys
D. decrease the osmotic pressure of blood
E. increase the osmotic pressure of blood

85. UME/93/49
Cold blooded animals are referred to as
A. Poikilothermic
B. homeothermic
C. Polythermic
D. homeostatic

86. PCE/98/10/TYPE: C
The group of animals that is poikilothermic is
A. birds, fishes and amphibians
B. reptiles, mammals and birds
C. mammals, amphibians and fishes
D. amphibians, fishes and reptiles

87. PCE/2001/30/TYPE: 4
Homeostasis is a feedback reaction which is largely
A. hormonal   B. voluntary
C. reflex   D. nervous

88. PCE/2003/39/TYPE: N
High body temperature in mammals can be regulated by
A. vasodilation and sweating
B. sweating and shivering
C. vasdilation and vasoconstriction
D. vasoconstriction and sweating

89. PCE/92/47
The temperature of a homoeothermic animal depends on the
A. temperature of the environment
B. heat obtained from the sun
C. heat it receives from the environment
D. in-built mechanisms that maintain a constant body temperature

90. PCE/00/26/TYPE: A
When the osmotic pressure of blood rises, the hormone released to the kidney causes the nephron to
A. decreases the reabsorption of water
B. increase the reabsorption of water
C. decrease the reabsorption of salt
D. increase the reabsorption of salt

MISCELLANEOUS QUESTIONS (14)

91. Which of the following is a homeostatic response in humans?
A. withdrawing the hand from a hot object
B. the mouth getting watery when food is sighted
C. yawning owing to tiredness
D. shivering in a cold environment
(UME/2005/31/TYPE: D)

92. The part of the brain that control heart beat and breathing is the
A. olfactory lobe   B. cerebellum
C. cerebral hemisphere   D. medulla oblongata
(UME/94/29)

93. The centre for controlling body temperature in the brain is the
A. cerebrum   B. cerebellum
C. medulla   D. hypothalamus
(UME/87/38)

94. Movement and positions of the head in man are detected by the
A. cochlea   B. malleus
C. utriculus   D. semicircular (UME/85/28)

95. The main function of the choroid is
A. protection of the eyeball
B. transmission of light
C. supply of nutrients to tissues of the eye
D. converging light   (UME/88/34)

96. What part of the central nervous system is concerned with answering an examination question?
A. Cerebrum   B. cerebellum
C. medulla oblongata   D. spinal cord (UME/88/35)

97. In the mammalian skin, mitosis occurs in the
A. cornified layer   B. granular layer
C. malpighian layer   D. dermal layer CE/99/26/TYPE: E)

98. The skin of a mammal will be rendered colourless by
A. a defective sebaceous gland
B. a defective malpighian layer
C. dead nerve endings
D. blocked sweat pores (PCE/2000/28/TYPE: E)

99. What would happen to a man whose pancreas has been surgically removed?
A. the level of blood sugar would increase
B. the glycogen content of the liver would increase
C. his blood pressure would decrease
D. his weight would increase appreciably

100. Which of the following growth activities in plants brought about by gibberellins?
A. rapid cell division   B. tropic response
C. cell elongation   D. main stem elongation
(UME/2000/33/TYPE: M)

101. One basic similarity between nervous and endocrine system is that they both
A. produce widespread effects
B. transmit very fast impulses
C. involve the use of chemical substances
D. electrostatic changes (UME 2006/6)

102. The unit of the nervous system is called:
A. The axons   B The myelin sheath
C. The neurons   D. The cell body
E. None of the above

103. The small gaps occurring between the nerve fibres are:
A. The juxtapositions   B. The bridges
C. The synapses   D. The nodes of Ranvier
E. None of the above

104. The function of the fatty myelin sheath is:
A. To insulate the axon and prevent the loss of impulses
B. To lubricate the axon and increase its conductivity
C. To replace damaged axons
D. To transmit impulses from the receptor to the effectors
E. None of the above

105. The nucleated sheath that covers the myelin sheath is:
A. The sheath nuclei   B. The neurilemma
C. The cytoplasm   D. The neurolemma
E. None of the above

106. The nodes of Ranvier are:
A. Areas in which the axons swell up
B. Areas where the neurilemma touches the axon
C. The points in which the axon is exposed
D. The contact points found over the non-myelinated nerve fibres
E. None of the above

107. Which of these actions is not reflex actions?
A. Jumping up when on a hot iron
B. The beating of the heart
C. The secretion of pepsin into the stomach
D. Ultra filtration of urea in the kidney
E. None of the above

108. What is the main difference between a reflex and a voluntary action?
A. A voluntary action is premeditated
B. A reflex action is initiated by the spinal cord while a voluntary action is the result of activity of the cerebral hemisphere of the brain
C. A reflex action has a longer period of execution
D. None of the above

109. What are the cranial nerves?
A. These are the nerves of the cranium
B. These are the nerves that control the brain
C. These are the shortest nerves of the body
D. These are the main nerves of the body
E. These are the nerves of the brain

110. While the ears and the eyes respond to the stimulus of sound and light respectively the taste buds and olfactory organs respectively responds to:
A. The stimulus of smell
B. The stimulus of bitter taste
C. The stimulus of chemicals
D. Bitterness, sourness and sweetness of substances
E. None of the above

111. The structures that respond to such sensations as sweetness, bitterness, sourness and saltiness are:
A. The taste-hairs   B. The tongues
C. The gustatory hairs   D. The taste-buds
E. None of the above

112. The vitreous humour differs from the aqueous humour in:
A. Being much more watery
B. Being more translucent
C. Having a less refractive index
D. Being more viscous
E. None of the above

113. Why is an area of the eye known as the blind spot?
A. This spot is dead
B. The spot is outside the receptive surfaces of the retina
C. The spot is covered by nerves from the retina
D. This spot being without visual cells is non receptive to light stimulus
E. None of the above

114. The function of the iris is:
A. To stir the aqueous humour and keep it always clean
B. To protect the lens
C. To regulate the size of the pupil
D. To direct the rays of light to the retina
E. To assist the ciliary muscles in varying the focal length of the lens

115. Which of these structures has practically no effect on the rays of light that enter the eye?
A. The cornea
B. The lens
C. The aquous humour
D. The vitreous humour
E. None of the above

116. A man discover that he cannot clearly see a far distant object. What is the defect of his eye?
A. Myopia
B. Astigmatism
C. Long-sightedness
D. Hypermetropia
E. A cataract of the eye

117. The watery fluid in the inner ear of mammals is known as
A. The endolymph
B. The perilymph
C. The labyrinth
D. None of the above

118. The most important structure of the utriculus of the inner ear responsible for balance is:
A. The cochlea
B. The ampulla
C. The semicircular canal
D. The sacculus
E. None of the above

119. Which of these is not an ossicle of the middle ear?
A. Malleous
B. Tympanic bone
C. Incus
D. Stapes
E. None of the above

120. If a student closes tightly his nostrils and mouth and blows from within him, he feels some unpleasantness on his tympanic membranes. Through which of these structures did air get into his ear?
A. The auditory meatus
B. The pharynx
C. The tympanic chamber
D. The lungs
E. The eustachian tube

121. The function of the pinnna of the ear is:
A. To locate the direction of sound
B. To distinguish between different types of sound reaching the ear.
C. To pick up the sound reaching the ear
D. To amplify sound reaching the ear
E. None of the above

122. Which of these is a function of the cochlea?
A. Hearing
B. To air the ear ossicles or bones
C. Detection of change in the angular disposition of the animal
D. Orientation or maintaining body posture in space
E. None of the above

123. Diabetes is an illness resulting from underactivity of the hormone
A Corticin
B. Thyroxin
C. Secretin
D. Insulin
E. Adrenalin

124. Which of these hormones is responsible for the growth and metamorphosis of tadpoles?
A. Adrenalin
B. Insulin
C. Pituitrine
D. Thyroxin
E. None of the above

125. A truant of a pupil escapping into a bush is confronted by a snake. Which of these hormones will be released into his blood-stream?
A. Insulin
B. Parathymin
C. Adrenalin
D. Thyroxin
E. Gastrin

CHAPTER 15

BASIC ECOLOGICAL CONCEPT

Understand the following after reading this chapter:

> Factors affecting the distribution of organisms
> Methods of measuring abiotic and biotic factor

Ecology is the study of living things in relation to their environment.

SOME BASIC ECOLOGICAL CONCEPTS
1. Autecology: The study of an individual organism and its environment.
2. Synecology: The study of two or more organism and their environments.
3. Environment: The totality of living and non living external factors which influence the life of an organism.
4. Habitat: The natural home of an organism.
There are two major types of habitats- **Terrestrial habitat** (living on land) and **Aquatic habitat** (living in water). There are three types of aquatic habitats:
i. Fresh water: A water body that has little or no salt concentration e.g. ponds, streams, rivers etc
ii. Marine or salt water: A water body with high salinity e.g. seas and oceans.
iii. Brackish or estuarine: A mix-up of fresh and marine waters with seasonally fluctuating salinity e.g. Deltas and Lagoons.
5. Population: Total number of individuals of the same species living in a particular area.
6. Community: A naturally occurring group of plants and animals living in a given area.
7. Niche: The smallest area within an habitat where an organism is found. It also means the role performed by an organism in the habitat.
8. Biome: A large land community with peculiar plants and animals.
9. Atmosphere: The open space above the earth surface which contains a mixture of gases and water vapour in different proportions.
10. Biosphere: The part of the earth that support the existence of living things.
11. Hydrosphere: Biosphere is made up of water e.g. river, stream etc.
12. Lithosphere: The solid part of the earth's crust e.g. rock, soil etc.

ECOLOGICAL OR ENVIRONMENTAL FACTORS
Ecological factors are broadly divided into two:
1. **Abiotic or non-living factors** i.e.
a. Climatic factors e.g. rainfall, light, wind, temperature, etc.
b. Topographic factors i.e. nature of soil surface e.g. slopes, valleys elevation, exposures etc.
c. Edaphic factors: soil types and properties i.e. moisture content, pH, organic matter content etc.
2. **Biotic or living factors**: activities of living things e.g. feeding, shading from light, competition dispersal, predator etc.

Ecological Factors in Aquatic Habitat
1. Salinity 2. Turbidity 3. Tides 4. Current

Ecological Instruments and their uses

Instrument	Uses
Anemometer-	For measuring the speed of wind
Barometer -	For measuring atmospheric pressure
Hygrometer -	For measuring relative humidity
Sweep net/ insect net-	For collecting flying insects
Min.Max. thermometer-	For measuring day temperature
Wind vane-	For determining the direction of wind
Rainguage-	For measuring amount of rain fall
Quadrant-	For determining the population of organism
Pooter-	For collecting small insects
Photometer/Light meter-	For measuring light intensity
Secchi disc-	For measuring turbidity of water
pH indicator-	For measuring water or soil acidity or alkalinity

Revision Questions

15.1. BASIC ECOLOGICAL CONCEPTS
1. UME/78/18
The study of the organism and the environment of an abandoned farmland is the ecology of
A. a community   B. a population
C. a species   D. a habitat
E. an ecosystem

2. UME/92/33
A population is defined as a collection of similar organisms that
A. behave in the same way
B. interbreed freely
C. are found in the same habitat
D. eat the same food

3. UME/93/35
What is the term used to describe the sum total of biotic and abiotic factors in the environment of the organism?
A. habitat   B. biome
C. ecosystem   D. ecological niche

Use the list of ecological constituents below to answer questions 4 and 5
1. mango   2. Spear grass   3. Goat   4. Sheep
5. Temperature 6. Beans   7. Rock   8. Water

4. UME/94/30
Item 1 – 4 can be regarded as
A. a population   B. a community
C. an ecosystem   D. a niche

5. UME/94/31
The physical factors are represented by
A. 1, 5, 6   B. 4, 5, 7
C. 5, 7, 8   D. 6, 7, 8

6. UME/2001/28/TYPE : T
The number of plant species obtained from a population study of a garden is as follows: Guinea grass (15), pornea spp (5), sida spp (7) and imperata spp (23). What is the percentage of occurrence of imperata spp.?
A. 35%   B. 16%
C. 46%   D. 13%

7. UME/2004/41/TYPE: 4
A caterpillar and an aphid living in different of the same habitat can be said to
A. be in similar microhabitats
B. occupy different ecological niches
C. occupy the same ecological niches
D. be in different habitats

8. PCE/96/26
The correct order of the ecological sequence in nature is
A. species → population → community → ecosystem
B. species → population → ecosystem → community
C. species → ecosystem → population → community
D. species → community → ecosystem → population

9. UME 2007/7
The highest level of ecological organization is the
A. ecosystem
B. population
C. biosphere
D. Niche

10. PCE/93/34
The functional position of an organism on the ecosystem is referred to as its
A. habitat   B. niche
C. community   D. micro-habitat

11. PCE/98/30/TYPE: C
A group of organisms of the same species living together in the same habitat is referred to as
A. community   B. a population
C. an ecosystem   D. an association

12. PCE/06/40/TYPE: A
A group of closely related organisms capable of interbreeding are termed
A. species   B. phylum
C. population   D. community

15.2. FACTORS AFFECTING THE DISTRIBUTION OF ORGANISMS
13. UME/84/44
Which of the following groups of factors is completely abiotic?
A. salinity, tide, plankton, turbidity
B. temperature, pH, soil, insect
C. wind, altitude, humidity, light
D. conifers, wind, pH, rainfall
E. soil, water, bacteria, salinity

14. UME/86/43
In an agricultural ecosystem, the biotic component consists of
A. crops, pest and beneficial insects
B. crops, temperature and humidity
C. pests, beneficial insects and water
D. crops, water and soil

15. UME/86/45
Which of the following ecological factors are common to both terrestrial and aquatic habitats?
A. rainfall, temperature, light and wind
B. salinity, rainfall, temperature and light
C. tides, wind, rainfall and altitude
D. pH, salinity, rainfall and humidity

16. UME/87/45
Which of the following relates to edaphic factors?
A. the structure of the earth's surface
B. the influence of living organisms on each other
C. Temperature, rainfall and humidity
D. the influence of soils on plants and animals

17. UME/88/42
Which of the following has the greatest influence on the distribution of animals in marine and fresh water habitats?
A. PH
B. salinity
C. water current
D. turbidity

18. UME/91/41
The most important factors which influence an organism's way of life in its habitat are
A. the physical and biotic environment
B. food and water availability
C. temperature, water, light and predator-prey relationship
D. competition for food and space

19. UME/93/37
The most important physical which affects an organism living in the intertidal zone of the seashore is
A. pH   B. salinity
C. wave action   D. temperature

20 UME/2003/33/TYPE: A
A density-dependent factor that regulates the population size of organisms is
A. sudden flood
B. disease
C. fire outbreak
D. drought

21. UME 2007/9
A biotic factor which affects the distribution and abundance of organisms in a terrestrial habitat is
A. temperature
B. competition
C. pH
D. light

22. UME 2008/34
An ecological factor that will have the most limiting effect on the abundance of phytoplankton in a turbid pond is
A. pH   B. oxygen
C. light   D. temperature

23. UME 2008/32
The main ecological problem facing intertidal organisms is
A. desiccation   B. floatation
C. salinity   D. humidity

24. UME 2009/32
In freshwater, marshes and swamps, the most important abiotic factor that organisms have to adapt to is
A. nature of substratum
B. high salinity
C. high temperature
D. low pH

25. UME 2010/37/TYPE: D
The activity of an organism which affects the survival of another organism in the same habitat constitutes
A. an abiotic factor
B. a biotic factor
C. physiographic factor
D. an edaphic factor

15.3.   METHODS OF MEASURING ABIOTIC/BIOTIC FACTORS
26. UME/84/42
An anemometer is an instrument for measuring
A. relative humidity   B. altitude
C. wind speed   D. turbidity
E. salinity

27. UME/84/33
The transect method can be used in ecology to show the
A. number of plants and animals in a habitat
B. population of a particular plant species
C. distribution of organisms along a line
D. heights of trees in a section of a forest
E. number of young plants across a forest

28. UME/88/38
The turbidity of a pond can be measured using the
A. anemometer   B. secchi disc
C. theodolite   D. hydrometer

29. UME/93/40
Which of the following instruments is NOT used in measuring abiotic factors in any habitat?
A. microscope   B. thermometer
C. hygrometer   D. wind vane

30. UME/95/31
The depth of illumination in a water body can be measured with a
A. photometer   B. secchi disc
C. hydrometer   D. anemometer

31. UME/97/33
The speed of wind can be measured with an instrument called
A. hygrometer   B. secchi disc
C. anemometer   D. wind vane

32. UME/2001/36/TYPE: T
The hygrometer is used for measuring
A. relative humidity   B. specific gravity
C. rainfall   D. salinity

33. UME/97/29
The groups of instruments that can be used in the measurement of ecological factors in a habitat is
A. secchi disc, microscope and pooter
B. potentiometer, hygrometer and chemical balance
C. anemometer, photometer and hygrometer
D. telescope, quadrat and wind vane

34. UME/2000/30/TYPE: A
A pooter is used in ecological studies to collect
A. flying insects   B. small insects
C. worms   D. rodents

35. UME/2003/22/TYPE: N
A maximum and minimum thermometer is suitable for measuring the
A. temperature of the soil
B. temperature of water at different depth
C. highest and lowest body temperatures of an animal
D. highest and lowest temperatures of the day

36. PCE/93/36
The estimation of the population of Tilapia in a pond is best done by
A. the quadrant method
B. counting
C. capture and recapture
D. transect method

MISCELLANEOUS QUESTIONS (15)

37. All members of a species inhabiting a given area make up the
A. community
B. population
C. biomass
D. ecosystem   (PCE/91/37)

38. In an ecological study, the biotic factors are
A. rainfall, light and temperature
B. general atmospheric conditions
C. the effects of living things on the ecosystem
D. the structures of the earth's surface (PCE/91/36)

39. The functional position of an organism in the ecosystem is referred to as its
A. habitat
B. niche
C. community
D. micro-habitat (PCE/94/34)

40. The habitat of bernacles and limpets is
A. freshwater
B. arboreal
C. marine
D. terrestrial (PCE/94/35)

41. The ecological factors common to both terrestrial and aquatic habitats are
A. tides, current, salinity and rainfall
B. salinity, rainfall, temperature and light
C. pH, salinity, rainfall and humidity
D. rainfall, temperature, light and wind
(PCE/2001/32/TYPE: 4)

42. The most important abiotic factor in the desert biome is
A. temperature
B. light
C. rainfall
D. wind (PCE/2002/15/TYPE:5)

CHAPTER 16

SYMBIOTIC INTERACTIONS OF PLANTS AND ANIMALS

Understand the following after reading this chapter:

> Food chains, food webs and trophic levels
> Energy flow in the ecosystem
> Nutrient cycling in nature

SYMBIOTIC INTERACTIONS OF PLANTS AND ANIMALS

Symbiosis: Symbiosis is a close association between two organisms in which both of them benefit from each other. Examples are:Algae and fungi in Lichen, Algae and plant roots in mycorrhiza, Protozoa in the intestine of termite, Nitrogen fixing bacteria in the root nodules of leguminous plants.

Commensalism: Commensalism is an association between two organisms living together in which only one (commensal) benefits from the association while the other neither benefited nor is harmed. Examples are: Remora fish and shark, Oyster and crab, Man and intestinal bacteria, Epiphytes growing on plants

Parasitism: Parasitism is a close association between two organisms in which one, known as the parasite, lives in or on and feeds at the expense of the other organism which is known as the host. Examples include: Man and the tapeworm, Mistletoe and flowering plant.

Predation: Predation is a type of association between two organisms in which the predator kills the other, called the prey and directly feeds on it. Examples are: Cat feeding on rats, Hawk and chicken.

FOOD CHAINS, FOOD WEBS AND TROPHIC LEVELS

Food chains is a linear feeding relationship in which an organism feeds on the one before it in the sequence e.g. Grass→Grasshoppers→Lizards→Snakes

Food web is a complex feeding relationship consisting of interconnected food chains in a community.

Trophic level refers to any step in the transfer of energy and matter in a community. This describes a specific level in a food chain. The term *trophic* refers to **nutrition**.

There are four important levels in most food chains:

- *Producers:* Organisms which convert some of the energy from the sun into stored chemical energy (usually plants).

- *Primary consumers:* Organisms that obtain energy by consuming producers. They are herbivores.

- *Secondary consumers:* Organisms which obtain energy by consuming primary consumers. They are carnivores.

- *Decomposers:* These organisms form the end point of every food chain. They are bacteria or fungi that obtain their energy by breaking down dead organisms from the other trophic levels.

Each description of a trophic level will describe an organisms role in the ecosystem. Organisms may occupy more than one trophic level, (e.g. when acting as omnivores).

ENERGY FLOW IN THE ECOSYSTEM

The sun is the ultimate source of energy on earth. The producers i.e. green plants produce their food through photosynthesis. The consumers i.e. animals and non-green plants depend on already made food by the producers. The decomposers i.e. microbes feed on dead plants and animals and in the process help to recycle nutrients.

Pyramids in ecology

Ecological pyramids are used as a tool to illustrate the feeding relationships of the organisms, which together make up a community.

Pyramid of numbers

This is the simplest way of illustrating the feeding relationships within a community. The commonest form shows that the numbers of organisms occupying each trophic level decreases from producers to secondary consumers and beyond

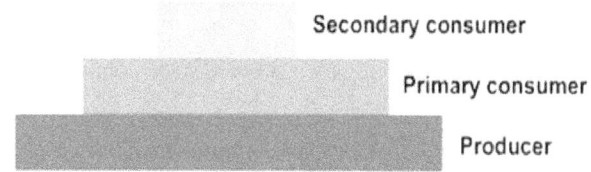

Two problems with this form of pyramid are that the numbers involved may be huge (in the hundreds of thousands) and some pyramids may be **inverted.**

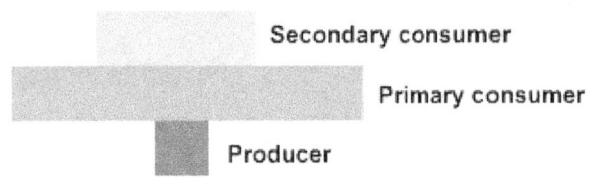

Pyramid of energy

This is the most accurate representation of the feeding relationship between the organisms at different trophic levels. It takes into account the energy gains and losses over a period of time.

Pyramid of biomass

This indicates the feeding relationship between organisms occupying different trophic levels with reference to their biomass.

Biomass can be measured as either wet mass or dry mass. Measuring the dry mass is more accurate as it does not include the variable water content of organisms.

The commonest form of the pyramid of biomass shows that the total biomass of organisms occupying each trophic level decreases from producers to secondary consumers and beyond.

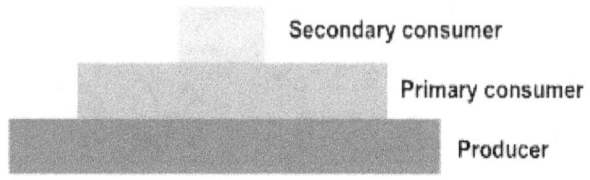

NUTRIENT CYCLING IN NATURE

CARBON CYCLE

Plants get their carbon from carbon dioxide in the atmosphere while animals obtain carbon through feeding on plants. Carbon dioxide is added to the atmosphere by: Respiration, combustion, decay; and volcanic eruption and removed by photosynthesis

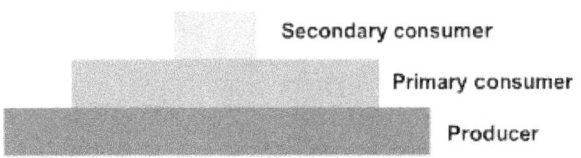

There is still the problem that a pyramid of biomass can be inverted and also it does not take account of changes over time. The sampling must all be carried out at one moment in time and therefore indicates the **standing crop** and not the **productivity**.

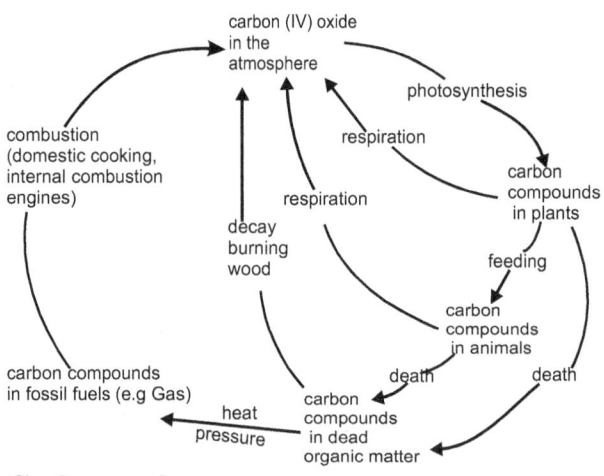

Carbon cycle

HYDROLOGICAL CYCLE

The hydrological cycle or water cycle is the continuous movement of water between the land, the sea and the air

Precipitation - All water released from clouds such as rain, snow, hail, sleet & snow
Surface Runoff - Water flowing across the surface of the land, whether in a channel or over the land
Interception - When trees or man made objects get in the way of rain reaching the ground surfac
Infiltration - When water soaks into the soil

Evaporation - When water is heated by the sun and rises into the sky as water vapour
Transpiration - When moisture from plants and leaves is lost to the atmosphere
Condensation - When water vapour is cooled and turns into water droplets to form clouds
Water Table - The level of saturated ground in the soil - it rises and falls depending on the amount of rain

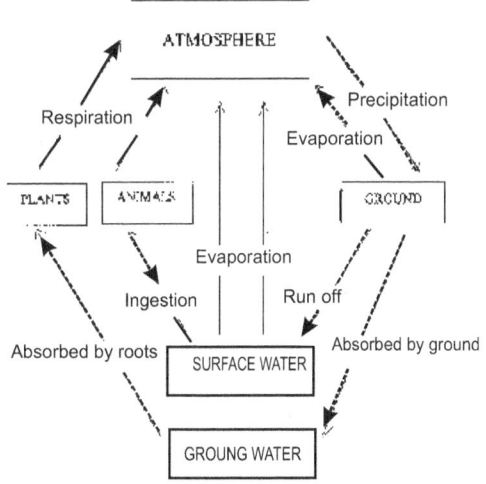

Water cycle

NITROGEN CYCLE

Nitrogen cycle involves a complex biochemical transformation which adds and removes nitrogen and its compound from the soil.

Nitrification (process that add nitrogen to the soil)

a. Nitrifying bacteria- Nitrosomonas and nitrobacter convert ammonia and nitrites to nitrates respectively.

b. Nitrogen fixation by;

i. Lightening or electrical discharge in the atmosphere

ii. Symbiotic nitrogen fixation by some bacteria e.g. Rhizobium

iii. Non-symbiotic nitrogen fixation by some free-living bacteria e.g. Clostridium and Azotobacter

iv. Application of nitrogen containing fertilizers e.g. NPK

Denitrification (process that remove nitrogen from the soil): uptake by plants, erosion, leaching, burning and denitrifying bacteria.

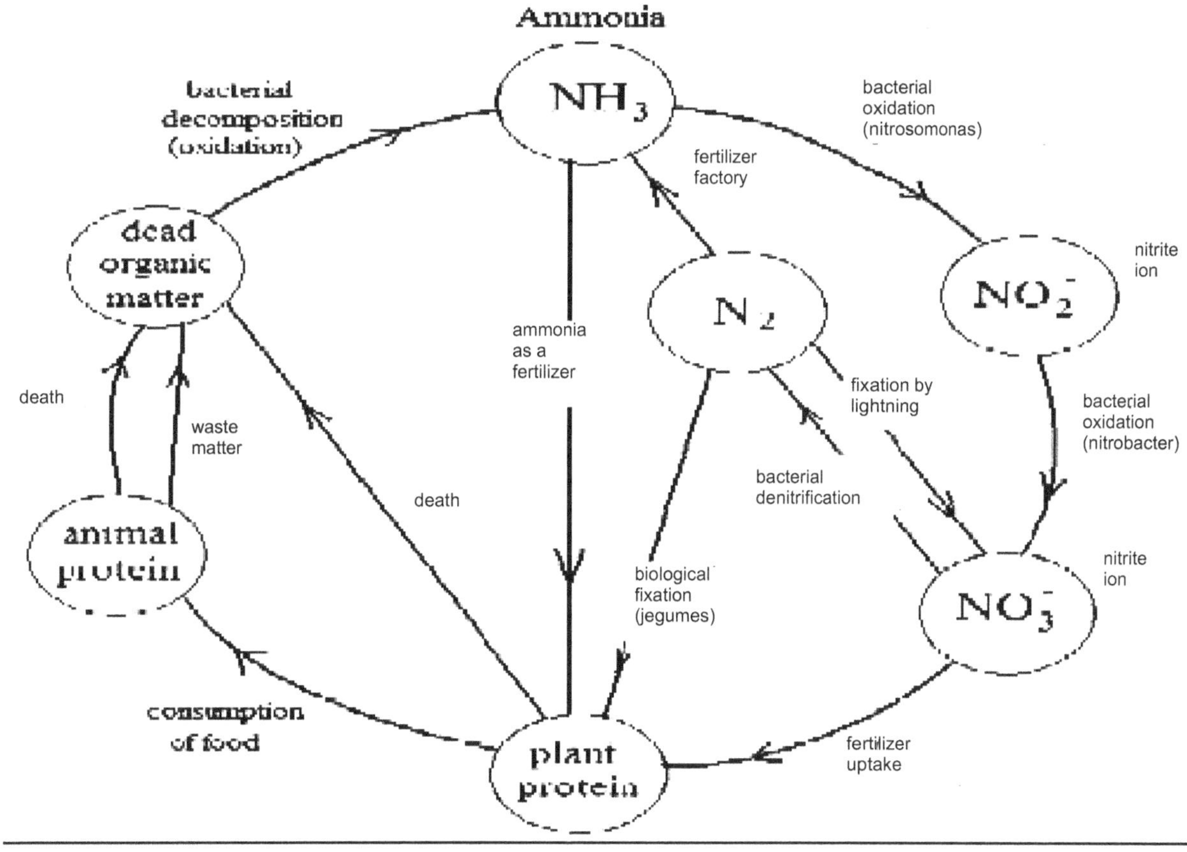

Nitrogen cycle

16.1. INTERACTIONS OF PLANTS AND ANIMALS

1. UME/79/19
In which of these associations is much harm done to one of the partners?
A. Symbiosis
B. Epiphytism
C. Commensalisms
D. Parasitism
E. Mutualism

2. UME/80/39
An organism X lives entirely on the waste products in another organism Y. In this association, X is a
A. Symbiont
B. Commensal
C. Saprophyte
D. Parasite
E. Epiphyte

3. UME/81/28
Which of the following statements is NOT true of symbiosis?
A. Symbionts must be living
B. It is an association of give and take
C. The association may involve two plants
D. Association between two similar species
E. symbionts derive mutual benefit

4. UME/87/46
Epiphytes growing on the branches of trees provide an example of the relationship known as
A. parasitism
B. commensalisms
C. saprophytism
D. holophytism

5. UME/88/41
Which of the following relationship involves only one organism?
A. saprophytism
B. commensalism
C. parasitism
D. symbiosis

6. UME/89/41
Lichen is an example of
A. a saprophytic organism
B. a symbiotic association
C. an epiphytic plant
D. a carnivorous plant

7. UME/91/50
Nitrogen fixing micro-organisms in leguminous plants live symbiotically in the
A. root nodules
B. tap roots
C. branch roots
D. root hairs

8. UME/2003/32/TYPE: A
Mycorrhiza is an association between fungi and
A. roots of higher plants
B. filamentous algae
C. bacteria
D. protozoans

9. UME 2007/4
Which of the following is an example of parasitism?
A. fungi growing on a dead tree branch
B. a squirrel living in an abandoned nest of a bird
C. mistletoe growing in an orange tree
D. cattle egrets taking ticks from the body of cattle

10. UME 2008/30
Which of the following associations is an example of mutualism?
A. *Hydra vividis* and *zoochlorellae*
B. Human and lice
C. Shark and remora fish
D. Bread and *Rhizopus stolonifer*

11. UME 2009/30
The association between bacteria residing in the caecum and the ruminant is
A. parasitism
B. predator
C. saprophytism
D. mutualism

12. UME 2010/29/TYPE: D
Mycorrhiza promote plant growth by
A. protecting it from infection
B. helping it to utilize atmospheric nitrogen
C. serving as a growth regulator
D. absorbing inorganic ions from the soil

16.2. FOOD CHAINS, FOOD WEBS AND TROPHIC LEVELS

13. UME 2010/28/TYPE: D
A food chain always begins with a
A. decomposer
B. producer
C. primary consumer
D. consumer

14. UME/79/43

Grasses→Grasshoppers→Lizards→Snakes→hawks
In the above food chain, the organisms which are the least in number are
A. grasses   B. grasshoppers
C. lizards   D. snake
E. hawks

15. UME/79/44

Which of these is NOT true? Grasses in the above food chain
A. trap all the sun energy
B. trap a small percent of the sun energy
C. are primary producers
D. are eaten by primary consumers

16. UME/81/14

Choose the sequence which represents the correct or der of organism in a   food chain
A. Grass→ snake→toad → grasshopper→ hawk
B. Grass→ grasshopper→ toad → snake → hawk
C. Grass→ grasshopper→ snake → toad → hawk
D. Grass→ snake → grasshopper→toad →hawk
E. Grass→ toad → snake → grasshopper→ hawk

17. UME/86/46

In a community, bacteria and fungi are referred to as
A. producers   B. decomposers
C. scavengers   D. consumers

18. UME/90/42

Which organism is an omnivores?
A. Praying mantis   B. Hawk
C. Mouse   D. Grasshopper

19. UME/90/43

Which of the organisms will have the lowest population in an ecosystem?
A. Hawk   B. Cowpea
C. Praying mantis   D. Mouse

20. UME/91/42.

Organisms in an ecosystem are usually grouped according to their tropic level as
A. carnivous and epiphytes
B. consumers and parasites
C. producers and consumers
D. producers and saprophytes

21. UME/93/38

At which tropic level would the highest accumulation of a non-biodegradable substance occur?
A. primary productions
B. tertiary consumers
C. primary consumers
D. secondary consumers

22. UME/94/32

Which of the following set is made up of decomposers?
A. rhizopus, earthworm and protozoa
B. mushroom, rhizopus and bacteria
C. bacteria, earthworm and nematodes
D. earthworm, sedges and platyhelminthes.

23. UME/2000/36/TYPE: M

In a food chain, each succeeding level in a forward direction, represents?
A. an increase in the number of individuals
B. a decrease in the number of individuals
C. an increase in the biomass of individuals
D. a gain in the total energy being transferred

16.3.   ENERGY FLOW IN THE ECOSYSTEM

24. UME/2001/32/TYPE: T

The highest percentages of energy in an ecosystem occurs at the level of the
A. secondary consumers   B. decomposers
C. producers   D. primary consumers

25. UME/82/36

A pyramid of numbers can be defined as
A. the number of plants and animals in an ecosystem
B. an arrangement of organisms according to their habitats

C. the numerical relationship of a food chain
D. the number of plants and animals in a population
E. the total number of species and genera in a community.

26. PCE/2002/21/TYPE: 5

The unidirectional flow of energy in an ecosystem is known as
A. feeding relationship   B. energy pyramid
C. food chain   D. food web

27. UME/89/40

In an ecosystem, the LEAST efficient transfer link is from
A. producers to primary consumers
B. sun to producers
C. primary consumers to secondary consumers
D. secondary consumers to decomposers

28. UME/92/34

In a typical predator food chain involving secondary and tertiary consumers, the organisms become progressively
A. smaller and more numerous along the food chain
B. equal in number and size along the food chain
C. larger and fewer along the food chain
D. parasitized along the food as consumers get bigger

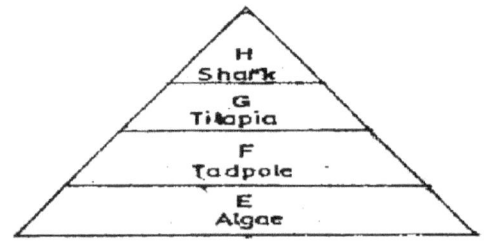

29. UME/92/35
Which level of the pyramid has the least total stored energy?
A. E   B. F   C. G   D. H

30. UME/92/36
Which organisms in the pyramid functions as a tertiary consumer?
A. Algae   B. Shark   C. Tadpole   D. Tilapia

31. UME/98/32
In a food chain involving a primary producer, a primary consumers as well are a secondary consumers, the sharing of tropic energy is in the form that the
A. primary consumer has more energy than the primary producers
B. secondary consumer takes up all the energy contents of the primary consumers
C. energy is shared equally between the three groups of organisms
D. secondary consumers gets only a small portion of the energy contained in the primary producer.

32. UME/98/29
Which of the following is the direct consequence of transferring energy from one tropic level to another?
A. an increase in biomass
B. a decrease in the efficiency of energy conversion
C. an increase in the total numbers of resulting individual
D. a decrease in the resulting biomass

16.4.   NUTRIENT CYCLING IN NATURE
16.4.1. CARBON CYCLE

33. UME/95/33
In spite of the removal of carbondioxide from the atmosphere, its amount remains more of less constant because
A. it is produced by green plants during photosynthesis
B. it is produced during respiration by animals
C. it is absorbed in ocean water
D. green plants release it during the day.

34. PCE/91/15
The carbon dioxide content of the atmosphere is kept constant by the processes of
A. decay, photosynthesis and transpiration
B. respiration and decay
C. photosynthesis and combustion
D. decay, respiration and photosynthesis

16.4.2. HYDROLOGICAL CYCLE

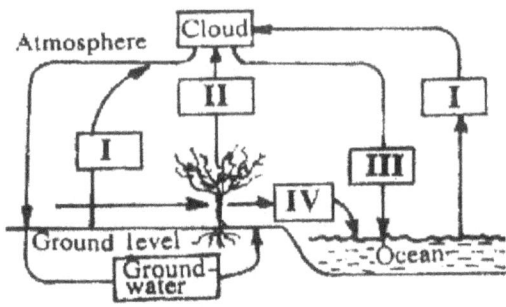

35. UME/2005/34/TYPE:D
Evaporation and transpiration are respectively represented by the components labelled
A. I and II
B. II and III
C. III and IV
D. IV and I

36. UME/2005/35/TYPE:D
The main reservoir of water in the cycle is the
A. cloud   B. groundwater
C. plant   D. ocean.

37. PCE/2003/29/TYPE: N
Water cycle is maintained in nature mainly by
A. evaporation and precipitation
B. precipitation and condensation
C. transpiration and evaporation
D. transpiration and precipitation

38. PCE/95/40
Which of the following is readily available to plants in the soil?
A. gravitational water
B. hygroscopic water
C. capillary water
D. run-off water

16.4.3. NITROGEN CYCLE

39. UME/80/44
During thunderstorms, the energy of lightning discharge cause
A. oxygen and nitrogen to combine
B. more carbon dioxide to be formed
C. nitrites to be converted to nitrates
D. nitrates to be converted to nitrogen
E. protect the microorganism in soil.

40. UME/2000/29/TYPE:M
Atmospheric nitrogen is converted to soil nitrogen for plant use by
A. nitrification and combustion
B. putrefaction and lightning
C. lightning and nitrification
D. combustion and putrefaction

41. UME/81/32
Nitrification means
A. conversion of nitrates to nitrogen
B. fixing nitrogen into plant
C. conversion of nitrates to nitrites
D. changing of ammonia to nitrites, then nitrates
E. nitrogen cycle.

42. UME/80/43
Leguminous plants, mucosa, are usually planted in cultivated farmlands because they
A. enrich the soil with phosphates
B. provides animals with food
C. enrich the soil with organic nitrogen
D. protect the soil from being overheated
E. protect the micro-organisms in the soil.

43. UME/85/49
Atmospheric nitrogen is directly replenished in nature through
A. activities of denitrifying bacteria
B. the breakdown of ammonium salts in the soil
C. the activities of nitrifying bacteria
D. the activities of nitrogen- fixing bacteria in root nodules.
E. egesting, death and decays.

44. UME/87/49
Nitrifying bacteria are important because they
A. release nitrogen to the atmosphere
B. convert atmospheric nitrogen to ammonia
C. combine ammonia with nitrogen.
D. oxidize ammonium salts to nitrates

45. UME/88/50
Dead plants and animal are decomposed by bacteria and fungi into
A. nitrates   B. nitrites
C. amino acids   D. ammonia

46. UME/94/15
The nitrifying bacteria, nitrosomonas, convert ammonia to
A. nitrites   B. nitric acid
C. nitrates   D. nitrous oxide.

47. UME/98/30
The condition that encourages denitrification is
A. low soil oxygen
B. high soil nitrogen
C. absence of soil bacteria
D. lightning and thunderstorm

48. UME/84/48
Denitrifying bacteria nature liberates gaseous nitrogen directly from
A. ammonium salt   B. soil nitrates
C. thunderstorms   D. soil nitrites
E. plant and animal proteins.

49. One of the ways in which the nitrogen of the atmosphere is converted into nitrates for plants is by the action of:
A. Putrefying bacteria   B. Lightning
C. Denitrifying bacteria   D. Decay
E. None of the above

50. The most important bacteria that convert nitrogen into nitrates are:
A. Nitrobacter   B. Nitrosomonas
C. Azotobacter   D. Denitricians
E. Bacterium radicicola

51. The symbiont that lives in the root nodules of leguminous plants is:
A. Azotobacter   B. Bacteria
C. Nitrosomonas   D. Nitrogen
E. None of the above

52. While the leguminous plant gains nitrates from the bacteria in its root nodules, it supplies the bacteria with:
A. Protein   B. Starch
C. Carbohydrates   D. Air
E. Fats

53. The nitrates absorbed from the soil by plants are built up into:
A. Plant nitrites   B. Plant nitrates
C. Animal proteins   D. Ammonium compounds
E. Plant proteins

54. The first product of putrefaction of plants and animal protein is:
A. Nitrites   B. Nitrates
C. Ammonium compounds
D. Nitrogen of atmosphere   E. Animal nitrate

MISCELLANEOUS QUESTIONS (16)

55. Two organisms of different species, living in close association but not dependent on each other are referred to as.
A. Parasites
B. commensals
C. Symbionts
D. autotrophs (UME/93/39)

56. An association between the root nodules of a leguminous plant and rhizobium sp is known as
A. commensalisms
B. mycorrhiza
C. parasitism
D. symbiosis (UME/2001/1/TYPE:T)

57. Bacteria which reduce nitrates in the soil into gaseous nitrogen are called
A. putrefying bacteria
B. nitrifying bacteria
C. saprophytic bacteria
D. denitrifying bacteria (PCE/2005/39/TYPE:P)

58. The presence of bacteria in the caecum of many herbivores is an example of
A. parasitism
B. commensalism
C. saprophytic
D. symbiosis (PCE/95/32)

59. Two organisms of different species living in close association but not totally dependent on each other are referred to as
A. parasites
B. commensals
C. symbionts
D. autotrophs (PCE/2001/31/TYPE:A)

60. A typical food chain, at which trophic level would the highest amount of energy occur?
A. producer
B. primary consumer
C. secondary consumer
D. decomposer (PCE/98/33/TYPE:C)

61. A correct food chain in a aquatic habitat is illustrated as
A. fish à worm à tadpole
B. phytoplankton à tadpole à fish
C. zooplankton à spirogyra à tadpole
D. algae à fish à tadpole (PCE/98/32/TYPE:C)

62. The release of inorganic salts from the decomposition of dead and animals in the soil is called
A. mineralization
B. volitilization
C. composting
D. humification (PCE/99/28/TYPE:E)

63. Which of the following organisms contribute to the process of denitrification?
A. Nitrosomonas
B. Nitrobacter
C. Rhizobium
D. Pseudomonas (PCE/99/30/TYPE:E)

64. The progressive loss of energy at each level in a food chain leads to
A. a decrease in biomass at each successive level
B. an increase in the number of organisms at each successive level
C. an increase in the total weight of living matter at each successive level
D. an increase in biomass at each successive level (UME/2004/42/TYPE 4)

156

CHAPTER 17

NATURAL HABITATS

Understand the following after reading this chapter:

Aquatic
Terrestrial

Habitat is any environment in which an organism lives naturally. It is an area where physical and chemical constituents required by organism are met. There are two main types of habitats:
1. Aquatic habitat : the habitat related to water.
2. Terrestrial habitat: the habitat that relate to land

AQUATIC HABITAT
Aquatic habitat comprises marine, estuarine and fresh water.

Marine Habitat: It is either ocean or high sea water bodies.
Characteristics of Marine Habitat
1. It has a high concentration of salt.
2. Its pressure increases with depth up to 1000 atmosphere.
3. Its surface current is driven by wind.
4. Tides occur due to alternative rise and fall in water level.
5. The high sea is relatively alkaline with a pH range of 8.0 and 8.4 at the surface.
6. Light can only penetrate up to about 200m or more.
Major Zones

They are (A) **Continental shelf**   (B) **Abyssal region** (C) **Oceanic region**
A. Continental shelf comprises of;
i. **Splash Zone**: Is just above the high tide mark. It is wetted with the spray of water from breaking waves.
ii. **Intertidal Zone**: Water stays in this area temporarily., especially when there is high tide, and gets exposed to air again when the tide goes down.
iii. **Littoral Zone**: It is found between high tide and low tide zones. It is about 200m deep.
iv. **Subtidal Zone**: Extends beyond 200m.
B. Abyssal region: This zone is more than 2000 metres to to the ocean floor.
C. Oceanic region: It is beyond the egde of continental shelf over the deep waters.

Estuarine Habitat: It is a transition between a river and the sea.

Characteristics of Estuarine Habitat
1. The salinity is always flunctuating.
2. Low species diversity: species are fewer in terms of diversity.
3. Turbidity increases especially during rainy season.
4. Very low oxygen content.
5. Very high productivity because of high nutrient level.

Fresh Water Habitat: It is known to have low salt content and can either be running (i.e. lotic) or stagnant (lentic). Example are rivers and ponds respectively.

Characteristics of Fresh Water Habitat
1. The salt content is very low.
2. Stratification of temperature is rare especially in running water.
3. Oxygen at the surface of the water is more than at the bottom.
4. Current can affect the distribution of gases, salts and small organisms e.g. in rivers or stream.
5. The waters are shallow.
6. Light penetrate more in running water than in stagnant water.
7. Low density of water.
8. The water bodies are always small.

TERRESTRIAL HABITAT
Soil and rain influence the nature of terrestrial habitat. They are divided into the following:
1. Marsh 2. Rain Forest 3. Grass Land
4. Arid Land

Marsh
1. It is a low land soil that is usually flooded.
2. It stands between aquatic and terrestrial habitats. Examples can be found in dried ponds or lakes as well.
Examples of plants are sedges, water drop-wart, marsh pennywort, balsam, hornwort (ceratophum), water lilies and duck weed.
Examples of animals are fishes, salamanders, frogs, toads, lizards, snakes, birds and mammals.

Rain Forest
Plants:
1. Scattered very tall trees of about 40m or above.
2. Continuous layer of trees of about 30m.
3. Possession of draw-out tips by trees for easy dripping.
4. The search for sun make the trees to be tall.

5. Possession of thin bark by trees so as to fasten gas exchange as well as transpiration

6. Plants that have the ability to climb. They do so in search of light e.g. Wild yam.

7. Possession of buttress root for strong anchorage e.g. silk cotton.

8. The broadness of the leaves enhances the rate of transpiration.

9. Clasping plants are used for support.

Animals:

1. Animals are protected by camouflage.

2. Some that are adapted to climbing do so with grasping pads tails to hold on to tree branches e.g. monkeys

3. Others have grasping scales e.g. snakes

4. The birds make use of their feathers.

Grassland

They are divided into these groups namely:

1. **Temperate grassland**

2. **Savanna or Tropical grassland** (southern guinea, northern guinea and sahel savanna as found in Nigeria)

3. **Meadow grassland**

Characteristics

1. Presence of intensive sunshine and high temperature

2. Rainfall is unevenly distributed

3. Humidity is always low

4. Deciduous plants are prominent

5. Trees which are fire resistant are common since there is constant bush burning

6. The habitat harbors a whole range of herbivores such as squirrel, grass cutters, antelopes and rodents.

7. Some carnivores which prey on herbivores are common e.g. Leopards and Lions.

Arid Land

They are divided into:

1. Hot arid lands (temperature as high as 77°C) e.g. hot deserts and semi desert.

2. Cold arid (temperature never up to 0°C).

Characteristics

1. Rainfall is always very low (below 25.4cm per annum)

2. There is high evaporative stress

3. Ecosystem is unstable

4. Extremely hot in the day and extremely cold at night.

5. Soil may be rocky or sandy

6. Extremely windy all year round.

7. Mobile sand dunes are common.

17.1.   AQUATIC

1. UME/82/50
The absence of stomata shows that a leaf may be
A. from a floating plant   B. from a submerged plant
C. variegated   D. from a terrestrial plant
E. from a parasitic green plant

2. UME/88/43
Which of these groups of animals is likely to be found in fresh water?
A. blood worm, pondskater and scorpion
B. bloodworm, pondskater and dragon fly larva
C. pondskater, scorpion and dragonfly larva
D. pond skater, bloodworm and ant – lion

3. UME/90/44
The salinity of a brackish environment
A. increases immediately after rain
B. increases at the end of the rainy season
C. decrease with increase in micro-organism
D. increase during the dry season

4. UME/93/41
Plants adapted for life in salty marsh are called
A. hydrophytes   B. xerophytes
C. halophytes   D. epiphytes

5. UME/98/31
A fresh water plant such as water lily can solve the problem of buoyance by the possession of
A. aerenchymatous tissues
B. dissected leaves
C. thin cell walls of the epidermis
D. water-repelling epidermis

6. UME/2001/45/TYPE: M
The stem of a typical aquatic plant usually has many
A. air cavities   B. intercellular space
C. water cavities   D. water conducting cells

7. PCE/93/40
The roots commonly found in red mangroves are the
A. buttress roots   B. breathing roots
C. clasping roots   D. stilt roots

8. PCE/96/31
Production of the viviparous seedlings is an adaptive feature of
A. mangrove forest plants
B. rain forest plants
C. savanna plants
D. desert plants

9. PCE/97/32
Pneumatophores are adaptations for
A. support
B. breathing
C. feeding
D. rainforest

10. PCE/97/33
Species diversity is highest in the
A. savanna
B. mangrove swamp
C. desert
D. rainforest

11. UME 2008/31
In a typical freshwater habitat, the edge of the stream or pond constitutes the
A. tidal zone
B. intertidal zone
C. littoral zone
D. euphotic zone

12. UME 2010/32/TYPE: D
Which of the following organisms is mainly found in the marine habitat?
A. Tilapia   B. Dog fish
C. Tortoise   D. Achatina

17.2.   TERRESTRIAL

13. UME/90/41
In a savanna ecosystem, the abiotic factors include
A. legumes, temperature and sandy soil
B. water, temperature and soil
C. minerals, oxygen and reptiles
D. water, soil and grasses

14. PCE/96/34
Which of the following plants is most adapted to the grassland ecosystem?
A. mahogany
B. pine
C. date- palm
D. baobab

15.   PCE/2000/34/TYPE: A
The most important adaptation required by terrestrial organisms is the ability to
A. search for food
B. escape from predators
C. conserve water
D. procure space

16. PCE/2003/26/TYPE: N
The most limiting factor in a terrestrial habitat is
A. high solar radiator
B. high wind action
C. the availability of water
D. fluctuating temperatures

17. UME 2009/29
The most important ecological factor in a terrestrial environment is
A. rainfall
B. humidity
C. wind
D. soil

18. PCE/98/31/TYPE: C
The greatest hazard faced by terrestrial organisms is
A. predation
B. dehydration
C. temperature changes
D. food scarcity

MISCELLANEOUS QUESTIONS (17)

19. One adaptation shown by terrestrial plants for combating drought is the
A. poor development of conducting tissues
B. possession of shorter roots which are less branched
C. the reduction in the number of stomata on the leaves
D. reduction in the intake of water from soil by roots (PCE 91/38)

20. In a forest habitat, emergent are best described as the
A. layer of shrubs
B. ground layer of shade – loving plants
C. layer of talls trees towering above others at intervals
D. layer of small trees than 10 metres tall (PCE/95/38)

21. A major difference between the marine and fresh water habitats is the
A. abundance of fishes
B. colour of the water
C. abundance of phytoplankton
D. concentration of salts   (PCE/98/36/TYPE: C)

22. The most widespread arboreal mammal is the
A. monkey
B. squirrel
C. pangolin
D. bat   (PCE/99/37/TYPE: E)

CHAPTER 18

LOCAL (NIGERIAN) BIOMES

Understand the following after reading this chapter:

Mangrove swamps
Tropical rain forest
Guinea savanna (Southern and Northern)
Sahel (Sudan) savanna
Desert

LOCAL (NIGERIA) BIOMES

In Nigeria there are three distinct biomes viz: Mangrove swamps or estuarine, Tropical rainforest and Savanna.

MANGROVE SWAMPS

In Nigeria, they are found in the delta regions of Lagos, Edo, Rivers, Cross Rivers and Delta states. The coastal areas of these states are muddy and water logged. The climate is hot and wet throughout the year. The areas support forests of small, evergreen broad-leaved trees growing in the shallow brackish water or muddy soil.

Characteristics of Mangrove Vegetation

1. Plants of the mangrove swamp possess specialised breathing roots called pneumatophores (in White mangrove) which grow into the air above the water surface. On it are tiny openings called lenticels which serve for gaseous exchange.
2. Many of the plants that float on the water surface have air-spaces in their leaves, stems and roots e.g. water lettuce and water lettuce.
3. Many mangrove trees have prop roots for support e.g. Red mangrove (*Rhizophora*).

Animals in Mangrove Swamp

The animals include worms (e.g. earthworms), crabs, frogs, snakes, insects, many fish including some species of Tilapia and the mud-skipper which can move rapidly over wet mud. Birds include herons, palm nut vultures while mammals include bats and monkeys.

TROPICAL RAINFORESTS

In Nigeria, the tropical rainforests cover parts of Rivers, Cross Rivers, AkwaIbom, Imo, Abia, Edo, Delta, Ogun, Ondo and Oyo states.

Characteristics

1. The forest is dense and contains tall trees which branch at the top to fprm a canopy. This cuts off light so that the forest interior has low light intensity, high humidity, still air and damp floor.

2. Most of the trees are broad-leaved and evergreen i.e. the leaves are green throughout the year, and they are gradually dropped and new ones grown to replace them.
3. The trees have tall, un-branched stems which are supported by large buttress roots at the base.
4. They have thin barks and they flower and form fruits in the wet season.
5. They are rich in epiphytes (plants that depend on others for support) and woody climbers known as lianas.
6. The vegetation on the forest floor is sparse because not enough light reaches it. Only small plants such as liverworts, mosses and ferns are found on the forest floor.

Notable rainforest trees in Nigeria include Obeche (*Triplochintonscleroxylon*), Iroko (*Chlorophora excels*); Mahogany timber trees (*Khayaivorensis*); Silk-cotton tree (*Ceibapetandra*); Teak (*Naucleadiderichii*); Afara (*Terminalia species*) etc.

N.B:
The trees are arranged in layers called storeys. There are five storeys or layers of vegetation, each consisting of plants with different light requirements.

1. **Upper storey:** This is the topmost layer with the tallest trees also called emergent. The trees may be 60 metres or more in height. They have an open canopy i.e. their canopies do not touch and form a continuous column.

2. **Middle storey:** This is made up of trees with dense and closed canopy i.e. their canopies touch and form a rather continuous column. The trees may be between 20-30 metres in height.

3. **Lower storey:** This layer is made up of trees of about 10-20 metres in height. The trees are scattered and do not form any canopy.

4. **Shrub layer:** This is made up of small plants below 10 metres in height (about 5 metres).

5. **Ground flora:** Only very little light reaches the ground flora as most of it is filtered off by the leaves of the trees and shrubs. Therefore only the smallest plants are found on the forest floor.

Animals of the Tropical Rainforest

1. There are 'arboreal' mammals including monkeys, squirrels, pottos and tree-hyrax. These mammals have long limbs and tail that enable them to climb and jump from tree to tree.
2. There are large forest birds including fruit-pigeons, hornbills, parrots and turacos.
3. Bats are the only flying mammals. They feed on fruits and insects. Their skin is modified to form wings which connect to fore and hind limbs.
4. There are insects including plant-bugs, beetles and the caterpillars of moths and butterflies.
5. Big mammals which move about in herds to protect themselves from predators include apes and elephants.
6. Many invertebrates which live in the litter or upper soil layer to prevent desiccation include earthworms and snails.

SAVANNA

Most parts of Nigeria outside the rainforest belt is covered by tropical Savanna vegetation. This is divided into three zones:

Guinea Savanna (South and North)

This borders the rainforest belt. In Nigeria, it is found in parts of Kaduna, Benue, Kogi and Kwara states. Because this zone is moist, the grasses grow tall during the rainy season. The trees occur quite close together especially on the fringes of the rainforest. There is the habit of burning the grasses during the dry season, hence Savanna trees are fire-resistant.

Sudan Savanna

This occurs to the North of the Guinea Savanna. It is therefore drier than the Guinea savannah and has shorter grasses, fewer and more scattered trees. It covers Kano and parts of Sokoto, Borno, Bauchi and Niger states.

Sahel Savanna

This borders the Sahara desert. The grasses here are generally very short and in clumps or tussocks. The trees and shrubs are few and sparsely distributed. Bush fire is a major problem here. During the dry season, when the grasses are burnt, the underground parts survive and grow again during the rainy season. Sahel occurs around the lake Chad Basin e.g. Borno, Yobe.

Animals of the Savanna Region

The Savanna supports an abundance of herbivores (plant eaters) and carnivores (flesh eaters). Examples of herbivores are insects (termites and grasshoppers), birds, rodents, antelopes, goats, cows, camels, elephant etc. Some of these graze the grass and browse on shrubs and trees. The carnivores include leopards, lions, tigers, cheetah etc. The scavengers are mostly vultures and hyenas. These eat up or clean up the remnants of animals killed by the carnivores.

DESERT

A desert is an area with less than 25.5cm rainfall a year or no rain fall at all for many years. Because of this, trees are scarce or completely absent i.e. the vegetation is very poor and consequently, the animal population is also poor.

There are two types of deserts;

Hot and dry desert: the temperature is very high (e.g. 800C or 1780F) e.g. Sahara desert.

Cold and frozen desert: the temperature here is usually below 00C during the winter and the top soil is frozen all year round e.g. Gobi desert in China.

There is no desert in Nigeria but the Sahel Savanna which is the narrow, most northern strip, gradually gives way outside the northern Nigerian border to the Sahara desert. The Sahara desert is the only desert in the West Africa.

Tropical rainforest: This has tall trees and climbers which form canopies. It is characterized by high rain fall and high temperature. It is usually stratified as follows:

A or uppermost layer, B or middle layer, C or lower layer, Shrub layer and Floor layer.

Common plants are: iroko, oil palm, rubber, obeche, wall nut etc. Common animals are snails, squirrel, snakes, monkeys etc.

Rain forest is found in Southern Nigeria i.e. Oyo, Lagos, Ogun, Edo, Ondo, Delta, Rivers, Enugu, Abia, Anambra, Imo, Cross Rivers and Bayelsa states.

Savanna: The savanna has more grasses than trees and shrubs. It is characterized by low rainfall, high temperature and a long dry season. It is further subdivided into:

a. **Southern Guinea savanna**: Niger, Benue, Kwara States.
b. **Northern Guinea savanna**: Kano, Sokoto States.
c. **Sahel savanna** found in Borno ,Yobe state

Swamp or Estuarine: It is solid if saturated and lacks oxygen. Common plants include mangrove, Raphia and coconut. It is found in the Niger Delta, Calabar, Lagos

.PCE\99\36\TYPE: E
In which of the following local biomes is Makurdi, Kaduna and Enugu located?
A. Sudan savanna
B. Sahel savanna
C. Guinea savanna
D. Rainforest

2. UME\98\32
The sequence of the biomes in Nigeria from Port Harcourt to Damaturu is
A. estuarine → rainforest→ Guinea savanna → sahel savanna
B. rainforest→ savanna→ rainforest → sahel savanna
C. estuarine→ Guinea savanna→ rainforest → sahel savanna
D. rainforest →estuarine→ Guinea savanna→ desert

3. UME\2003\36\TYPE: A
The correct sequence of biomes from northern to southern Nigeria is
A. estuarine → tropical rainforest→ guinea savanna→ sahel savanna
B. sahel savanna sudan savanna→ guinea savanna→ tropical rainforest→ estuarine
C. sahel savanna→ tropical rainforest→ estuarine→ guinea savanna
D. guinea savanna sudan savanna→tropical rainforest→sahel savanna → estuarine.

18.1.   TROPICAL RAINFOREST
4. UME\95\34.
In a tropical rainforset, non- epiphytic ferns and fern allies occurs as
A. middle storey species
B. upper storey species
C. shade- loving species
D. emergent species.

5. PCE\98\35\TYPE:C
Epiphytes and climbers are mostly found in the
A. sahel savanna
B. sudan savanna
C. rainforest
D. mangrove swamp

18.2.   GUINEA SAVANNA (SOUTHERN AND NORTHERN)
6. UME\88\44
One of the characteristics of plants in the savanna is the
A. possession of thin, smooth barks
B. possession of large tap roots
C. production of seedlings on mother plant.
D. possession of thick flaky barks.

7. UME/97/37
The southern Guinea savanna differs from the Northern Guinea savanna in that it has
A. lower rainfall and shorter grasses
B. less grasses and scattered tress
C. more rainfall and taller grasses
D. less arboreal and burrowing animal

8. PCE/2004/38/TYPE 9
Grasses are better adapted than trees in the savanna because of their possession of
A. corms   B. bulbs
C. rhizomes   D. tubers

9. UME/2005/37/TYPE: D
In Nigeria, the Guinea savanna belt borders the
A. magroove swamps and the sahel savanna
B. rainforests and the sudan savanna
C. deserts and the sudan savanna
D. rainforests and the deserts

Use the list of biome below to answer questions 10 and11
1. Deserts
2. Rainforest
3. Southern Guinea Savanna
4. Northern Guinea Savanna

10. UME/94/35
A biome with a low annual rainfall, few scattered trees within dense layer of grasses an found in Kano and Katsina states is
A. 1   B. 2   C. 3   D. 4

11. UME/94/36
A biome where there are many cacti and the small mammals undergo aestivation for long period is
A. 1   B. 2   C. 3   D. 4

12. UME 2006/42
Grasses recover quickly from bush fires in the savanna because of their
A. perennating organs   B. rapid growth rate
C. fibrous roots   D. succulent stems

18.3.   SAHEL (SUDAN) SAVANNA
13. UME/2002/21/TYPE: Y
In a population study using the transect method a student is likely to record the highest number of specie in
A. a tropical rainforest   B. a guinea savanna
C. a sahel savanna   D. an estuarine swamp

14. PCE/92/34
Sokoto state is located in the
A. rainforest   B. southern guinea savanna
C. sahel savanna   D. desert

15. UME 2006/50
The main purpose of establishing shelter belts in the Sahel region is to
A. check desert encroachment   B. provide wood fuel
C. beautify the region
D. break the harmattan wind

18.4.   DESERT

16.   UME/97/36
Adaptive feature of plant to desert condition include
A. thick barks, succulent stems and sunken stomata
B. thin barks, succulent stems and sunken stomata
C. thick barks, airfloats on stems and sunken stomata
D. air spaces in tissues, adventitious root and thick barks

Ecological Zone	Temperature(^{0}C)	Rainfall (mm)
I	45	300
II	32	2000
III	30	2200
IV	15	800

17. UME/2004/44/TYPE: 4
Which of the zones is likely to be a desert?
A. III   B. IV   C. I   D. II

18. UME/2004/45/TYPE: 4
High relative humidity will be expected in zones
A. II and III   B. II and IV
C. I and IV   D. III

19.UME/2004/45/TYPE :4
A state in Nigeria that is most susceptible to desert encroachment is
A. Kaduna   B. Katsina
C. Kwara   D. Taraba

MISCELLANEOUS QUESTIONS (18)
20. The least adaptive feature for arboreal life is the
A. possession of four limbs
B. possession of claws
C. development of long tail
D. counter shading of coat colour
(UME/99/46/TYPE:D)

21. In which of the following zones of the marine habitat is the intensity of light at its minimum?
A. littoral   B. intertidal
C. benthic   D. splash
(PCE/96/32)

22.The guinea savanna is mainly characterized by
A. grass and thorny shrubs
B. tall grasses and sparse trees
C. tall grasses and shrubs
D. short and sparse grasses (PCE/2003/31/TYPE:N)

23. In which of the following local biomes is Makurdi, Kaduna and Enugu located?
A. Sudan savanna
B. Sahel savanna
C. Guinea savanna
D. rain forest (PCE/99/36/TYPE: E)

24. The typical features of the grassland of the Obudu and Jos Plateaux in Nigeria are
A. heavy rainfall, rocky highlands and high temperature
B. low temperature, thick forests and low rainfall
C. rocky highlands and low temperature
D. thick forests and grassland (PCE/06/47/TYPE: A)

25. The cactus plant is structurally adapted to survive by the possession of
A. swollen leaves that store food and water
B. swollen roots that store food and water
C. thick corky bark that reduces loss of water D. swollen stems that store food and water.
(PCE/2002/26/TYPE: 5)

26. The mambilla plateau is a unigue Nigerian biome located in
A. Plateau state   B. Borno state
C. Taraba state   D. Benue state.
(UME/2005/36/TYPE:D)

27. The mangrove swamp in Nigeria is restricted to the A. Sahel savanna
B. Sudan savanna
C. Guinea savanna
D. Tropical rain forest (UME 2007/8)

28. UME 2007/10
In which of the following Nigerian States can Montana vegetation be found?
A. Taraba   B. Enugu
C. Bauchi   D. Plateau

29. Which of the following biomass could be characterized by very low rainfall, cold night, hot days and fast blooming plants?
A. Northern Guinea Savanna
B. Southern Guinea Savanna
C. Tropical desert
D. Montana forest (UME 2009/33)

30. Low annual rainfall, sparse vegetation, high diurnal temperatures and cold nights are characteristic features of the
A. desert
B. montana forest
C. guinea savanna
D. tropical rainforest.(UME 2010/35/TYPE:D)

CHAPTER 19

THE ECOLOGY OF POPULATIONS

Understand the following after reading this chapter:

> Patterns of growth in population size
> Factors affecting population sizes
> Ecological succession

CHARACTERISTICS OF A POPULATION

Population density or size: This is the number of individuals occupying a given habitat.

Natality or birth rate: This is the number of new individuals produced per unit time.

Mortality or death rate: This is the number of individuals dying per unit time.

Growth rate: This is the increase in number of organisms in a community.

Dominance: This refers to the species which is dominant over others in terms of cover or density or both.

Age distribution: This is the proportion of individuals in different age range within a population.

Dispersability: This is the ability of members of a population to disperse or spread.

ECOLOGICAL SUCCESSION

This is a long term, gradual or progressive series of changes occurring in the structure, composition, variety or diversity and number of species in an area till a stable or climax community is established.

Pioneer species to climax communities

Pioneer species: These are the first species to occupy a new habitat, starting new communities. They have rapid reproductive strategies, enabling them to quickly occupy an uninhabited area. Many have an asexual stage to their reproduction.

Seres: These are the various stages that follow on from the pioneer species.

Climax community: This is the stable community that is reached, beyond which, no further succession occurs.

TYPES OF SUCCESSION

Primary succession

This occurs when the starting point is a bare ecosystem, (e.g, following a volcanic eruption or a landslide). The pioneer species are usually lichen, then by moss or algae. They are able to penetrate the bare surface, trap organic material and begin to form humus.

Over several generations soil begins to form. The soil can be used by a more diverse range of plants with deeper root systems. Gradually larger and larger plants occupy the ecosystem along with a diversity of animals.

Finally a climax community is reached and the species present do not change unless the environment changes in some way.

An example of primary succession forming oak woodland:
1. Bare rock is colonised by lichen and then mosses.
2. Small plants, ferns and grasses take over.
3. Larger plants with deeper roots appear.
4. Bushes and shrubs replace non-woody plants.
5. Fast growing trees form a dense, low wood.
6. Larger, slow growing oak trees create the oak woodland.

Secondary succession

This occurs when the starting point is bare, existing soil, (e.g, following a fire, flood or human intervention). This type of succession proceeds in the same way as primary succession except that the pioneer species tend to be grasses and fast growing plants.

An example of secondary succession forming an oak woodland:
1. Bare soil is colonised by grasses and pioneer plants.
2. Grasses begin to predominate with time.
3. Shrubs replace the grasses.
4. Fast growing trees appear.
5. Slow growing oaks create the climax community.

FACTORS AFFECTING A POPULATION

These are divided into:

Density dependent factors e.g. predation, food, competition, diseases etc.

Density independent factors e.g. flood, bush fires, earthquake, erosion etc.

PROBLEMS AFFECTING THE POPULATION
1. Overcrowding   2. Food shortage

Causes of Overcrowding
1. Lack of space   2. Increase in birth rate
3. Absence of predators 4. Immigration

Effects of Overcrowding
1. Decrease in natality 2. Stress and death

3. Spread of disease 4. Shortage of food
5. Competition
Causes of Food shortage
1. Drought 2. Pests
3. Diseases 4. Over population
Effects of Food shortage
1. Competition 2. Emigration
3. Decline in natality

FAMILY PLANNING AND BIRTH CONTROL

Family planning is a device by which couples determine the number of children they want and when they want them. Birth control is the used to prevent a woman from becoming pregnant, for as long as she wishes.
Birth control methods include:
1. Withdrawal method
2. Rhythm method or safe period
3. Use of condom (sheath) in males
4. Cap (Diaphragm) in females
4. Spermicidal cream or tablet
5. Intra-uterine device (IUD)
6. Contraceptive pill
7. Injection
8. Sterilization

Revision Questions

19.1. PATTERN OF GROWTH IN POPULATION SIZE

1. UME/92/33
A population is defined as a collection of similar organisms that
A. behave in the same way B. interbreed freely
C. are found in the same habitat D. eat the same food

2. PCE/2004/34/TYPE: 4
A combination of density-dependent factors that can limit the size of a population is
A. predation, behavioural adaptations and competition
B. parasitism, physiological adaptations and competition
C. predation, parasitism and competition
D. competition, behavioural adaptations and physiological adaptations.

3. UME/2000/30/TYPE : M
I. High birth rate and high immigration rate
II. Low mortality rate and low emigration rate
III. Low birth rate and high immigration rate
IV. High mortality rate and high emigration rate. Which combination above can cause rapid overcrowding in climax biotic communities and human settlements?
A. II and III B. I and III
C. I and IV D. I and II

4. UME/2002/14/TYPE: Y
Plants tend to prevent overcrowding by means of efficient
A. water uptake B. seed germination
C. pollination D. seed dispersal

5. UME/2003/33/TYPE: A
A density-dependent factor that regulates the population size of organisms is
A. sudden flood B. disease
C. fire outbreak D. drought

6. PCE/93/36
The estimation of the population of Tilapia in a pond is best done by
A. the quadrant method
B. counting
C. capture and recapture method
D. transect method

7. PCE/99/31/TYPE: E
The combination of factors most likely to cause over crowding is
A. competition, immigration and high death rate
B. immigration, natality and high survival rate
C. natality, competition and migration
D. high survival rates, scarce resources and migration

8. PCE/99/32/TYPE: E
The factor that affects distribution of organisms in an aquatic habitat is
A. humidity B. rainfall
C. wind D. turbidity

9. UME 2006/49
A population that doubles in size at constant intervals is an indication of
A. sigmoid growth B. exponential growth
C. population explosion D. rapid growth

10. UME 2010/36/TYPE: D
The average number of individuals of a species per unit area of the habitat is the
A. population frequency
B. population size
C. population distribution
D. population density

11. PCE/07/32
A population is said to have reached equilibrium if it
A. no longer faces competition
B. has attained its carrying capacity
C. is declining with time
D. has exceeded its carrying capacity

19.2.   FACTORS AFFECTING POPULATION SIZES

12. UME/2004/47/TYPE: 4

The scarcity of food causes a sudden decrease in populations in size by

A. minimizing the rate of competition
B. raising the mortality rate
C. bringing about immigration
D. decreasing the reproductive rate

13. PCE/2004/32/TYPE: 9

The study of changes in population size over time is known as

A. population census   B. population dynamics
C. population control   D. population estimate

19.3.   ECOLOGICAL SUCCESSION

14. UME/93/42

Which group of plants would be the first colonizers in an ecological succession changing rocks to soil?

A. Mosses   B. Ferns
C. Lichen   D. Grasses

15. UME/95/35

Colonization of a bare rock surface is termed

A. evolution   B. speciation
C. primary succession   D. secondary succession

16. UME/98/43

The biological association that contributes directly to succession in a community is

A. competition   B. predation
C. parasitism   D. commensalism

17.   UME/97/38

After the heavy rainfall and the formation of a large pond, the most likely sequence of changes in the vegetation of the food is

A. Euglena→Waterlily → Spirogyra
B. Waterlily→ Spirogyra Grass → Euglena
C. Spirogyra →Euglena → Grass→ Waterlily
D. Euglena→ Spirogyra→ Waterlily→grass

18. UME 2006/43

Climax communities in a biotic succession are usually characterized by

A. a constant change in the appearance of the communities
B. different species that are constantly changing
C. a stable composition of plant and animal species
D. rapid change in the plant and animal species

MISCELLANEOUS QUESTIONS (19)

19. Secondary succession is much faster than primary succession because

A. pioneer coloniers are more in number
B. soil is already present
C. secondary succession requires less nutrients
D. species competition is increased(UME/2005/39/TYPE:D)

20. In ecological succession, older communities are characterized by greater productivity as a result of increasing

A. Stability   B. biomass
C. population   D. species turnover
(PCE/2005/35/TYPE:P)

21. Which of the following is true of a climax community?

A. species diversity increases
B. species diversity decreases
C. the community is almost in a state of equilibrium
D. the community keeps changing(PCE/92/46)

22. Succession that takes place on an abandoned farmland is known as

A. primary succession   B. climax
C. animal succession   D. secondary succession
(PCE/93/41)

23. One effective means of reducing intra-specific competition in a population is to

A. decrease the resources competed for
B. increase the resources competed for
C. increase the number of competitors
D. increase the number of species in the community
(PCE/95/35)

24. In a normal succession, species diversity

A. decreases   B. increases
C. is constant   D. is zero (PCE/96/35)

25. Species diversity is highest in the

A. savanna   B. mangrove swamp
C. desert   D. rainforest (PCE/97/33)

26. In a bare rock environment, which of these plants is likely to be the pioneer occupant?

A. lichens   B. ferns
C. spermatophytes   D. bryophytes (PCE/92/33)

27. The correct formula for determining the space by organisms in a habitat is

A.   Number of organisms
         Area of habitat
B.   Area of habitat
         Number of organisms
C.   Number of organisms
         Mass of organisms
D.   Density
         Number of habitats   PCE/99/35/TYPE:E

28. In nature, organisms prevent overcrowding by

A. Increase in the rate of reproduction
B. Territorial behaviour   C. Immigration
D. Avoiding predators   (PCE/01/35/TYPE: A)

CHAPTER 20

SOIL

Understand the following after reading this chapter:

> Components and characteristics of different types of soil
> Loss of soil fertility
> Renewal and maintenance of soil fertility

The soil gives plant support, in addition to harbouring animals, supply of mineral salts, and production of vegetation.

Types of soil are as follows:

Sandy Soil: Their particles are coarse, loosely bound and have poor nutrient and water-holding capacity. They have very low capillarity and are well drained.

Loamy Soil: They are mixtures of clay and sand particles, with a higher proportion of organic matter. They are more moist, riches in terms of plant nutrients; and drain lesser than sandy soil.

Clay Soil: They are made up of smallest and finest particles. They have water-retaining capacity. They reduce and form lumps when dry. They are damp and sticky to touch when wet.

Laterites: They are earthy, granular or concretionary mass, chiefly of iron and aluminium oxides and hydroxides, occurring as a layer or as scattered nodules in tropical soils. They are impervious to water and contain little nutrients.

To determine the particle sizes in a soil, shake up about 100g of soil with same water in 500ml measuring cylinder and allow settling. The particles settle down in the following order: gravel, sand, silt and clay; clay particles are also suspended in the water. The humus floats on the surface.

SOIL FERTILITY

A fertile soil supports the growth of plants adequately. It is usually a loamy soil with a good humus content, good drainage and aeration and adequate nutrient content. Soil fertility can be increase by adding fertilizer.
Soil can lose its fertility through loss of organic matter (humus), soil erosion and leaching.

EXPERIMENTS ON SOIL

Experiment: 1 To determine the amount of air in the soil.

Materials required. A plastic container, two measuring cylinders of $500cm^3$ each, water, glass rod.

Method. The internal volume of the plastic container is first determined. This is done by pouring water into the container until it is filled to the brim. The water is poured from the container into one of the cylinders. Fresh soil is then packed into the container. The soil is made tight to the top of the container. Next, the soil is levelled off at the top. The volume of the soil and its component air, is taken to be **x.** The soil is scraped into the second measuring cylinder which contains 200ml of water. The soil and water are then stirred continuously with a glass rod to drive away any gas in the soil. The volume of the soil and the water in the cylinder is then noted. This volume is known as **y.** The amount of air in the soil is calculated on a percentage basis as shown below:

Volume of soil + Air	$= x$
Volume of soil – Air	$= y - 200$
∴ Volume of air	$= x - (y - 200)$
∴ Percentage of soil air	$= x - (y - 200) \times 100/1$
	$= x - y + 200 \times 100 /x$

Experiment: 2 To determine the amount of water in the soil.

Materials required. A small dish, small amount of soil, an oven.

Method. A small dish is weighed and the weight is recorded as **a** gram. The dish and some amount of soil in it again is weighed and this weight is also noted and it is taken as **b** gram. Next, the dish with the soil is placed in an oven whose temperature is 100oC, to dry the soil. After some hours, say 12 hours, the dish is removed from the oven. It is then allowed to cool down in a dessicator. Next, thedish and the dried soil are weighed and recorded. Drying, cooling as before and weighing are continued until a more or less constant weight is obtained. This weight is taken as **c** gram.

Observations. There is a loss in weight of the oven dried dish and soil.
Conclusion. The loss in weight is due to water given off by evaporation. This shows that water is present in the soil.
Calculation. The amount of water in the soil can be calculated on a percentage basis as follows:

Weight of dish = a gm
Weight of dish + soil = b gm
Weight of soil only = (b-a) gm
Weight of dish + soil from oven = c gm
∴ Loss in weight = Weight of water = (b - c) gm
∴ Percentage of water in the soil = b – c / b – a x 100/1

Experiment:3 To determine the amount of humus in the soil.

Materials required. A lump of soil, a platinum crucible.
Method. A platinum crucible weighing **b** gram is filled with soil. The soil is then dried at 100oC as in experiment 21 until all the water in the soil is evaporated. The weight is taken to be **a** gram. Heat is then applied to the platinum and the soil is stirred with a wire. Smokes comes out of the soil but heating is continued until the smoke stops and the soil turns red. The soil is then allowed to cool in a desiccator. The red soil with the crucible is then weighed to be **c** gram.

Observations. There is a loss in weight of the soil which received very strong heat until it turned red.

Conclusion. The loss in weight is due to humus (organic matter) burnt during heating. This shows that humus is present in the soil.
Calculation. The amount of humus in the soil can be calculated on a percentage basis as follows:

Weight of crucible   = b gm
Weight of crucible + dry soil containing humus = a gm
Weight of dry soil with humus   = (a-b) gm
Weight of crucible + dry soil without humus   = c gm
∴ Weight of humus lost   = (a - b) – (c - b) = a - c
∴ Percentage of humus lost   = a – c / a – b x 100/1

Experiment: 4 To compare the porosity of different samples of soil (sandy, clay, and loamy).

Materials required. Three measuring cylinders, sandy soil, clay soil, loamy soil, 3 filter funnels, water, labels.
Method. Three measuring cylinders of size 200cc are taken. They are labelled A, B, and C. a filter funnel is placed in the mouth of each. Next, soil is put in the funnel of each cylinder; sandy soil in A, clay soil in B and loam soil in C. A known volume of water, say 70cc is then poured onto the soil in each funnel at the same time. The time it takes the water to pass through in the different samples is recorded.
Observations. The water drains through the soil at different rates, being fastest in sandy soil, followed by loam and least in clay soil.
Conclusion. There is difference in porosity in different samples of soil.

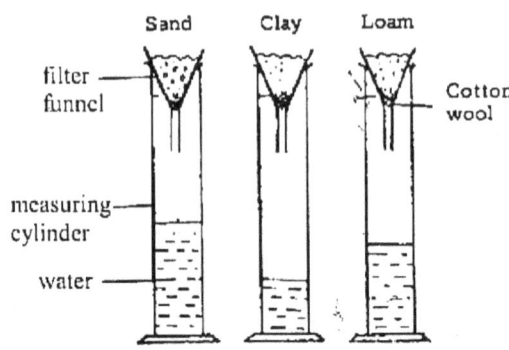

Experiment :5 To compare the capillarity of different samples of soil.

Materials required. Three long glass tubes, cotton wool, labels, sandy soil, loamy soil, clay soil, a trough of water, and maize grains.
Method. Three long glass tubes are collected and at one end of each, some cotton wool is placed. The tubes are labelled A, B, and C. the tubes are filled with soil (about ¾ up) as follows: A to contain sandy soil. B to contain loamy soil. C to contain clay soil. The tubes are stood in the trough of water as shown in Fig. 40.12. Next, two or three maize grains are sowed in each. The tubes are then left for some days.
Observations. The grain in A germinates first, after a few days. This is followed by grains in B. Those in C germinates last.

Conclusion. The observations made as above are so because water first got to the grains in A before B and C. this shows that capillarity varies with different soil samples.

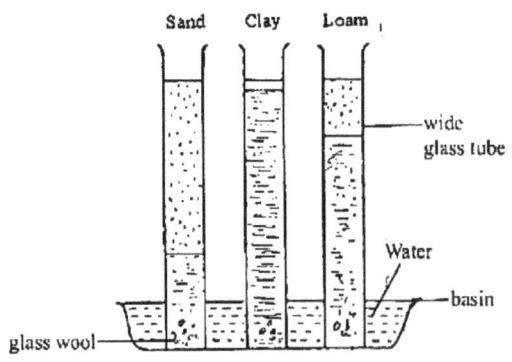

0.1. COMPONENTS AND CHARACTERISTICS OF DIFFERENT TYPES OF SOIL

1. UME/92/38
Soil with the finest particles is called
A. silt   B. clay
C. sand   D. gravel

2. UME/79/26
Which of these is NOT a type of soil?
A. sand   B. granite
C. loam   D. clay
E. sandy loam

3. UME/85/48
The origin of mineral particles in the soil is
A. humus   B. water
C. micro-organisms   D. weathered rock
E. organic matter

4. UME/97/39.
In a soil, the relative amounts of the different soil particles can best be determined by the process of
A. filtration   B. centrifugation
C. precipitations D. sedimentation

5. UME/2004/43/TYPE: 4
The soil type that is difficult to plough in a wet season is one that is
A. sandy   B. loamy
C. salty   D. clayey

6. UME/80/41.
If three 30cm lengths of glass tubing are tightly packed with clay, sand and loamy soil respectively and then stood in a beaker of water for one week. The level of water will be
A. lowest in the tubes with clay
B. the same in all the tubes
C. lowest in the tube with loamy soil
D. highest in the tube with sandy soil
E. lowest in the tube in the with sandy soil

7. UME/80/45.
A few grams of dried soil were first heated until red and hot and then further heated until no more smoke was released. This experiment was to determine the
A. amount of water in soil
B. percentage of water in soil
C. presence of humus in soil
D. resistance of laterites to heat
E. release of smoke from the soil

8. UME/84/47.
If a handful of soil is shaken with water and left to settle, the soil particles will settle from light to heavy particles as follows
A. humus, clay, silt, sand, stones
B. humus, silt, clay, sand, stones
C. humus, silt clay, stones, sand
D. humus, sand, silt, clay, stones
E. clay, humus, silt, sand, stones,

9. UME/85/47
Water retention is highest in soils which are rich in
A. sand, poor in humus and devoid of clay
B. clay and sand, but poor in humus
C. clay and humus but poor in sand
D. clay, poor in humus and devoid of sand
E. sand and humus, but poor in clay

10. UME/86/48
A large percentage of tropical soils tend to be acidic because they
A. contain large quantities of potash
B. contain large quantities of lime
C. lose a high proportion of their organic matter to running water
D. lose lime and potash from the top soil through rain action

11. UME/94/39
The water-retention capacity of a soil indicates its
A. Fertility
B. capillarity
C. aeration level
D. loss of organic matter due to exposure to direct sunlight
E. washing out of chalk and limestone from upper layers of soil by heavy rains

12. UME/88/48
The mineral nutrient that easily gets leached out of the soil is
A. phosphorus   B. calcium
C. magnesium   D. nitrate

13. UME/89/47
Most irrigated lands often become unproductive in later years because
A. loss of fertility   B. increase in salinity
C. soil erosion   D. loss of water

14. UME/2004/50/TYPE: 4
A farm practice that results in the loss of soil fertility is
A. continuous cropping   B. mixed farming
C. bush fallowing   D. shifting cultivation

15. PCE/2001/36/TYPE: 4
Leaching of soil nutrients is mainly due to water
A. flocculation   B. percolation
C. erosion   D. capillarity

16. UME 2007/2
The increasing order of the particular size in the following soil type is
A. clay → sand→silt→ gravel
B. silt→sand→ clay→ gravel
C. clay→ silt→sand→ gravel
D. silt→ clay →sand→gravel

17. UME/2007/6
A crucible of 5gm weighed 10gm after filling with fresh soil. It is then heated in an oven at 100°C for 1 hour. After scaling in desiccators, the weight was 8gm. The percentage of water in the soil is
A. 80%   B. 60%   C. 20%   D. 40%

18. UME/2008/37
In an experiment to determine the percentage of air in a soil sample, the following readings were recorded:
Vol. of water in a measuring cylinder = 500cm³
Volume of soil added to water = 350cm³
Volume of water and soil after stirring = 800cm³
A. 6.25%   B. 10.36%
C. 14.28%   D. 43.28%

19. UME/2009/36
Mass of a crucible = 10g
Mass of a crucible & soil before heating = 29%
Mass of a crucible & soil after heating = 18g
From the information above, determine the percentage of water in the given soil sample?
A. 20%   B. 25%
C. 40%   D. 50%

20. The soil of any locality is normally richer in
A. Humus   B. Rocks
C. Carbonates   D. Ammonia
E. Bacteria

21. In an experiment to determine the proportion of air in sandy and clay soil, it isdiscoveredthat this former has much more air in the same volume of soil as the latter.This shows that:
A. Clay particles being smaller are impervious to air
B. Sand is better for the growth of crops than clay
C. Sand can also retain much more water than clay
D. Sand particles being coarser pack less tightly than the clay, leaving much room for air
E. None of the above

22.Sand, though usually well drained and well aerated, is regarded as being "warm" This is due to:
A. its poor porosity   B. its good capillarity
C. Its poor imbibitory power D. its low specific heat
E. none of the above

23. Crops grow on clay soils are usually poor and ripen late. This is due to the fact that:
A. clay soils are badly aerated and water-logged
B. clay soils are too warm
C. clay soils have low capillarity power
D. clay soils allow water to run through easily
E. Clay soils have fewer nutrients

24.The temperature of the soil is partly affected by:
A. The amount of water that evaporates from the soil
B. The amount of humus in the soil
C. The amount of carbon compounds in the soil
D. The texture of the soil E. None of the above

20.2. RENEWAL AND MAINTENANCE OF SOIL FERTILITY

25. UME/84/46
Erosion can be reduced along a slope by
A. ridging across slope   B. ridging up slope
C. ridging down slope   D. crop rotation
E. bush fallowing system

26. UME/86/49
The following are methods of soil conservation EXCEPT
A. contour terracing   B. strip cropping
C. contour ploughing   D. mixed grazing

27. UME/87/50
The process by which lime is added to soils is known as
A. sedimentation   B. flocculation
C. leaching   D. manuring

28. UME/90/48
An acidic soil can be improved upon by
A. sedimentation   B. leaching
C. flocculation   D. watering

29. UME/98/33
Soil micro-organisms are beneficial because of their involvement in
A.photosynthesis   B. translocation
C. cycling of nutrients   D. respiration using soil air

30. UME/2002/29/TYPE:7
The addition of lime to clayey soil serves to
A. aid water retention   B. close up the texture
C. prevent water-logging   D. improve capillary action

31. UME/2003/35/TYPE:A
Soil fertility can best be conserved and renewed by the activities of
A. microbes   B. earthworms
C. man   D. rodents

32. PCE/03/21/TYPE: N

Excessive application of inorganic fertilizers on a farmland leads to an

A. increase in soil fertility   B. alteration in soil pH

C. improvement in soil texture

D. increase in number of soil organisms

33. PCE/07/33

The organisms that help to increase soil fertility include

A. termite, butterfly, larvae, earthworms and bacteria

B. bacteria, termites, domestic fowls and fungi

C. earthworms, termites, bacteria and fungi

D. earthworms, spiders, bacteria and fungi

34. UME/2010/38/TYPE: D

The loss of soil through erosion can be reduced by

A. crop rotation   B. manuring

C. irrigation   D. watering

MISCELLANEOUS QUESTIONS (20)

35. In a community bacteria and fungi are referred to as

A. producers   B. decomposers

C. scavengers   D. consumers (UME/86/46)

36. The importance of practicing crop rotation in agriculture is to

A. maintain soil fertility

B. improve the maturational value of crops

C. control soil erosion

D. ensure the growth of crops

(UME/2005/41/Type: D)

37. Which of the following types of soil possesses the least water retaining capacity

A. clayey soil   B. sandy soil

C. humus soil   D. loamy soil (PCE/93/42)

38. A soil sample maybe described as humus soil if it

A. it is black in appearance

B. has a greater proportion of sand than clay

C. allows water to run through it slowly

D. contains inorganic and organic materials

(PCE/94/41)

39. Which of the following is readily available to plant in the soil ?

A. gravitational water   B. hygroscopic water

C. capillary water   D. run-off water (PCE/95/40)

40. The group that accurately describes the composition of the soil is

A. mineral matter, humus, water, air and organisms

B. silica, humus, magnesium and water

C. air, water organisms and carbondioxide

D. organisms, water ,oxygen ,humus and minerals

(PCE/97/35)

41. The best method of preventing leaching is

A. the planting of cover crop   B. mulching

C. terracing

D. strip-cropping (PCE/96/37)

42. A soil with large particle size, little humus and coarse texture is a

A. clayey soil   B. loamy soil

C. sandy soil   D. humus soil

(PCE/2002/16/TYPE:5)

43. The most important function of soil water is to

A. facilitate absorption of nutrients

B. dissolve inorganic material

C. enhance plant growth

D. give a cooling effect (PCE/2000/37/TYPE:A)

44. Planting of leguminous cover crops increases soil fertility by

A. checking the growth of parasites

B. increasing soil nitrogen content

C. increasing soil colloids

D. adding humus to the soil (PCE/2001/37/TYPE:4)

45. In an experiment, the following results were obtained:

I. Original weight of soil = xg

II. weight of oven dried soil = yg

III. weight humus in soil = zg

The percentage of water in the soil is therefore given by the expression

A. $(x-y/z) \times (100/1)$ B. $(x-y/x) \times (100/1)$

C. $(z/x-y) \times (100/1)$ D. $(x/x-y) \times (100/1)$

(PCE/96/36)

46. A dry soil (x grams) is heated in a known weight of crucible until red hot without smoke. If the final weight of the soil is y grams, the percentage weight of humus in the soil is

A. $100 (y-x)$ grams   B. $100 (y-x)/y$ grams

C. $100 (x-y)/x$ grams   D. $100 (x-y)$ grams

PCE/06/39/TYPE: A

47. An adverse effects of excessive application of inorganic fertilizer on the soil is an increase in its

A. Microflora   B. aeration

C. Alkalinity   D. acidity PCE/06/48/TYPE: A

48. In an experiment to determine the percentage of water in a soil sample, the following date were obtained:

Mass of crucible $= X_1$ g

Mass of crucible and soil $= X_2$ g

Mass of crucible and soil after heating $= X_3$ g

What is the amount of soil water?

A. (x_2-x_1) g   B. (x_3-x_1) g

C. (x_3-x_1) g   D. (x_2-x_3) g PCE/07/34

CHAPTER 21

HUMANS AND THEIR ENVIRONMENT

Understand the following after reading this chapter:

> Diseases
> Pollution and its control
> Conservation of natural resources

DISEASES
Disease simply means an illness affecting humans, animals or plants often caused by infections.
Diseases caused by micro-organism:
1a. **Animal diseases caused by viruses**: poliomyelitis, infective hepatitis, measles, common cold, influenza, small pox, chicken pox, yellow fever, rabies, Acquired Immune Deficiency Syndrome (AIDS)
1b. **Plant diseases caused by viruses**: rosette disease, cassava mosaic, maize streak, yam mosaic, cowpea mosaic.
2a. **Animal diseases caused by bacteria**: tuberculosis, leprosy, tetanus, typhoid, dysentery, cholera, pneumonia, anthrax, diphtheria, gonorrhea, syphilis.
2b. **Plant diseases caused by bacteria**: tomato rot, onion rot, banana wilt.
3a. **Animal diseases caused by fungi**: ringworm (tinea), aspergilosis, athletic foot, thrush.
3b. **Plant diseases caused by fungi**: maize smut, maize rust, rice blight, leaf spot, cocoa black pod
4a. **Animal diseases caused by protozoa**: malaria, trypanosomiasis, trichomoniasis, coccidiosis.

CHOLERA
Cholera is caused by a bacteria called *Vibrio cholerae*. It is found in water contaminated with human faeces; no animal reservoirs. It is an important cause of severe infection in developing countries where portable water and sewage systems are poor.
It causes vomiting and frequent watery stools (diarrhoea) and may lead to death.

SCHISTOSOMIASIS (BILHARZIA)
Schistosomiasis is caused by Schistosomes, a fluke worm. Schistosome include: *Schistosomamansoni, S. japonicum, S. haematobium*.
Schistosomes have seperate male and female worms. Infective cercariae penetrate the skin, enter the circulation and develop into male and female worms in the intestine (*S. mansoni, S. japonicum*) or bladder (*S. haematobium*). The female worm produces eggs which penetrate the intestinal or bladder wall and are passed in feaces or urine. Ova reach fresh water, hatch into larvae and infect snails.

ONCHOCERSIASIS (RIVER BLINDNESS)
Onchocersiasis is caused by a roundworm, *Onchocerca volvulus*, through the bite of the blackfly, *Simuliumdamnosum*, which breeds in fast flowing streams.

AFRICAN TRYPANOSOMIASIS (SLEEPING SICKNESS)
African trypanosomiasis is caused by *Trypanosomabruceigambiense and T. bruceirhodesiense*, and is transmitted by tsetse fly. The infective stage (trypomastigote) is present in salivary glands of the tsetse fly. Trypomastigotes enters the host via the insect bite and reach the lymphatic system, blood and CNS.

MALARIA
Malaria is caused by protozoan parasite, *Plasmodium spp.*, through the bite of female |Anopheles mosquito bites a human and infectious plasmodia (sporozoites) are introduced into the blood stream.

POLLUTION AND ITS CONTROL
Pollution is the discharge of harmful substances into the environment in amount that can cause damage to man animals and crops. There are three types of pollution: air, water and land.

Causes of Air Pollution: It is caused by discharge of gases and particles from factories, domestic fires, vehicle exhaust pipes, dust, storms, fog, noise and radioactive elements.
Effects of Air Pollution: Bronchitis, silicosis, eye and nose irritation, green house effect (excess CO_2 in atmosphere).

Control of Air Pollution:
1. Industries should be sited away from residential areas.
2. Tall factory chimneys should be built.
3. Banning of old cars.
4. The use of lead-free fuels.

Causes of Water Pollution: Discharge of untreated sewage, chemical wastes, fertilizers, heat etc.
Effects of Water Pollution: Contamination of drinking water, death of aquatic life, out-break of epidemic diseases like cholera and diarrhea.

Control of Water pollution:
1. Proper treatment of sewage before discharge.

2. Biological control should replace chemical control to avoid pollution.
3. Minimizing oil-spillage.

Causes of Land Pollution: Indiscriminate dumping of refuse, indiscriminate dumping of dilapidated vehicles and oil-spillage.
Effects of Land Pollution: Poor crop yield, breeding of pest.
Control of Land Pollution:
1. Recycling of wastes.
2. Refuse should be incinerated.
3. Public enlightenment campaign.

CONSERVATION OF NATURAL RESOURCES
Conservation of natural resources can be defined as the process of making full use of natural resources in order to ensure their continuous availability and to keep the quality or original nature of the environment safe from decaying. Natural resources can be renewable and non renewable.
Renewable natural resources: These are natural resources that are recoverable. Examples are plants and animals.
Non renewable natural resources: these are resources which when exhausted cannot be replaced. Examples are mainly minerals, coal, oil, gas.

The Need for Conservation
1. To maintain and to promote the conservation of natural resources.
2. To prevent the destruction of natural environment
3. To prevent the destruction of natural ecosystem; this will allow the organisms in the ecosystem to survive
4. To promote the recycling of some scarce mineral resources eg water
5. To prevent naturally beautiful sceneries for their authentic value Natural resources that need to be conserved are: wildlife, fresh water, forest, mineral resources and soil.

Method of Conserving of Wildlife
1. Establishment of game or forest reserves
2. Establishment of zoological gardens
3. Control of hunting to prevent extinction of some animal species
4. Prohibiting the killing of animals in game reserves.
5. Prohibition of bush burning as this may lead to migration or displacement of wildlife
6. Prohibition of deforestation and encouragement of a forestation or reforestation
7. Creation of awareness on the value of wildlife
8. Prevention of pollution to prevent the destruction of aquatic life.

Methods of Conserving Fresh Water Supply.
1. Use fresh water economically and prevent unnecessary wastage.
2. Reduce water pollution.
3. Prevent deforestation.
4. Prevent draining of marches and swamp so as to hold rain water and release it slowly to under ground water supplies, rivers and lakes.
5. Reduce unproductive irrigation schemes which causes Salinization of the land.

Methods of Conserving Soil
1. Prevention of over grazing which may cause soil erosion.
2. Prevention of bush burning which may expose the soil to erosion.
3. Prevention of indiscriminate falling of trees.
4. Prevention of pollution of land so as not to destroy useful soil organism.
5. Avoidance of clean clearing which may expose the soil to erosion.

Methods of conserving Mineral resources
Mineral resources, unlike other resources are non renewable resources because once they are exhausted, they cannot be replaced, hence the need to conserve. The method of conserving mineral resources are:
1. There should be legislation against indiscriminate mining of mineral resources.
2. There should be proper pricing of mineral resources and their product to ensure maximum value for the mineral products.
3. Effective and efficient extraction method of mining should be adopted to prevent wastage.

Revision Questions

1.1. DISEASES
1. UME/78/42
A farmer X working in a swamp did not eat any food nor drink any water. Which of these diseases can he not contract?
A. Cholera   B. Bilharzia
C. River blindness   D. Malaria
E. Sleeping sickness

2. UME/80/47
Below are some groups of diseases. Which group is caused by bacteria?
A. tuberculosis, small pox   B. gonorrhea, measles
C. tuberculosis, polio
D. sleeping sickness, measles   E. syphilis, gonorrhea

3. UME/84/45
Which of the following lists of diseases, their causes and transmission is CORRECT?
A. cholera, virus, severe diarrhea, infected water
B. malaria, protozoan, high fever, contact with infected person
C. syphilis, virus, veneral disease, sexual intercourse
D. smallpox, virus, skin with blister, close contact with infected person
E. sleeping sickness, bacteria, tiredness, headache and dozing, tsetsefly bite.

4. UME/85/43
The primary and secondary hosts respectively of bilharzia are
A. fish and man　　　　B. man and dog
C. snail and man　　　　D. man and snail
E. fish and snail

5. UME/86/44
Which of these diseases CANNOT be prevented by immunization?
A. Poliomyelitis　　　　B. Tuberculosis
C. Cholera　　　　　　D. Onchocerciasis

6. UME/86/47
The swollen shoot disease of cocoa tree is caused by a
A. virus　　　　　　　B. fungus
C. bacterium　　　　　D. protozoan

7. UME/88/45
Which of the following diseases can be contracted in areas with fast flowing rivers?
A. schistosomiasis　　　B. elephantiasis
C. syphilis　　　　　　D. onchocerciasis

8. UME/90/46
Which set of diseases is spread mainly by insect vectors?
A. Cholera, tenia and gonorrhea
B. Cholera, malaria and tuberculosis
C. Poliomyelitis, tuberculosis and syphilis
D. Malaria, cholera and river blindness

9. UME/95/36
Which of the following is a measure for the control of bilharzia?
A. cutting low bushes around homes
B. application of molluscides in water bodies
C. screening windows and doors with mosquito net
D. application of herbicides in water bodies

10. UME/97/41
One of the most effective ways of controlling guinea worm is by
A. treating　　　　B. public enlightenment campaigns
C. accelerating rural development
D. provision of portable drinking water

11. UME/98/34
Which of the following groups of diseases are associated with water ?
(i) Onchocerciasis (ii) Schitosomiasis
(iii) Dracunculiasis (iv) Elephantiasis (v) Taeniasis
A. i, ii and iii　　　　B. ii, iv and v
C. ii, iii and iv　　　　D. i, ii and v.

12. UME /99/41/TYPE: D
Vaccinations is carried out in order to
A. check the production of poison
B. increase the activity of white blood cells
C. increase the number of red blood cells
D. stimulate the production of antibodies

13. UME /99/42/ TYPE: D
The construction of dams may lead to an increase in the prevalence of
A. typhoid fever, measles and yellow fever
B. tuberculosis, leprosy and trypanosomiasis
C. guinea worm, malaria and tuberculosis
D. malaria, bilharziasis and onchocersiasis

14. UME/2006/41
The causative agent of typhoid fever is
A. *Salmonella*　　　　B. *Entamoeba*
C. *Escherichia*　　　　D. *Shigella*

15. UME/2007/3
The causative agent of bird flu is a
A. protozoan　　　　　B. virus
C. bacterium　　　　　D. fungus

16.　　　UME/2008/38
A boy who is found of swimming in a pond finds himself passing urine with traces of blood. He is likely to have contracted
A. schistosomiasis　　　B. onchocerciasis
C. poliomyelitis　　　　D. salmonellosis

17. PCE/93/43
The vector of river blindness is
A. *Musca domestica*　　B. *Simulium damnosum*
C. *Anopheles gambiae*　D. *Culex pipiens*

18. PCE/94/42
Which of the following diseases is caused by a virus?
A. Poliomyelitis　　　　B. Cholera
C. Biharzia　　　　　　D. Malaria

19. PCE/95/42
The goal of immunization is to
A. cure diseases
B. increase protection against certain diseases
C. enable a person to travel abroad
D. eradicate deadly diseases

20. PCE/97/38
The urinary bladder can become damaged when an individual is infected by
A. *Schistosoma mansoni*   B. *Schistosoma japonicum*
C. *Schistosoma intercalatum*
D. *Schistosoma haematobium*

21. PCE/98/39
The infective stage of malaria parasite is
A. trophozoites   B. schizozoites
C. gametocytes   D. sporozoites

22. PCE/99/33/TYPE: E
An example of an obligate parasite is
A. pediculu   B. boophilus
C. mallophagus   D. Taenia

23. PCE/02/18/TYPE: 9
Diseases that are caused by faecal pollution of water include
A. dysentery, cholera and typhoid fever
B. dysentery, malaria and poliomyelitis
C. river-blindness, malaria and cholera
D. tuberculosis, typhoid fever and cholera

24. PCE/07/35
The spread of tuberculosis is favoured by
A. malnutrition   B. contaminated water
C. air pollution   D. overcrowding

25. UME/2009/37
I. Onchocerciasis   II. Schistosomiasis
III. Salmonellosis   IV. Meningitis
Which of the diseases listed above are associated with water?
A. I and II only   B. II, III and IV
C. I, II and III   D. II and IV

26. UME/2010/39/TYPE: D
The vector for yellow fever is
A. anopheles mosquito   B. tsetse fly
C. blackfly   D. aedes mosquito

27. UME/2010/40/TYPE: D
The protozoan plasmodium falciparum is transmitted by
A. female aedes mosquitoes
B. female culex mosquitoes   C. female blackfly
D. female anopheles mosquitoes

21.2.   POLLUTION AND ITS CONTROL
28. UME/87/48.
One of the function of UNICEF is to
A. prevent and control major diseases
B. prevent disease outbreak by administering vaccines
C. improve the health and nutrition of children and nursing mothers
D. monitor environmental pollution

29. UME /88/46
Which of the following causes pollution?
A. consumption of canned drinks
B. the addition of fertilizers to farmland
C. respiration of living organisms
D. burning of refuse.

30. UME/89/43.
Which of the ways of controlling bilharzia can result in pollution ?
A. clearing waterweed on which the snails feed
B. treating infected people with drugs
C. preventing contamination of the water by infected urine and faeces
D. applying chemicals to kill the snail

31. UME /90/47
Which of the following constitutes pollution?
A. droppings from birds
B. loud disco music
C. a pack of cigarettes
D. refuse in an incinerator

32 UME /93/45
Carbon monoxide poisons tissues by
A. constricting the blood vessels B. killing the cells
C. combining with haemoglobin
D. rupturing the blood vessels

33. UME/94 /40
People who suck petrol with their mouth run the risk of increasing in their blood the concentration of
A. iron   B. lead
C. calcium   D. magnesium

34. UME/95/38
Environmental pollution which can work through the media of water ,soil and air include
A. carbon monoxide   B. noise
C. sulphur (iv) oxide   D. smoke

35.   UME/97/40
Which is the most important pollutant of the marine environment in Nigeria?
A. insecticides   B. sewage
C. oil   D. inorganic fertilizers

36.   UME/2000/41/TYPE:M
The pollution that contribute to the depletion of the ozone layer in the atmosphere are
A. radioactive materials   B. oxides of sulphur
C. oxides of carbon   D. chlorofluorocarbons

37. UME/2002/20/TYPE:Y
The excessive use of agro-chemicals could lead to the pollution of
A. the lithosphere   B. the atmosphere
C. Freshwater   D. space

38. PCE/92/40
Which of the following may result in air pollution?
A. incomplete combustion of fuel
B. old vehicle
C. sewage
D. noise

39. PCE/94/43
Which of the following pollutions is carcinogenic?
A. Detergents   B. smoke
C. carbon monoxide   D. radioactive materials

40. PCE/97/36
An smoke from car exhausts is a poisonous air pollutant because it combines readily with
A. haemoglobin   B. antibodies
C. leucocytes   D. blood plasma

41. PCE/06/46/TYPE: A
A major effect of deforestation is
A. a reduction in greenhouse effect
B. an increase in greenhouse effect
C. an increase in the ozone layer of the atmosphere
D. a depletion of atmospheric carbon (iv) oxide

42. PCE/07/28
Excess carbon (iv) oxide in the atmosphere could lead to
A. global warming   B. high rainfall
C. algal bloom   D. condensation

43. UME/2008/39
The easiest way to establish the level of pollution in a local stream is to measure the level of
A. Oxygen   B. carbon (iv) oxide
C. ammonia   D. alkalinity

44. UME/2009/38
The major cause of global warming is the
A. burning of fossil fuel
B. construction of dams
C. use of electricity
D. exploration of space

21.3. CONSERVATION OF NATURAL RESOURCES
45. UME/2001/33/TYPE:T
The greatest influence on a stable ecosystem in nature is exerted by
A. man   B. pollution
C. animal   D. rainfall

46. UME/2003/34/TYPE:A
The most effective method of dealing with non –biodegradable pollution is by
A. burying   B. dumping
C. incineration   D. recycling.

47. PCE /95/39
Conservation of wildlife in Nigeria is carried out by
A. cattle grazing   B. annual bush burning
C. changing mining techniques
D. establishment of forest reserves

48. PCE/2000/40/TYPE: A
Conservation of wildlife in Nigeria is carried out by the
A. World Wildlife Fund for Nature
B. Nigeria Game Reserve Authority
C. Federal Environmental Protection Agency
D. Forestry Research Institute of Nigeria

49. PCE/98/38/TYPE: C
The most important resources in conservation is
A. water   B. wildlife
C. soil   D. forest

50. PCE/03/27/TYPE: N
Mineral resources need to be conserved because they are
A. not renewable   B. very expensive
C. not abundant   D. aesthetic in nature

51. PCE/07/36
Conservation of wildlife can be enhanced by
A. intensified poultry farming
B. legislating against poaching
C. house-to-house-campaign
D. planting various species of plants

52. UME/2008/29
The major consequence of bush burning in an ecosystem is
A. the loss of water absorbing ability of the soil
B. the loss of biological diversity
C. a decrease in animal population
D. an increase in soil fertility

MISCELLANEOUS QUESTIONS (21)
53. The greatest impact on a natural ecosystem is made by
A. bacteria   B. man
C. viruses   D. green plants
(PCE/2005/38/TYPE:P)

54. The blackfly is a vector of
A. malaria   B. trypanosomiasis
C. onchocerciasis   D. yellow-fever
(PCE/2005/11/TYPE: O)

55. One disease NOT caused directly by a bacteria is
A. malaria   B. tuberculosis
C. pneumonia   D. tetanus
E. cholera (UME/83/27)

56. Which of the following disease could be exclusively associated with a river basin?
A. malaria   B. syphilis
C. onchocerciasis   D. cholera
E. poliomyelitis (UME/83/33)

57. The recycling method of solid waste disposal is unsuitable for
A. organic matter   B. glass
C. Plastic   D. metal scraps
(UME/2005/42/TYPE:D)

58. Sources of air pollutants are
A. industrial chimneys, burning fossil oils and river dams
B. sulphur dioxide, acid rain and pesticides
C. sulphur mines, vehicle exhausts and aerosols
D. sewage, smoke and old vehicles   (UME/91/48)

59. A renewal natural resources is
A. wildlife   B. gold
C. petroleum   D. coal
(PCE/2004/40/TYPE:9)

60. A non-renewable alternative source of energy is
A. a wind generators   B. solar panels
C. nuclear energy   D. hydroelectric power
(UME/2005/43/TYPE:D)

61. Acid rain is due mainly to the release into the atmosphere of the oxides of
A. carbon and phosphorus
B. sulphur and nitrogen
C. carbon and sulphur
D. phosphorus and nitrogen (PCE/2004/41/TYPE:9)

62. A causative agent of severe acute respiratory syndrome (SARS) is a
A. protozoan
B. bacterium
C. fungus
D. virus   (PCE/2005/32/TYPE:P)

63. In the human body, blood fluke inhabits the
A. rectum   B. liver
C. stomach   D. bladder (PCE/96/39)

64. The clearing away of plants from water banks and draining of water collections are control measures targeted towards the eradication of
A. Malaria
B. schistosomiasis
C. Onchocerciasis
D. draconiasis
(PCE/96/40)

65. The stage of a parasite harboured by a definitive host is the
A. egg   B. cyst
C. adult   D. larva PCE/97/37

66. The introduction of snails that feed on mosquito larvae in stagnant water is a
A. cultural control
B. chemical control
C. biological control
D. mechanical control (PCE/01/38/TYPE: 4)

67. Which of these diseases can be controlled by immunization?
A. bilharzias
B. polio
C. AIDS
D. diarrhoea (PCE/01/39/TYPE:4)

68. The best way to eradicate poliomyelitis is to
A. isolate the affected individuals
B. inoculate children from zero to five years
C. keep a clean environment
D. treat patients at the emergence of symptoms
(PCE/06/50)

69. Which of the following types of immunity do babies obtain from their mothers?
A. Artificial passive immunity
B. Artificial active immunity
C. Natural active immunity
D. Natural passive immunity PCE/98/37

CHAPTER 22

VARIATION IN POPULATION

Understand the following after reading this chapter:

> Meaning of variation
> Types of variation
> Causes of variation
> Application of variation

CAUSES OF VARIATION

There are many causes of variation as this chart shows:

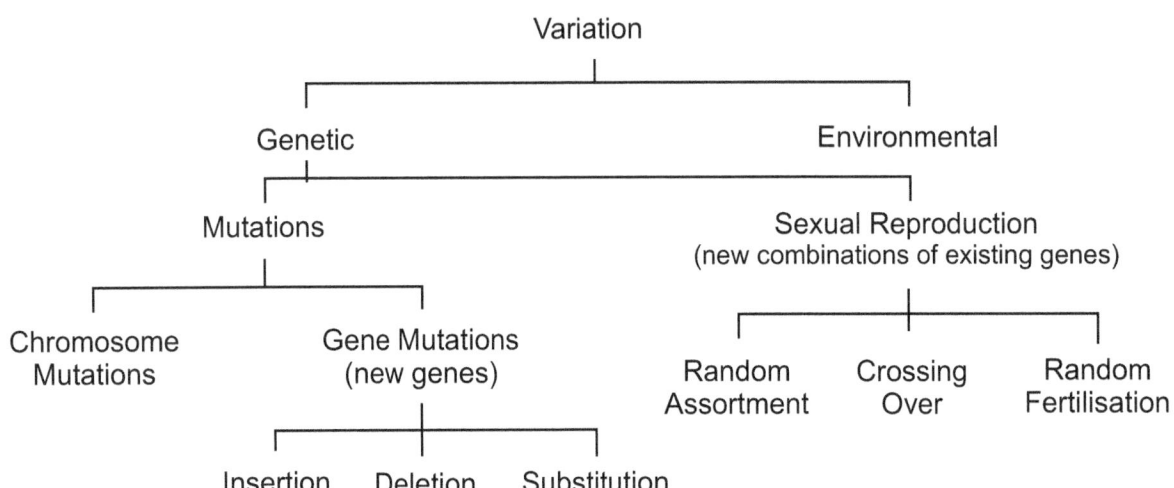

VARIATION

Variation means the differences in characteristics (phenotype) within a species.
Variation consists of differences <u>between</u> species as well as differences within the <u>same</u> species. Each individual is influenced by the environment, so this is another source of variation.

genotype + environment = phenotype

TYPES OF VARIATION

For each characteristic, the population may show either

discontinuous or *continuous* variation.
Discontinuous or physiological variation: limited or no intermediate forms. You either have the characteristic or you don't. Blood groups are a good example: you are either one blood group or another - you can't be in between. Such

data is called discrete (or categorical) data.

The characteristics of discontinuous variation:
- have distinct categories into which individuals can be placed
- tend to be qualitative, with no overlap between categories
- are controlled by one gene, or a small number of genes
- are largely unaffected by the environment

Discontinuous characteristics are rare in humans and other animals, but are more common in plants. Some examples are human blood group, ability to roll tongue, ability to detect phenylthiocarbamide (PTC), heart rate, muscle efficiency, intelligence, growth rate, rate of photosynthesis, etc.

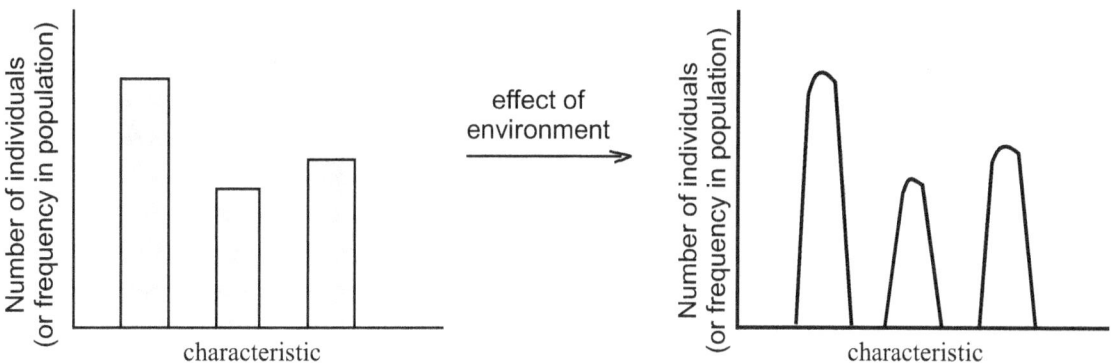

effect of environment →

Continuous or morphological variation: prevalence of many intermediates between the two extremes
The characteristics of continuous variation:

- have no distinct categories into which individuals can be placed
- tend to be quantitative, with overlaps between categories
- are controlled by a large number of genes (polygenic)
- are significantly affected by the environment

Continuous characteristics are very common in humans and other animals. Some examples are height, weight, shoe size, hand span, hair colour, milk yield in cow.
Milk yield in cows, for example, is determined not only by their genetic make-up but is also significantly affected by environmental factors such as pasture quality and diet, weather, and the comfort of their surroundings. When plotted as a histogram, these data show a typical bell-shaped normal distribution curve, with the mean (= average), mode (= biggest value) and median (= central value) all being the same.

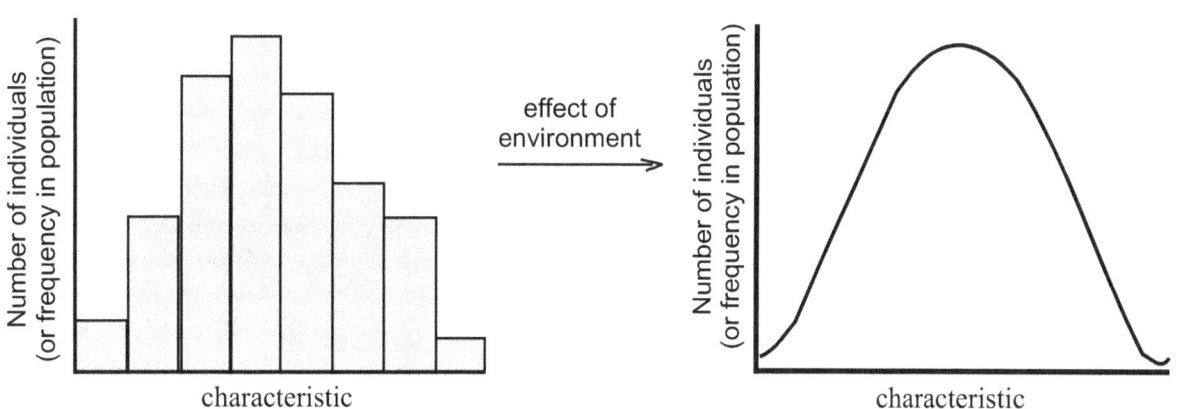

effect of environment →

APPLICATION OF VARIATION

1. Crime detection though analysis of chromosomes and genes.
2. Crime detection through the use of finger prints
3. Personal identification.
4. In the determination of the paternity of a disputed child.
5. In the determination of compatibility of the blood groups of prospective donors and recipients before transfusion.

ABO BLOOD GROUPS IN HUMANS AND BLOOD TRANSFUSION

There are four blood groups based on the presence or absence of antigens in the red blood cells. The blood plasma also contains antibodies which are sensitive to the antigens. Therefore, before blood transfusion, a donor's and the recipient's blood must be tested for compatibility else there will be aggulatination or clotting of blood in the recipient vessels which can lead to death.

Blood Group	Antigens on RBC	Antibodies (serum)	Can donate blood to	Can receive blood from
A	A	Anti-B	A and AB	A and O
B	B	Anti-A	B and AB	B and O
AB	A and B	None	AB	All groups
O	None	Anti-A and Anti-B	All groups	O

From the table it is obvious that blood group AB is the universal acceptor and blood group O the universal donor.

RHESUS GROUPINGS

The **Rh (Rhesus) blood group system** (including the **Rh factor**) is one of thirty current human blood group systems. Clinically, it is the most important blood group system after ABO. Persons who had this antigen are termed as Rh positive (Rh+) and the persons in whom the Rh antigen is absent are designated as Rh negative (Rh-) people. This is called Rh Grouping. During blood transfusion, Rh factor/Rh antigen is also taken into account. It was discovered by Landsteiner and Wiener in 1940 after immunizing the rabbits with the blood of a monkey - *Macaca rhesus.*

Revision Questions

2.1. VARIATION IN POPULATION
1. UME/98/39
After one week of life, the weights of five chicks of the same sex hatched simultaneously from the eggs of the same hen and fed on the same diet were 45g, 40g, 35g, 33g and 30g. This is an example of
A. growth rate
B. natural selection
C. Variation
D. mutation

2. UME/2006/22
One advantage of variation in a species population is that individuals
A. with favoured traits become dominant
B. easily reach their reproductive age
C. are easily recognized by mates
D. are better adapted to changes

22.2. TYPES OF VARIATION

3. PCE/2004/43/TYPE:9
Which of the following groups are physiological variations?
A. height, tongue rolling and blood group
B. fingerprint, blood group and colour blindness
C. blood group, tongue rolling and PTC tasting
D. PTC tasting, skin colour and height

4. UME/93/46
Which of the following is an example of discontinuous variation observed in man?
A. skin colours
B. tongue rolling
C. body weight
D. height

5. UME/86/39
Which of the following is an example of discontinuous variation?
A. skin coloration
B. left-handedness
C. body weight
D. height

6. UME/2002/46/TYPE: Y
In a population of living things, the parameters of size, height, weight and colour are examples of
A. discontinuous variations
B. continuous variations
C. non-heritable variations
D. physiological variarions

7. PCE/2003/7/TYPE: N
Arches, loops, whorls and compounds are types of variation in
A. finger prints
B. hair colour
C. blood group
D. eye colour

8. PCE/2000/41/TYPE: A
The ability in plants and animals to resist certain diseases is
A. a physiological variation
B. a morphological variation
C. a continuous variation
D. an acquired variation

9. UME/2010/41/TYPE: D
A dilute solution of phenylthiocarbamide tastes bitter to some people and is tasteless to others. This is an example of
A. discontinuous variation
B. morphological variation
C. continous variation
D. taste bud variation

10. UME/2009/40
A phenotypic character with that intermediate forms that can be graded from one extreme to the other is referred to as
A. discontinuous variation
B. continous variation
C. a mutant
D. a genome

22.3. CAUSES OF VARIATION

11. UME/81/27
Normally any character shown by an organism is due to the effects of
A. hormones and chromosomes   B. chromosomes
C. mutations   D. hormones and genes
E. genes and environment

12. UME/95/39
Human height is an example of a feature which depends on both
A. genotype and phenotype
B. mother's genotype and environmental factors
C. genotype and environmental factors
D. phenotypic and environmental factors.

13. UME/2006/18
Hassan and Hussain are identical twins but Hassan grows taller and fatter than Hussain. This is probably because
A. Hassan inherits genes for tallness and fatness from the father
B. they have dissimilar genotypes
C. Hassan is endowed with genes for shortness and thinness
D. they are raised in different environment

14. UME/2008/40
Which of the following is a major cause of variation among organisms?
A. inbreeding   B. backcrossing
C. sexual reproduction   D. gene dominance

22.4. APPLICATION OF VARIATION
22.4.1. CRIME DETECTION

15. PCE/95/45
One of the best applications of variation in the detection of crime is found in the use of
A. size of the criminal
B. fingerprints of the criminal
C. colour of the eyes of the criminal
D. skin colour of the criminal

22.4.2. BLOOD TRANSFUSION

16. UME/81/29
A person whose blood can be donated to all other people must have the blood group
A. O   B. AB
C. B   D. A
E. none of the above

17. UME/98/36
In blood transfusion, agglutination occurs when
A. white blood cells from two individuals meet
B. two different antibodies meet
C. two different antigens meet
D. contrasting antigens and antibodies meet

18. UME/2008/41
The Rhesus factor of blood was first identified in a category of
A. monkeys
B. human females
C. human males
D. chimpanzees

19. PCE/01/91/TYPE: 4
The possible blood groups of offspring that can be produced from a man in blood group O and a woman in blood group AB are
A. A and O
B. A and AB
C. O. and AB
D. A and B

20. PCE/06/18/TYPE: A
What is the number of recognizable groups in the ABO blood system?
A. 3   B. 4
C. 1   D. 2

22.4.3. DETERMINATION OF PATERNITY

21. PCE/2001/42/TYPE: 4
If a baby with blood group B is claimed to be born to parents with blood group O, it can be implied that
A. the paternity of the body is questionable
B. the paternity of the baby is certain
C. blood group B is dominant over O
D. B and O exhibit co-dominance

22. UME/2004/23/TYPE: 4
Paternity disputes can most accurately be resolved through the use of
A. DNA analysis   B. finger printing
C. tongue rolling   D. blood group typing

23. UME/2009/36
An accurate identification of a rapist can be carried out by conducting a
A. RNA analysis   B. DNA analysis
C. blood group test   D. behavioural traits test

MISCELLANEOUS QUESTIONS (22)

24. Characteristics that could have intermediate forms such that they could be graded from one extreme to the other is referred to as
A. intermediate variation
B. discrete variation
C. discontinuous variation
D. continous variation (PCE/92/44)

25. The difference in height, weight, skin colour and texture of hair between two individuals is an example of
A. genetic variation
B. physiological variation
C. morphological variation
D. anatomical variation   (PCE/97/40)

26. Discontinuous variation is evidenced in
A. variation in the body weight
B. variation in height
C. the ability to roll the tongue
D. the ability to determine colours (PCE/98/41/TYPE:C)

27. Which of the following indicates physiological variation?
A. appearance of different colours in flowers of the same species
B. differences in the eye colours of students in a class
C. differences in the ability to taste phenylthiocarbamide
D. differences in the uniqueness of finger prints (PCE/99/43/TYPE:E)

28. A person with type O blood can donate to a patient with type A because the donor's blood
A. lacks antigens
B. anti-A antibodies
C. lacks anti-B antibodies
D. has both anti-A and anti-B antibodies (UME/95/40)

29. If a baby with blood group B is claimed to be born to parents with blood group O, it can be implied that
A. the paternity of the baby is questionable
B. the paternity of the baby is certain
C. blood group B is dominant over O
D. B and O exhibit co-dominance (PCE/2001/42/TYPE:4)

30. The finger prints of man can be classified into
A. arches, simple, complex and compound
B. arches, loops, whorls and compound
C. arches, loops, simple and compound
D. arches, simple. whorls and compound (PCE/97/41)

CHAPTER 23

HEREDITY (GENETICS)

Understand the following after reading this chapter:

Inheritance of characters in organisms
Chromosome- the basis of heredity
Probability in genetics
Applications of the principle of heredity

DEFINITION OF TERMS IN GENETICS

- A -

albinism the genetically inherited condition in which there is a marked deficiency of pigmentation in skin, hair, and eyes. An individual with these traits is an "albino." Since the gene for albinism is recessive, it only shows up in the phenotype of homozygous recessive people. This is a pleiotropic trait.

alleles alternate forms or varieties of a gene. The alleles for a trait occupy the same locus or position on homologous chromosomes and thus govern the same trait. However, because they are different, their action may result in different expressions of that trait.

amino acids small molecules that are the components of proteins. There are 20 different kinds of amino acids in living things. Proteins are composed of different combinations of amino acids assembled in chain-like molecules. Amino acids are primarily composed of carbon, oxygen, hydrogen, and nitrogen.

autosome a chromosome not involved in sex determination. The diploid human genome consists of 46 chromosomes, 22 pairs of autosomes, and 1 pair of sex chromosomes (the X and Y chromosomes).

- B –

back cross this is the crossing of an organism with the homozygous recessive organisms from the original parent generation. Test cross and back cross is used to determine the genotype of organisms showing dominant phenotype.

- C -

carrier an individual who is heterozygous for a trait that only shows up in the phenotype of those who are homozygous recessive. Carriers often do not show any signs of the trait but can pass it on to their offspring. This is the case with haemophilia.

chromosomes thread-like, gene-carrying bodies in the nucleus of a cell. Chromosomes are composed primarily of DNA and protein. They are visible only under magnification during certain stages of cell division. Humans have 46 chromosomes in each somatic cell and 23 in each sex cell.

codominance the situation in which two different alleles for a trait are expressed unblended in the phenotype of heterozygous individuals. Neither allele is dominant or recessive, so that both appear in the phenotype or influence it. Type AB blood is an example. Such traits are said to be codominant.

- D –

diploid a full set of genetic material, consisting of paired chromosomes one chromosome from each parental set. Most animal cells except the gametes have a diploid set of chromosomes. The diploid human genome has 46 chromosomes.

dominant allele an allele that masks the presence of a recessive allele in the phenotype. Dominant alleles for a trait are usually expressed if an individual is homozygous dominant or heterozygous.

DNA (deoxyribonucleic acid) a large organic molecule that stores the genetic code for the synthesis of proteins. DNA is composed of sugars, phosphates and bases arranged in a double helix shaped molecular structure. Segments of DNA in chromosomes correspond to specific genes.

- E -

evolution genetic change in a population of organisms that occurs over time. The term is also frequently used to refer to the appearance of a new species.

- F -

f1 generation the first offspring (or filial) generation. The next and subsequent generations are referred to as f2, f3, etc.

- G -

gene flow the transference of genes from one population to another, usually as a result of migration. The loss or addition of individuals can easily change the gene pool frequencies of both the recipient and donor populations--

that is, they can evolve.

gene pool all of the genes in all of the individuals in a breeding population. More precisely, it is the collective genotype of a population.

genes units of inheritance usually occurring at specific locations, or loci, on a chromosome. Physically, a gene is a sequence of DNA bases that specify the order of amino acids in an entire protein or, in some cases, a portion of a protein. A gene may be made up of hundreds of thousands of DNA bases. Genes are responsible for the hereditary traits in plants and animals.

genetic drift evolution, or change in gene pool frequencies, resulting from random chance. Genetic drift occurs most rapidly in small populations. In large populations, random deviations in allele frequencies in one direction are more likely to be cancelled out by random changes in the opposite direction.

genetics the study of gene structure and action and the patterns of inheritance of traits from parent to offspring. Genetic mechanisms are the underlying foundation for evolutionary change. Genetics is the branch of science that deals with the inheritance of biological characteristics.

genome the full genetic complement of an individual (or of a species). In humans, it is estimated that each individual possesses approximately 2.9 billion base units in his or her DNA.

genome imprinting an inheritance pattern in which a gene will have a different effect depending on the gender of the parent from whom it is inherited. Genome imprinting is also known as genetic imprinting.

genotype the genetic makeup of an individual. Genotype can refer to an organism's entire genetic makeup or the alleles at a particular locus. See phenotype.

- H –

haploid a single set of chromosomes (half the full set of genetic material), present in the egg and sperm cells of animals and in the egg and pollen cells of plants. Human beings have 23 chromosomes in their reproductive cells.

haemophilia an X-linked genetically inherited recessive disease in which one or more of the normal blood clotting factors is not produced. This results in prolonged bleeding from even minor cuts and injuries. Swollen joints caused by internal bleeding is a common problem for hemophiliacs. Hemophilia most often afflicts males.

heterozygous a genotype consisting of two different alleles of a gene for a particular trait (Aa). Individuals who are heterozygous for a trait are referred to as heterozygotes.

homologous chromosomes chromosomes that are paired during the production of sex cells in meiosis. Such chromosomes are alike with regard to size and also position of the centromere. They also have the same genes, but not necessarily the same alleles, at the same locus or location.

homozygous having the same allele at the same locus on both members of a pair of homologous chromosomes. Homozygous also refers to a genotype consisting of two identical alleles of a gene for a particular trait. An individual may be homozygous dominant (AA) or homozygous recessive (aa). Individuals who are homozygous for a trait are referred to as homozygotes.

hybrids offspring that are the result of mating between two genetically different kinds of parents--the opposite of purebred.

- K -

karyotype photomicrograph of an individual's chromosomes arranged in a standard format showing the number, size, and shape of each chromosome type; used in low-resolution physical mapping to correlate gross chromosomal abnormalities with the characteristics of specific diseases.

- L -

Locus the location of a gene or transcription unit on a chromosome.

- M -

meiosis cell division in specialized tissues of ovaries and testes which results in the production of sperm or ova. Meiosis involves two divisions and results in four daughter cells, each containing only half the original number of chromosomes--23 in the case of humans.

Mendelian genetics inheritance patterns which can be explained by simple rules of dominance and recessiveness of genes.

modifying gene a gene that can alter the expression of another gene in the phenotype of an individual.

monozygotic twins (identical twins) twins that come from the same zygote are essentially the same genetically. Differences between monozygotic twins later in life are virtually always the result of environmental influences rather than genetic inheritance. Fraternal twins may look similar but are not genetically identical.

multiple-allele series a situation in which a gene has more than two alleles. The ABO blood type system is an example. Multiple-allele series only partly follow simple Mendelian genetics.

mutation an alteration of genetic material such that a new variation is produced. For instance, a trait that has only one allele (A) can mutate to a new form (a). This is the only mechanism of evolution that can produce new alleles of a gene.

- O -

ovum (plural ova) a female sex cell or gamete.

phenotype the observable or detectable characteristics of an individual organism--the detectable expression of a genotype.

pleiotropy the situation in which a single gene is responsible for a variety of traits. The collective group of symptoms known as sickle-cell trait is an example.

polygenic trait

an inherited trait that is determined by genes at two or more loci. Simple Mendelian rules of dominance do not apply to the complex interaction of these genes. As a result, phenotypes may appear as apparent blends or intermediate expressions. Human skin and hair color are polygenic traits. Many polygenic traits are also influenced by environmental factors.

probability the likelihood that a specific event will occur. Probability is usually expressed as the ratio of the number of actual occurrences to the number of possible occurrences.

proteins any of a large number of complex organic molecules that are composed of one or more chains of amino acids. Proteins can serve a wide variety of functions through their ability to bind to other molecules. Proteins may be enzymes, hormones, antibodies, structural components, or gas-transporting molecules.

proteome the full complement of proteins produced by an individual (or a species). It is estimated that each human produces approximately 90,000 types of proteins.

Punnett square a simple graphical method of showing all of the potential combinations of offspring genotypes that can occur and their probability given the parent genotypes. See example below. Punnett squares are commonly used by genetics counselors to predict the odds of a couple passing on particular inherited traits.

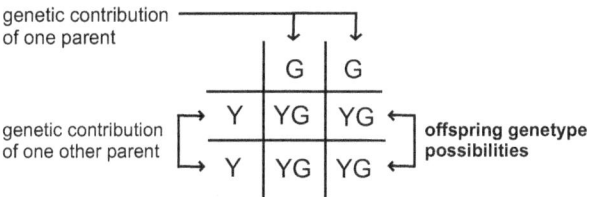

purebred offspring that are the result of mating between genetically similar kinds of parents--the opposite of hybrid. Purebred is the same as true breeding.

- R -

recessive allele an allele that is masked in the phenotype by the presence of a dominant allele. Recessive alleles are expressed in the phenotype when the genotype is homozygous recessive (aa).

- S -

sex cell a gamete, either a sperm or an ovum. Sex cells are produced by the meiosis process.

sex-controlled gene a gene that can be inherited by both genders but is usually expressed differently in males and females.

sex-limited gene a gene that can be inherited by both genders but is usually expressed in only males or females.

sickle-cell trait a genetically inherited recessive condition in which red blood cells are distorted resulting in severe anemia and related symptoms that are often fatal in childhood. Sickle-cell trait is the result of a pleiotropic gene. Sickle-cell trait is also known as sickle-cell anemia.

somatic cell any cell in the body except those directly involved with reproduction. Most cells in multicellular plants and animals are somatic cells. They reproduce by mitosis.

sperm a male sex cell or gamete.

- T -

test cross this is the crossing of an organism with the homozygous recessive organisms e.g. crossing genotype Tt against tt.

- X -

X-linked referring to a gene that is carried by an X sex chromosome.

- Z -

zygote a "fertilized" ovum. More precisely, this is a cell that is formed when a sperm and an ovum combine their chromosomes at conception. A zygote contains the full complement of chromosomes (in humans 46) and has the potential of developing into an entire organism.

INHERITANCE OF CHARACTERS IN ORGANISMS (CLASSICAL GENETICS)

Classical genetics is the study of inheritance of characteristics at the whole organism level. It is also known as transmission genetics or Mendelian genetics, since it was pioneered by Gregor Mendel.

Gregor Mendel

Mendel (1822-1884) investigated inheritance in pea plants and published his results in 1866. They were ignored at the time, but were rediscovered in 1900, and Mendel is now recognised as the "Father of Genetics". His experiments succeeded where other had failed because:

- Mendel investigated simple well-defined characteristics (or traits), such as flower colour or seed shape, and he varied one trait at a time. Previous investigators had tried to study many complex traits, such as human height or intelligence.
- Mendel use an organism whose sexual reproduction he could easily control by carefully pollinating stigmas with pollen using a brush.

Peas can also be self-pollinated, allowing self crosses to be performed. This is not possible with animals.

- Mendel repeated his crosses hundreds of times and applied statistical tests to his results.
- Mendel studied two generations of peas at a time.

A typical experiment looked like this:

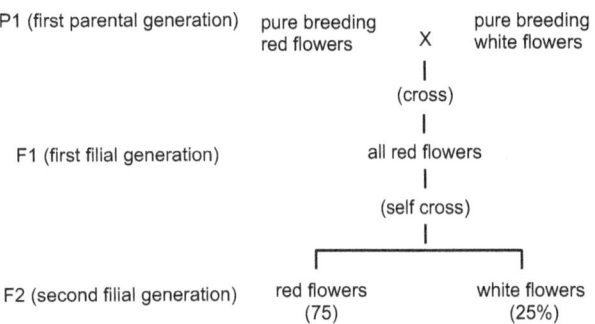

P1 (first parental generation) pure breeding red flowers   X   pure breeding white flowers
|
(cross)
|
F1 (first filial generation)   all red flowers
|
(self cross)
|
F2 (second filial generation)   red flowers (75)   white flowers (25%)

Mendel made several conclusions from these experiments:

1. There are no mixed colours (e.g. pink), so this disproved the widely-held blending theories of inheritance that characteristics gradually mixed over time.
2. A characteristic can disappear for a generation, but then reappear the following generation, looking exactly the same. So a characteristic can be present but hidden.
3. The outward appearance (the phenotype) is not necessarily the same as the inherited factors (the genotype) For example the P1 red plants are not the same as the F1 red plants.
4. One form of a characteristic can mask the other. The two forms are called dominant and recessive respectively.
5. The F2 ratio is always close to 3:1. Mendel was able to explain this by supposing that each individual has two versions of each inherited factor, one received from each parent. We'll look at his logic in a minute.
6. Mendel's factors are now called genes and the two alternative forms are called alleles. So in the example above we would say that there is a gene for flower colour and its two alleles are "red" and "white". One allele comes from each parent, and the two alleles are found on the same position (or locus) on the homologous chromosomes. With two alleles there are three possible combinations of alleles (or genotypes) and two possible appearances (or phenotypes):

Genotype	Name	Phenotype
RR	homozygous dominant	Red
rr	homozygous recessive	White
Rr, rR	Heterozygous	Red

The Monohybrid Cross

A simple breeding experiment involving just a single characteristic, like Mendel's experiment, is called a monohybrid cross. We can now explain Mendel's monohybrid cross in detail.

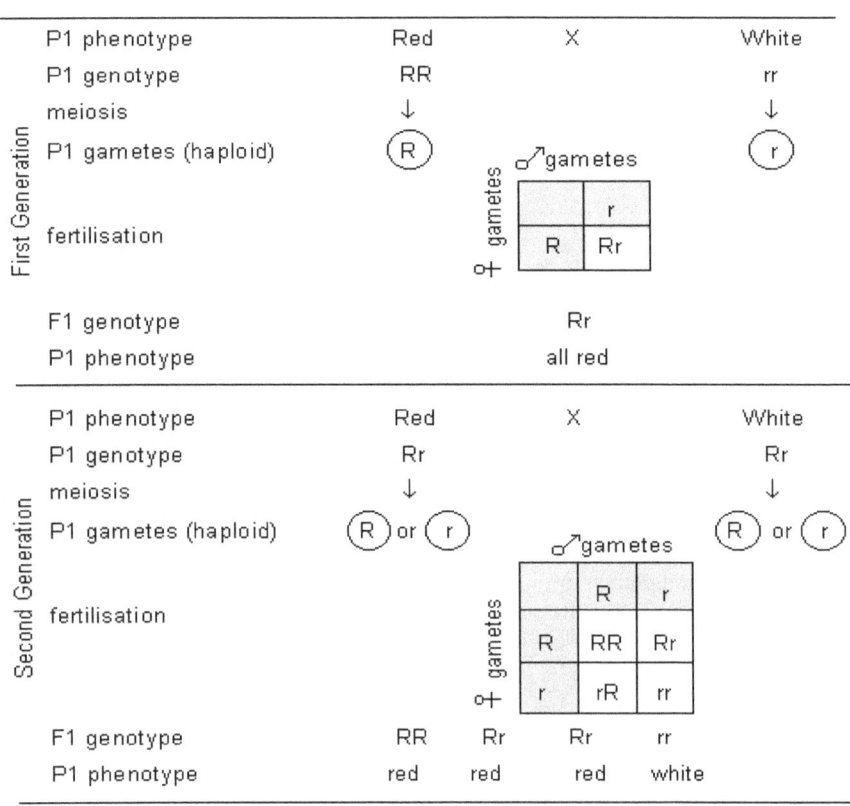

186

At fertilisation any male gamete can fertilise any female gamete at random. The possible results of a fertilisation can most easily be worked out using a <u>Punnett Square</u> as shown in the diagram. Each of the possible outcomes has an equal chance of happening, so this explains the 3:1 ratio (phenotypes) observed by Mendel.

This is summarised in <u>Mendel's First Law</u>, which states that individuals carry two discrete hereditary factors (alleles) controlling each characteristic. The two alleles <u>segregate</u> (or separate) during meiosis, so each gamete carries only one of the two alleles.

The Test Cross

You can see an individual's phenotype, but you can't see its genotype. If an individual shows the recessive trait (white flowers in the above example) then they must be homozygous recessive as it's the only genotype that will give that phenotype. If they show the dominant trait then they could be homozygous dominant or heterozygous. You can find out which by performing a test cross with a pure-breeding homozygous recessive. This gives two possible results:

- If the offspring all show the dominant trait then the parent must be homozygous dominant.
- If the offspring are a mixture of phenotypes in a 1:1 ratio, then the parent must be heterozygous.

The Dihybrid Cross

Mendel also studied the inheritance of two different characteristics at a time in pea plants, so we'll look at one of his dihybrid crosses. The two traits are seed shape and seed colour. Round seeds (R) are dominant to wrinkled seeds (r), and yellow seeds (Y) are dominant to green seeds (y). With these two genes there are 4 possible phenotypes:

Genotypes	Phenotype
RRYY, RRYy, RrYY, RrYy	round yellow
RRyy, Rryy	round green
rrYY, rrYy	wrinkled yellow
rryy	wrinkled green

Mendel's dihybrid cross looked like this:

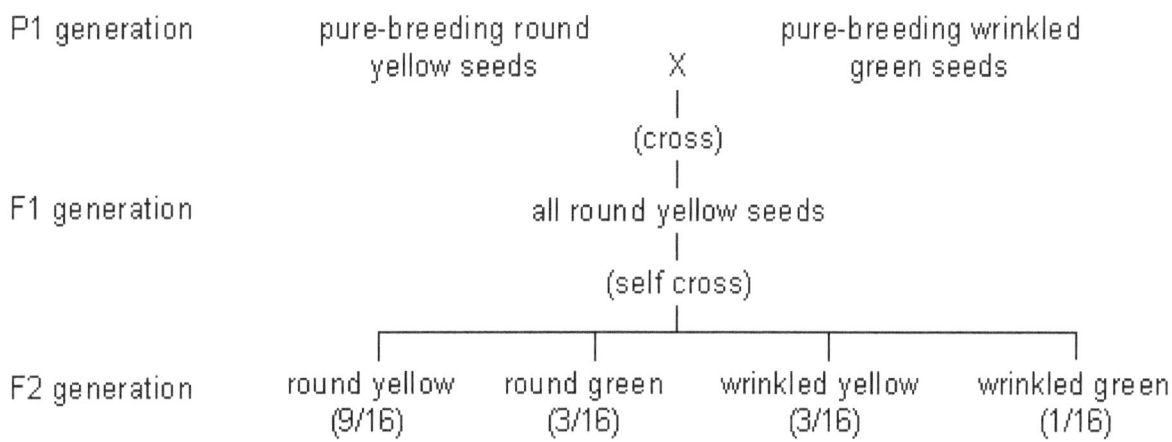

All 4 possible phenotypes are produced, but always in the ratio 9:3:3:1. Mendel was able to explain this ratio if the factors (genes) that control the two characteristics are inherited independently; in other words one gene does not affect the other.

This is summarised in Mendel's second law (or the law of independent assortment), which states that alleles of different genes are inherited independently.

We can now explain the dihybrid cross in detail.

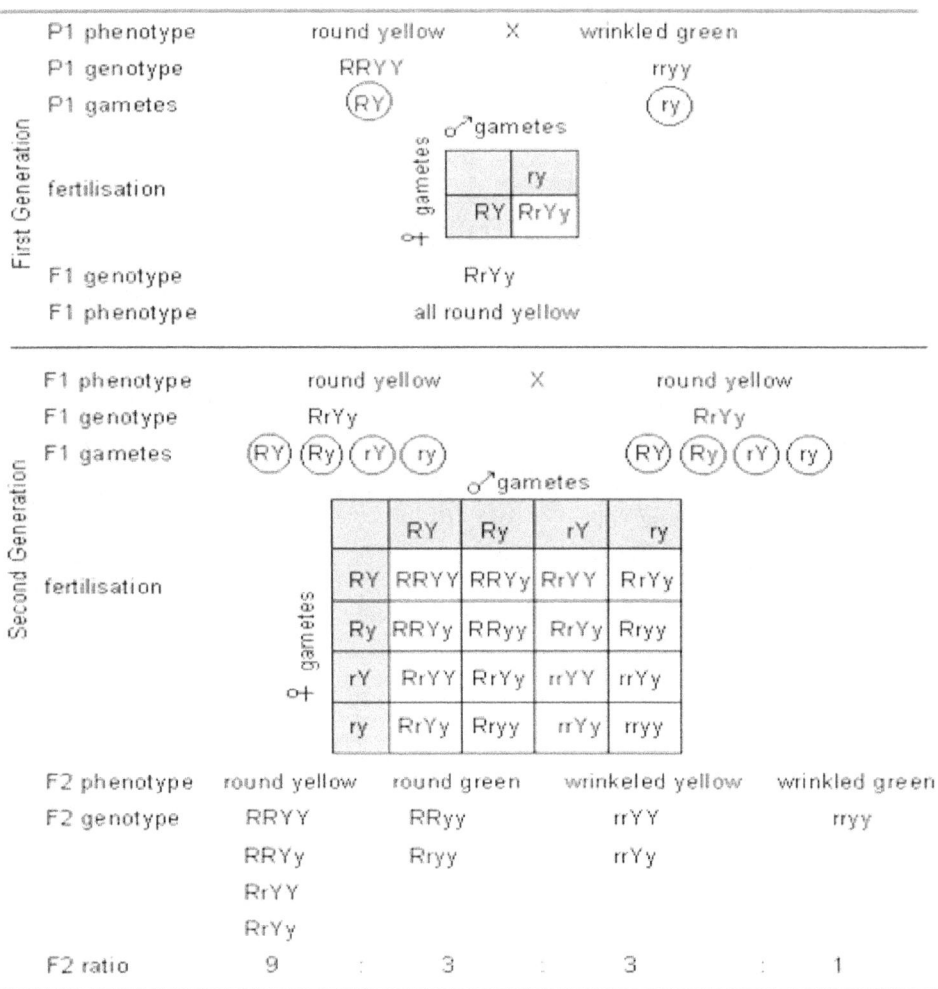

The gametes have one allele of each gene, and that allele can end up with either allele of the other gene. This gives 4 different gametes for the second generation, and 16 possible genotype outcomes.

Dihybrid Test Cross

There are 4 genotypes that all give the same round yellow phenotype. Just like we saw with the monohybrid cross, these four genotypes can be distinguished by crossing with a double recessive phenotype. This gives 4 different results:

Original genotype	Result of test cross
RRYY	all round yellow
RRYy	1 round yellow : 1 round green
RrYY	1 round yellow : 1 wrinkled yellow
RrYy	1 round yellow : 1 round green: 1 wrinkled yellow: 1 wrinkled green

CHROMOSOME- THE BASIS OF HEREDITY (MOLECULAR GENETICS)

Molecular Genetics (or Molecular Biology), which is the study of heredity at the molecular level, and so is mainly concerned with the molecule DNA. It also includes genetic engineering and cloning, and is very trendy.

Structure of DNA

The three-dimensional structure of DNA was discovered in the 1950's by Watson and Crick. The main features of the structure are:

- DNA is double-stranded, so there are two polynucleotide stands alongside each other. The strands are antiparallel, i.e. they run in opposite directions (5' 3' and 3'5')
- The two strands are wound round each other to form a double helix.
- The two strands are joined together by hydrogen bonds between the bases. The bases therefore form base pairs, which are like rungs of a ladder.

- The base pairs are specific. A only binds to T (and T with A), and C only binds to G (and G with C). These are called underlined complementary base pairs. This means that whatever the sequence of bases along one strand, the sequence of bases on the other strand must be complementary to it. (Incidentally, complementary, which means matching, is different from complimentary, which means being nice.)

Function of DNA

DNA is the genetic material, and genes are made of DNA.

DNA therefore has two essential functions: replication and expression.

? Replication means that the DNA, with all its genes, must be copied every time a cell divides.

? Expression means that the genes on DNA must control characteristics. A gene was traditionally defined as a factor that controls a particular characteristic (such as flower colour), but a much more precise definition is that a gene is a section of DNA that codes for a particular protein. Characteristics are controlled by genes through the proteins they code for, like this:

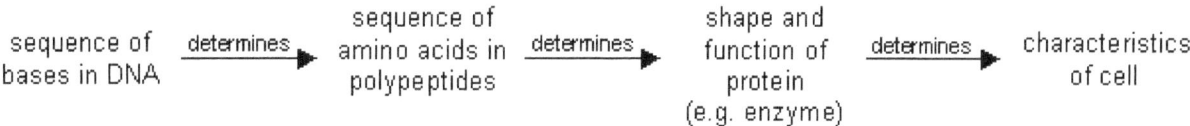

Expression can be split into two parts: transcription (making RNA) and translation (making proteins). These two functions are summarised in this diagram (called the central dogma of genetics).

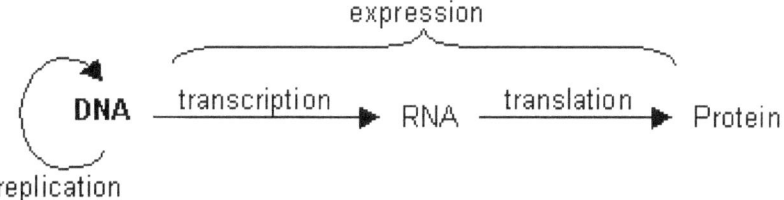

No one knows exactly how many genes we humans have to control all our characteristics, the latest estimates are 60-80,000. The sum total of all the genes in an organism is called the genome.

The table shows the estimated number of genes in different organisms:

RNA

RNA is a nucleic acid like DNA, but with 4 differences:
- RNA has the sugar ribose instead of deoxyribose
- RNA has the base uracil instead of thymine
- RNA is usually single stranded
- RNA is usually shorter than DNA

Messenger RNA (mRNA)

mRNA carries the "message" that codes for a particular protein from the nucleus (where the DNA master copy is) to the cytoplasm (where proteins are synthesised). It is single stranded and just long enough to contain one gene only. It has a short lifetime and is degraded soon after it is used.

Ribosomal RNA (rRNA)

rRNA, together with proteins, form ribosomes, which are the site of mRNA translation and protein synthesis. Ribosomes have two subunits, small and large, and are assembled in the nucleolus of the nucleus and exported into the cytoplasm.

Transfer RNA (tRNA)

tRNA is an "adapter" that matches amino acids to their codon. tRNA is only about 80 nucleotides long, and it folds up by complementary base pairing to form a looped clover-leaf structure. At one end of the molecule there is always the base sequence ACC, where the amino acid binds. On the middle loop there is a triplet nucleotide sequence called the anticodon. There are 64 different tRNA molecules, each with a different anticodon sequence complementary to the 64 different codons. The amino acids are attached to their tRNA molecule by specific enzymes. These are highly specific, so that each amino acid is attached to a tRNA adapter with the appropriate anticodon.

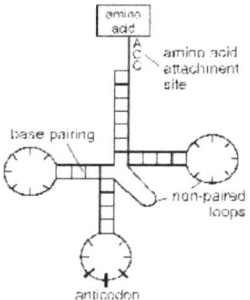

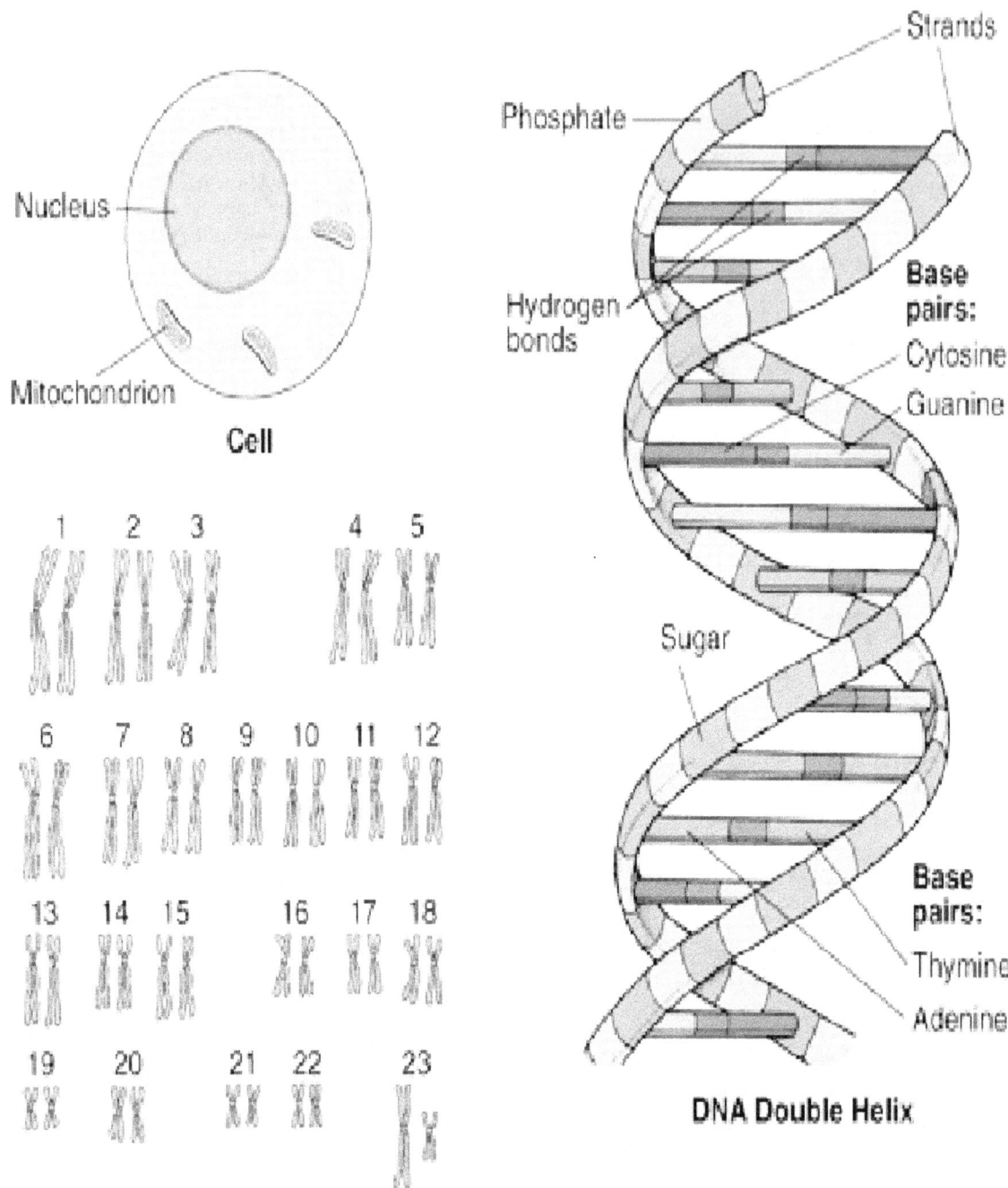

Nucleus

Mitochondrion

Cell

1 2 3   4 5

6 7 8 9 10 11 12

13 14 15   16 17 18

19 20   21 22   23

X Y

**Pairs of Chromosomes
in a Human Cell**

Strands

Phosphate

**Hydrogen
bonds**

**Base
pairs:**

Cytosine

Guanine

Sugar

**Base
pairs:**

Thymine

Adenine

DNA Double Helix

Chromosome and DNA

PROBABILITY AND GENETICS

Probability is the study of the likelihood of the occurrence of a particular event or offspring. The chance or probability that an event will take place can be expressed as a fraction (1/4), ratio (1:4) or % (25%).

Probability = # of chances for an event
                    # of possible combinations

THE RULE OF INDEPENDENT EVENTS: previous events have no impact on future events. The chance of having a girl is 1/2. If you already have one girl the chance that your next baby will be a girl is still 1/2. Each event is regarded as an individual event.

THE PRODUCT RULE: the chance that independent events will occur together is the product of their individual probabilities. Thus the chance of having 3 girls in a row is: 1/2 x 1/2 x 1/2 = 1/8 or 12.5%.

These principles only predict theoretical possibilities and there is no certainty that the event will occur.

EXAMPLE:

Rr x Rr (heterozygous monohybrid cross)
Probability of RR is 1/2 from mom
1/2 from dad
thus 1/2 x 1/2 = 1/4
Probability of rr is 1/2 from mom
1/2 from dad
thus 1/2 x 1/2 = 1/4
Probability of Rr is R: 1/2 from mom and 1/2 from dad 1/2 x 1/2 = 1/4
r: 1/2 from mom and 1/2 from dad 1/2 x 1/2 = 1/4
Thus 1/4 + 1/4 = 2/4 or 1/2
Our phenotypic ratio of 3:1 is met, 3 dominant to 1 recessive.
As is our genotypic ratio of 1:2:1, 1 RR to 2Rr to 1 rr, as seen in a punnet.
Dihybrid crosses are based on the numbers of the heterozygous monohybrid cross, for example:
DdQq x DdQq (heterozygous dihybrid cross)
Probabilities are based on the monohybrid cross:

	D	d
Q	DQ	dQ
q	Dq	Dq

Thus in the monohybrid cross, 3/4 have dominant alleles, 1/4 are recessive.
Now apply these numbers to the dihybrid cross DdQq x DdQq.
Probability of being D Q is 3/4 x 3/4 = 9/16
Probability of being D q is 3/4 x 1/4 = 3/16
Probability of being d Q is 1/4 x 3/4 = 3/16
Probability of being ddqq is 1/4 x 1/4 = 1/16
This matches our 9:3:3:1 phenotypic ratio.

(See page 188 for details).

APPLICATION OF GENETICS TO AGRICULTURE

Plants and animals breeding have the following advantages
1. Production of quality fruits and seeds
2. Production of strains resistant to diseases and climatic factors.
3. Production of high yielding varieties
4. Development of early maturing varieties of crops and animals.
5. Production of improved qualities of animals for sports and recreation.

APPLICATION OF GENETICS TO MEDICINES

1. Diagnosis and Management of certain diseases which tend to run in families.
2. Management of inheritable diseases and genetic counselling.

Revision Questions

23.1. HERITABLE AND NON-HERITABLE CHARACTERS

1. UME/82/11
If a 26 year old blind man married a young one-eyed woman and they had four children how many of them would be blind like their father?
A. All   B. 3
C. 2   D. 1
E. none

2. UME/87/41
An example of monohybrid inheritance in man is
A. astigmatism   B. cretinism
C. hyperthyroidism   D. albinism

3. UME/94/43
In the gene locus for eye colour in humans, the allele for brown-eyes is dominant over the allele for blue eyes. If a homozygous brown-eyed girl has a brother with blue eye, what are the likely phenotypes of their parents eye colour? *(Eye colour is not a sex-linked trait)*
A. both parents have blue eyes
B. their father has blue eyes and their mother has brown eyes
C. both parents have brown eyes
D. their mother has blue eyes and their father has brown eye.

4. UME/2000/40/TYPE: M
A cross between an albino female and a genetically normal male will result in offspring that are
A. all albino
B. all phenotypically normal
C. all genetically normal
D. half albino and half normal.

5. UME/2009/41
A farmer's assumption that the seed from a good harvest will produce a good yield is explained by the theory of
A. evolution   B. adaptation
C. variation   D. heredity

23.2.   MENDELS WORK IN GENETICS
6. UME/91/37
The greatest contribution to genetic studies was made by
A. Thomas morgan   B. Gregor Mendel
C. Charles Darwin   D. Robert Hooke

7. PCE/99/44/TYPE: E
The segregation of genes during gamete formation and the recombination of genes at fertilization are explained in Mendelian
A. first and second laws respectively
B. first law only
C. second law only
D. second and first law respectively.

8. UME/2006/23
Mendel's second law of inheritance states that
A. alleles combine randomly
B. alleles segregate independently
C. alleles separate predictably
D. chromosomes segregate independently

9. UME/2003/39/TYPE: A
An organism that has been extensively used to test the chromosome theory of heredity is
A. *Homo sapiens*   B. *Drosophilia melanogaster*
C. *Zea mays*   D. *Musca domestica*

10. UME/79/33
If you cross-breed a tall variety (Tt) of maize with a short variety (tt), the ratio of tall to short plants in the offspring will be
A. 2:1   B. 1:2   C. 1:1   D. 1:3   E. 3:1

11. UME/81/5
What is the genetic ratio of a cross between a homozygous tall plant and a homozygous dwarf plant?
A. 0 tall   : 4 short
B. 3 tall   : 1 short
C. 2 short: 2 tall
D. 1 tall   : 1 short
E. 4 tall   : 0 short

12. UME/83/49
What is the genetic ratio of the F_1 generation if members of F_1 generation are allowed to self pollinate?
A. 1 tall   : 3 short   B. 3 tall   : 1 short
C. 1 tall   : 1 short   D. 4 short : 0 tall
E. 4 tall   : 0 short

13. UME/84/20
A man with a normal haemoglobin (AA) marries a woman who has sickle-cell haemoglobin (SS). They have a child who has a sickle-cell trait. Which of the following genotypes could be associated with the child's haemoglobin?
A. AA   B. OO
C. AO   D. AS
E. SS

14. UME/87/42
If a plant, homozygous for round and yellow (RR:YY), is crossed with a wrinkled green type (rr:yy) all of the resulting seeds will be
A. blue and wrinkled
B. round and yellow
C. wrinkled and yellow
D. round and greenish yellow

15. UME/92/42
If R and r denote the genes for a character, the offsprings of the cross between RR and Rr
A. RR, 2Rr, rr   B. 2RR, 2rr
C. 2RR, 2Rr   D. 4Rr

16. UME/94/44
If a woman's genotype is Tt Qq Rr what will be gene content of her eggs?
A. TQr, tqr   B. TQR, tqr
C. TqR, tQr   D. tQr, TQR

17. UME/99/44/TYPE: D
If the offspring of a cross between a brown mouse (bb) and a black mouse (BB) are allowed to interbreed, how many different genotypes would result?
A. 2   B. 3   C. 4   D. 5

18. UME/88/37
A red-coloured flower when crossed with a white-coloured one produced pink flowers. This is an example of
A. complete dominance   B. blending inheritance
C. interaction of genes   D. back crossing

19. UME/2004/25/TYPE : 4
In a Mendelian cross of red and white varieties of the four O'clock plant, the F_1 generation expresses incomplete dominance by having flowers which are
A. multicolored   B. pink
C. red   D. white

20. UME/2009/42
In Mendelian inheritance, discontinuous characters are controlled by the
A. centromeres   B. alleles
C. chromosomes   D. chromatids

21. UME/2001/38/TYPE: T
Both recessive and dominant characters are found
A. on different chromosomes in the cell
B. at the same locus of a homologous chromosome
C. at different loci of a homologous chromosome
D. on the same chromatid in a chromosome

22. PCE/96/41
An example of incomplete dominance in the inheritance of blood group in human population is
A. OO   B. OA   C. OB   D. AB

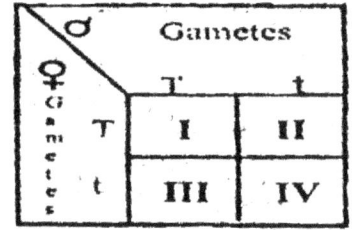

23. UME/2008/42
Which of the four offspring of the cross will be short if T the gene for tallness is dominant to t
A. I   B. II
C. III   D. IV

24. UME/2008/43
The genotype ratio of the offspring of the cross is
A. 1 : 2   B. 1:2:1
C. 1:1:1:1   D. 3:1

23.3.   TERMS USED IN GENETICS
25. UME/83/24
An organism having one pair of identical genes is
A. heterozygote   B. a hybrid
C. an allelomorph   D. a homozygote
E. a diploid

26. UME/84/8
The character-producing factors in living organisms are
A. chromomeres   B. alleles
C. chromatids   D. chromosomes
E. genes

27.   UME/85/34
The characters by which an organism is recognized are termed its
A. phenotype   B. genotype
C. morphology   D. anatomy   E. physiology

28. UME/87/40
A gene which expresses itself only in the homozygous condition is
A. a mutant   B. dominant
C. recessive   D. lethal

29. UME/89/37
When the two alleles present in an organism are of the same type, the genotype is described as
A. heterozygous   B. heterogamous
C. homozygous   D. homologous

30. UME/2010/45/TYPE: D
A pair of genes that control a trait is referred to as
A. Recessive   B. dominant
C. an hybrid   D. an allele

23.4. CHROMOSOMES-THE BASIS OF HEREDITY
31. UME/98/41.
The correct increasing order for the cell components responsible for heredity is
A. Chromosome, DNA, nucleus, gene
B. DNA, gene, chromosome, nucleus
C. Chromosome, nucleus, DNA, gene
D. DNA, gene, nucleus, chromosome

32. UME/85/53
The hereditary material in cell is known as
A. ADP   B. CNS
C. RNA   D. ATP
E. DNA

33. UME/89/36
During cell division, two strands of chromosomes are joined at a point called
A. spindle   B. chromate
C. centromere   D. aster

34. UME/90/36
In an organism were the 2n number of chromosomes is 16, the number of chromosomes in each gamete will be
A. 32   B. 16
C. 8   D. 4

35. UME/91/38.
The exchange of genes between homologous chromosomes is called
A. test cross   B. back cross
C. crossing over   D. mutation

36. UME/95/41
The DNA molecule is a chain of repeating
A. nucleosides   B. nitrogenous bases
C. sugar phosphates   D. nucleotides

37. UME/95/42

The specific number of chromosomes in each somatic cell is represented by

A. 2^N B. 23
C. 2N   D. n

38. UME/2003/43/TYPE:A

If a DNA strand has a base sequence TCA, its complementary strand must be

A. ATG   B. GAT
C. AGT   D. TAG

39. UME/2006/19

The two normal types of sex chromosomes are

A. xxy and xyy   B. xx and xyy
C. xy and xxy   D. xx and xy

40. UME/2010/46/TYPE: D

The chromosome number of a cell before and after the process of meiosis is conventionally represented as

A. $n \rightarrow n$ B. $n \rightarrow 2n$
C. $2n \rightarrow n$ D. $2n \rightarrow 2n$

41. The nitrogenous base found in the RNA but not in the DNA is

A. Thymine   B. Adenine
C. Guanine   D. Uracil
E. Cytosine

23.5. PROBABILITY IN GENETICS

42. UME/97/44

The homozygous condition HbS HbS results in sickle cell anaemia whereas HbA HbS has the sickening trait. What is the probability that a couple with the sickening trait will give birth to one normal child?

A. 1/2   B. 1/4
C. 1/8   D. 0

43. UME /2002/50/TYPE: Y

What proportion of the offspring of a cross between two heterozygous parents will exhibit the recessive condition phenotypically?

A. $\frac{1}{4}$ B. $\frac{1}{2}$
C. $\frac{3}{4}$ D. $\frac{4}{4}$

44. UME/2002/48/TYPES: Y

The first four children of a couple were all girls. The probability that fifth will also be a girl is

A. $\frac{1}{5}$ B. $\frac{1}{4}$
C. $\frac{1}{3}$ D. $\frac{1}{2}$

45. UME/95/44

The F1 of a cross between a tall and a dwarf plant was tall. The F1 was advanced to F2 how many of 120F2 plant will be dwarf?

A. 30   B. 60   C. 90   D. 120

46. UME/86/38

If a woman who is a carrier of sickle cell trait (AS) married a man who is a sickler (SS) and they had four children how many of them would be normal

A. three   B. two   C. one   D. none

47. UME/89/38

If parents with blood group AB and OO produce six children

A. three of them will have group B
B. two of them will have group A
C. all the offsprings will have O
D. none of them will have group A

48. UME/94/46

A man who has the trait for colour blindness marries a normal woman. What percentage of their children would be sufferers, carriers and normal respectively?

A. 25%, 25% and 50%   B. 25%, 50% and 25%
C. 50%, 25% and 25%   D. 25%, 37.5% and 37.5%

49. UME/2003/41/TYPE:A

A man and his wife are both heterozygous for the sickle cell trait. The likely percentage of their offspring that will be either carriers or sicklers is

A. 50%   B. 25%
C. 75%   D. 100%

23.6. APPLICATIONS OF THE PRINCIPLES OF HEREDITY

23.6.1. AGRICULTURE

50. PCE/2003/4/TYPE:N

The knowledge of genetic is applied in agriculture to improve

A. character variations and disease resistance
B. yield, quality and disease resistance
C. variation, yield and disease resistance
D. yield, character and quality

51. UME/92/44

Pawpaw seed collected from a tree with many desirable agronomic qualities

did not give rise to plants of desirable characters as the parent because

A. seeds are not reliable for propagating plant
B. uncontrolled out crossing can introduce unwanted variability
C. vegetative propagation is the best form of reproduction for all crops
D. seed were not physiologically mature at harvest

52. UME/2000/42/TYPE: M

The surest way to combine the best qualities of both parent in the offspring by

A. cross-breeding   B. in breeding
C. selective breeding   D. pure breeding

53. UME/2002/45/TYPE: Y
To select and retain the desirable trait of large body size which a farmer has observed in his herd, the farmer need to
A. feed the animals in the herd with more food
B. cross breed his animals with a different herd
C. inbreed the animal in his herd
D. prevent diseases in his herd.

54. UME/2008/44
Genetically modified food products have not become universally accepted because
A. they are not tasty as others produced by conventional means
B. their effects on human consumers is not yet fully understood
C. they are usually costlier than others produced by conventional means
D. the technology can be applied only in developed countries

23.6.2. MEDICINE
55. UME/99/43/TYPE: D
The biological factor that is unique to individual is the
A. DNA   B. eye colour
C. blood group   D. RNA

56. UME /2000/43/TYPE: M
Blood grouping is human being is derived from the combinations of
A. two different alleles   B. four different alleles
C. three different alleles   D. two different genes

57. UME /2002/49/ TYPE: Y
Genetic counseling is important when a marriage is planned between
A. Rh–woman and Rh+man
B. Rh–woman and Rh–man
C. Rh+ woman and Rh–man
D. Rh+ woman and Rh–man

58. UME/99/42/TYPE: E
Where a child with blood group O has parents with blood group A and B respectively the
A. child does not belong to the father biologically
B. parent are homozygous A and B respectively
C. blood group O is more favoured
D. parents are heterozygous A and B respectively

59. UME/2007/49
Which of the following is true of cloning?
A. it is welcome as an ethically and morally sound science
B. only one cell of the original organism is needed to initiate the process
C. the clone is similar to but not exactly like the original organism
D. it involves the asexual multiplication of the tissues of the original organism

60. UME/2009/44
A health condition that is known to have resulted from gene mutation
A. haemophilia   B. colour blindness
C. sickle cell anaemia   D. anaemia

MISCELLANEOUS QUESTIONS (23)
61. Which of the following is the best explanation for a child which is phenotypically short and born of two tall parents?
A. The father possesses a gene for shortness
B. The mother possesses a gene for shortness
C. Nature makes the child short
D. Both parents possess genes for shortness (UME/2005/44/TYPES: D)

62. A man who has the trait for colour blindness marries normal woman. What percentage of their children would be sufferers, carriers and normal respectively?
A. 25%. 25% and 50%
B. 25%, 50% and 25%
C. 50%, 25 and 25%
D. 25%, 37.5 % and 37.5% (UME/94/46)

63. A heterozygous brown eyed woman (Bb) marries a blue-eyed man (bb) and produces four children. The proportion of the eye colour of their children is
A. 25% brown, 75%blue
B. 20% brown, 80%blue
C. 75% brown, 25%blue
D. 50% brown, 50%blue (PCE/2005/25/TYPES: P)

64. A nucleotide of DNA consist of
A. 5-carbon sugar, inorganic base and phosphoric acid
B. 5-carbon sugar, nitrogenous base and phosphoric acid
C. 5-carbon sugar, ditrogenous base and phosphoric acid.
D. 4-carbon sugar, inorganic base and phosphoric acid (PCE/2005/25/TYPES: P)

65. Homozygous and heterozygous traits that show dominance can be distinguished by
A. a test cross   B. a progeny cross
C. inbreeding   D. interbreeding
(PCE/2005/21/TYPES: P)

66. One advantage of out breeding is that it increase
A. hybrid vigor   B. the rate of mutation
C. the rate of fertility   D. hybrid size
(PCE/2005/22/TYPES: P)

67. In DNA structure, the nitrogenous base Adenine complementary pairs with
A. Thymine   B. Uracil
C. Cytosine   D. Guanine
(PCE/2002/50/TYPES: E)

68. A pure stock of red and white flowered pea plants is crossed. If red colour is dominant over white, what will be the phenotype of flowers of the first filial generation?
A. 75% red and 25%, white
B. 59%, white and 50%.red
C. 100% red
D. 100% white   (PCE/2008/44/TYPES: N)

69. Gene flow indicates the
A. interbreeding is possible between members of different species
B. interbreeding is possible between of the same species
C. some genes may spread more than others.
D. random changes alter genetic information (PCE/2001/50/TYPES: 4)

70. A gene which can only express its characters in a homozygous condition is said to be
A. recessive
B. dominant
C. co-dominant
D. segregated (PCE/2000/42/TYPES: A)

71. What is the probability that two consecutive pregnancies in a newly wedding couple will result in male issues?
A. 1
B. ¾
C. ½
D. ¼ (PCE/2000/43/TYPES: A)

72. The genetic constitution of the children whose parents are both sickle cell trait carriers
A. AA, AS, AS, SS
B. AA, AS, AA, AS
C. AS, SS, SS, AS
D. AS, AA, SS, SS   (PCE/98/42/TYPES: C)

73. The number of chromosomes in the nucleus of a fertilized human ovum is
A. 23
B. 26
C. 46
D. 92 (PCE/97/42)

74. The abnormality that is caused when the body cell has more than the normal number of chromosomes is
A. mongolism
B. mutation
C. cretinism
D. haemophilia (PCE/97/4)

75. A gene which expresses itself only in the homozygous conditions is
A. recessive
B. an alleles
C. dominant
D. a mutant   (PCE/06/19/TYPE: N)

76. Genetic traits that are concerned primarily with forms and structures are said to be
A. morphological
B. ecological
C. phenotypical
D. physiological (PCE/06/19/TYPE: A)

77. A pure-breed black rabbit is crossed with a pure-breed white rabbit and all the F$_1$ offspring have black hair. The F$_1$ offspring are
A. hybrids
B. zygote
C. autosomes
D. homozygous (PCE/06/22/TYPE: A)

CHAPTER 24

SEX DETERMINATION AND SEX-LINKED CHARACTERS

SEX DETERMINATION

SEX is determined by the sex chromosomes (X and Y).They are called heterosomes or sex chromosomes, while the other 22 pairs are called autosomes.In humans the sex chromosomes are homologous in females (XX) and non-homologous in males (XY).Using monohybrid-cross to show that there will always be a 1:1 ratio of males to females.Sex is determined solely by the sperm.There are techniques for separating X and Y sperm, and this is used for planned sex determination in farm animals using In-Vitro Fertilization (IVF).

SEX-LINKED CHARACTERS

Characters controlled by genes located on the X chromosomes are usually described as sex-linked characters. The male has only one gene for any sex-linked character. Whatever genes, dominant or recessive, that are on the X chromosome in the male are therefore expressed in the phenotype.
Examples of sex-linked characters in man are:
1. **Baldness**: lack of hair on the head.
2. **Haemophilia**: a disease in which the blood takes an abnormally long time to clot thereby resulting in excessive bleeding even in a small cut;
3. **Colour blindness**: inability to recognize some colours usually red and green.
4. **Muscular dystrophy**: gradual weakening of the muscles.

The genes for these characters are recessive.
A female has two genes for each sex-linked trait, one on each of the sex chromosomes. If she is homozygous, the trait will be expressed. If she is heterozygous, she will be normal but she carries the abnormal trait involved. A female who is heterozygous for a recessive sex-linked trait is known as a carrier.

CHROMOSOME ABNORMALITY

A **chromosome anomaly**, **abnormality** or **aberration** reflects an atypical number of chromosomes or a structural abnormality in one or more chromosomes. A Karyotype refers to a full set of chromosomes from an individual which can be compared to a "normal" karyotype for the species via genetic testing. A chromosome anomaly may be detected or confirmed in this manner. Chromosome anomalies usually occur when there is an error in cell division following meiosis or mitosis. There are many types of chromosome anomalies. They can be organized into two basic groups, **numerical** (alteration in chromosome number) and **structural** (alteration in chromosome structure)anomalies.

ALTERATIONS IN CHROMOSOME NUMBER:

Nondisjunction occurs when either *homologues fail to separate* during anaphase I of meiosis, or *sister chromatids fail to separate* during anaphase II. The result is that one gamete has 2 copies of one chromosome and the other has no copy of that chromosome. (The other chromosomes are distributed normally).
If either of these gametes unites with another during fertilization, the result is **aneuploidy** (abnormal chromosome number)
A **trisomic** cell has one extra chromosome (2n +1) = example: trisomy 21. (**Polyploidy** refers to the condition of having **three** homologous chromosomes rather then two)
A **monosomic** cell has one missing chromosome (2n - 1) = usually lethal except for one known in humans: Turner's syndrome (monosomy XO).
The frequency of nondisjunction is quite high in humans, but the results are usually so devastating to the growing zygote that miscarriage occurs very early in the pregnancy. If the individual survives, he or she usually has a set of symptoms - a **syndrome** - caused by the abnormal dose of each gene product from that chromosome.

1. Human disorders due to chromosome alterations in autosomes (Chromosomes 1-22). There only 3 trisomies that result in a baby that can survive for a time after birth; the others are too devastating and the baby usually dies in utero.
A. Down syndrome(trisomy 21)
B. Patau syndrome (trisomy 13)
C. Edward's syndrome (trisomy 18)

2. Nondisjunction of the sex chromosomes (X or Y chromosome): Can be fatal, but many people have these karyotypes and are just fine!
a. Klinefelter syndrome: 47, XXY males.
b.47, XYY males.
c. Trisomy X: 47, XXX females.
d.Monosomy OX (Turner's syndrome)

ALTERATIONS IN CHROMOSOME STRUCTURE:
Sometimes, chromosomes **break**, leading to 4 types of changes in chromosome structure:
1. Deletion
2. Duplication
3.Translocation
4. Inversions.

1. UME/80/42
If a baby is a female, her mother's ovum must have been fertilized by a sperm carrying the chromosome
A. X
B. XY
C. YY
D. Y

2. UME/82/10
A married couple have 20 children and they are all girls, which of the following is the CORRECT explanation?
A. the woman is incapable of producing male children
B. the man's sperms are very weak
C. the man is not athletic enough
D. the Y component of the man's sex chromosomes was always involved
E. the X component of the man's chromosomes was always involved.

3. UME/87/44
In which of the following crosses will all the female offspring be colour blind?
A. colour blind mother X colour blind father
B. colour blind mother X normal father
C. carrier mother X colour blind father
D. carrier mother X normal father

4. UME/91/40
Which of the following characteristics is NOT sex-linked?
A. river blindness
B. baldness
C. haemophilia
D. colour blindness

5. UME/92/40
Which of the following is true of the children of a haemophilic man who marries a woman that is not haemophilic and does not carry the trait?
A. all their sons will be haemophilic
B. all their daughters will be haemophilic
C. all their daughters will be carriers
D. all their sons will be carriers.

6. UME/92/45
Women do not suffer from colour blindness because
A. the trait is sex-linked
B. only men are colour blind
C. the genes are recessive and sex-linked
D. the genes occur on both the X and Y chromosomes

7. UME/95/45
All the sons of a colour-blind woman will be colour blind regardless of the state of the father because
A. the egg determines the phenotype of the son
B. sons inherit the sex chromosomes of the mother
C. the father's sex chromosome is weaker in son
D. sex-linked traits express dominance in females.

8. UME/2004/24/TYPE:4
Sex-linked genes are located on
A. X- and Y- chromosomes
B. homologous chromosomes
C. X-chromosome
D. Y-chromosome

9. PCE/99/40/TYPE:E
A male child cannot inherit any of his father's sex-linked traits because
A. Y chromosomes is usually genetically empty
B. the male child cannot receive Y-chromosome from his father
C. X-chromosome is dominant over Y-chromosome
D. Y-chromosome is dominant over X-chromosomes from his father.

10. PCE/2003/5/TYPE: N
All the sons of a colour-blind woman will be colour-blind regardless of the their father because
A. sex-linked traits express dominance in females
B. the father's sex chromosome is weaker than that of the mother
C. the egg determine the phenotype of the sons
D. the sons inherit the gene for colour-blindness from their mother.

11. PCE/97/45
The pair of colours that cannot be distinguished by a person with traits of colour blindness is
A. white and blue
B. yellow and green
C. red and blue
D. red and green

12. UME/2002/47/TYPE: Y
If X^N is the dominant alleles for normal vision and X^n the recessive allele for colour-blindness, a boy with the genotype $X^N Y$ will
A. have normal vision
B. colour blind
C. be totally blind
D. be a carrier of colour-blindness

13. UME/2003/38/TYPE: A
If the pair of alleles for baldness is given as Bb, a female carrier will be denoted by
A. X^BX^b
B. X^BX^B
C. X^bY
D. X^BY

14. UME/2001/39/TYPE T
The probability of a baby being a boy or a girl depends on the contribution of the
A. father's sex cell
B. father's somatic chromosomes
C. mother's sex cell
D. mother's X-chromosomes

15. PCE/99/41/TYPE: E
In man, sex determination is based on the XY-chromosome system. An individual that carries only one X-chromosome with no allele (XO) will be a
A. sterile female
B. normal male
C. sterile male
D. normal female

16. UME/2007/50
An example of a sex-linked trait is the
A. ability to grow long hair in females
B. colour of the skin in humans
C. ability to roll the tongue
D. possession of facial hair in adult human

17. UME/2010/48/TYPE: D
At what stage in the life history of a mammal is the sex of an individual set?
A. at puberty
B. at birth
C. at conception
D. at adolescence

18. PCE/04/46/TYPE: 9
In humans, baldness is peculiar to the male because the trait is
A. carried by the x-chromosomes
B. carried by the y-chromosomes
C. Heterogametic
D. homogametic

19. PCE/07/47/TYPE: A
The result of mating between a colour-blind male (X^cY) and a normal female (X^cX^c) is
A. 3 carrier females   :   1 normal male
B. 1 normal female   :   3 carrier males
C. 2 normal female   :   2 normal males
D. 2 carrier females   :   2 normal males

20. PCE/07/43
Which of the following is responsible for contributing the chromosome that determines the sex of an unborn offspring?
A. The father
B. The mother
C. The grandfather
D. The grandmother

MISCELLANEOUS QUESTIONS (24)
21. Which of the following characteristics is NOT sex-linked?
A. albinism
B. baldness
C. haemophilia
D. colour blindness (UME/97/45)

22. Examples of water-borne and sex-linked diseases are
A. taeniasis and malaria
B. cholera and gonorrhoae
C. typhoid and syphilis
D. dracunculiasis and haemophilia (UME/2000/46/TYPE:M)

23. Colour blindness is a defect which occurs
A. more often in women than men
B. more often in men than women
C. only in aged people
D. more in young adolescents (PCE/96/45)

24. The pair of colours that cannot be distinguished by a person with traits of colour blindness is
A. white and blue
B. yellow and green
C. red and blue
D. red and green   (PCE/97/45)

25. Characters expressed by a gene located on only X-chromosomes are said to be
A. recessive
B. dominant
C. sex-linked
D. alleles (PCE/98/45/TYPE:C)

26. In man, sex determination is based on the XY-chromosome system. An individual that carries only one X-chromosome with no allele (XO) will be a
A. sterile female
B. normal female
C. sterile male
D. normal female (PCE/97/41/TYPE:E)

CHAPTER 25

ADAPTATION FOR SURVIVAL

Understand the following after reading this chapter:

> Structural adaptation
> Adaptive colouration and its function
> Behavioural adaptation in social animals

COMPETITION

Competition is the interaction between organisms in a community in order to obtain mutually required resources. There are two types:

Intra Specific Competition: This occurs between members of a population.

Inter Specific Competition: This occurs between members of different population.

ADAPTATIONS FOR SURVIVAL

There are three major types of adaptation viz; structural adaptation, adaptative colouration and behavioural adaptation.

Structural adaptation for protection and defense:

1. Stinging: This is used to inflict pains and ward off enemies by some animals e.g. bees, ants, scorpions.
2. Shells: These serve as hard protective cover for some animals e.g. snails and tortoise.
3. Claws: These are the pointed nails of some animals used for scratching and defence e.g. birds.

Structural adaptations for obtaining food:

1. Modification of mouth parts e.g. beaks for feeding in birds.
2. Trapdoor used by bladderwort to obtain food.

Structural adaptation for securing mates

1. Red head of male lizards to attract the female
2. Beautiful tail feathers of some birds e.g. turkey, peacock
3. Beautiful colouration of some insects e.g. butterfly.

Structural adaptation for water conservation in:

Plants

1. Possession of small leaves or spines e.g.Acacia
2. Possession of thick cuticle over the leaves e.g. citrus
3. Absence of leaves e.g. cactus
4. Possession of sunken stomata e.g. Oleanders
5. Presence of water storage tissues e.g. cactus.

Adaptations to dry habitats in:

Plants

Plants in different habitats are adapted to cope with different problems of water availability.

Mesophytes plants adapted to a habitat with adequate water

Xerophytes plants adapted to a dry habitat

Halophytes plants adapted to a salty habitat

Hydrophytes plants adapted to a freshwater habitat

Animals

1. Possession of exoskeleton e.g. chitin in arthropods
2. Possession of scales e.g. reptiles
3. Possession of shell e.g. snails
4. Possession of hairs e.g. Mammals

2. ADAPTIVE COLOURATION

1. **Bright colours** in insect pollinated flowers.
2. **Cryptic (concealing /camouflage)**

Protective colouration in some animals to deceive enemies or predators by blending their backgrong and be less visible. The different types are:

a. **Colour change** to blend with the surrounding e.g chameleon

b. **Disruptive colouration**: Possession of spots and stripes to break up the familiar shapes of their bodies , enabling them to go undetected e.g. leopard, tiger, giraffe.

c. **Counter shading**: Possession of lighter colour on the ventral surface and a darker colour on the dorsal surface e.g. fish

3. **Warning or aposematic colouration**: It consists of bright and conspicuous colours possessed by some animals advertising themselves as poisonous and distasteful.

3. BEHAVIOURAL ADAPTATIONS

Behavioural adaptations of predators

1. Hide and wait position
2. Chasing and overtaking the prey with greater speed
3. Setting up traps to catch the preyse.g spider

Behavioural Adaptations for protection from Predators

1. Swaying in the air
2. Feigning death
3. Secretion of offensive and repulsive odour
4. Retraction into a burrow when threatened by enemies
5. Escaping by running or flying away from enemies

Behavioural Adaptations against unfavourable Climatic Condition

1. **Dormancy**: Inactive state of an organism to withstand unfavourable climatic condition.
2. **Hibernation**: Inactive state or deep sleep in some animals during cold season (winter) e.g.   bat, fox.
3. **Aestivation**: Inactive state in some animals during hot weather (drought) e.g. snail, frog
4. **Migration**: Outright relocation to more suitable environment

Shape of Bird Foot	Type of Bird Foot	Adaptation and Lifestyle
	Climbing	Feet like these help birds, like woodpeckers, climb trees. Notice the sharp nails for digging into the wood, and the back toes so that the bird doesn't topple backward.
	Swimming	Webbed feet help birds, like ducks, paddle through the water more efficiently.
	Running	For running quickly, birds like emus, often have three toes, all of which face forward.
	Perching	Feet with four toes, one of which is in the back, are useful for perching on tree branches. Birds, like blue jays, wrap their toes around the branch to help balance.
	Grasping	Predatory birds, like hawks, have clawlike feet called talons for grabbing their prey.
	Scratching	Chickens, and other birds that scratch in the dirt for insects, usually have feet with four toes, all of which have strong nails for digging into the ground.

Feet of birds and their adaptation

SOCIAL ANIMALS BEES

The bee caste system include the:

1. Queen: This is the fertile female which lays eggs

2. Drones: These are the fertile males which mate with queen

3. Workers: These are the sterile females which perform several functions such as feeding older larvae with honey and pollen, and foraging. However, their functions varies with age and needs of the colony

TERMITES

The termite caste is composed of the;

1. Reproductive: Which consist of the king (male) queen (female) and winged reproductive whose function is mainly to reproduce

2. Workers: These perform among other functions searching for food, feeding other members of the colony, constructing and repairing the nest.

3. Soldiers: These consist of the nasute and mandibulate soldiers whose function is to defend and protect the colony

Economic importance of termites

They destroy wooden materials and products.

They attack some crop plants e.g. rubber trees, oil palm.

They help the process of decay and recycling of nutrients.

They aerate the soil by burrowing thereby increasing soil fertility. They can serve as food. Their termitaria (ant-hill) are used for surfacing tennis courts.

Revision Questions

1. UME/2001/47/TYPE: T

The ability of an organism to live successfully in an environment is known as

A. resistance

B. competition

C. succession

D. adaptation

2. UME/2001/50/TYPE: T.

For heterotrophic organisms, competition is least caused by the inadequacy of

A. mates

B. space

C. light

D. nutrients

3. PCE/92/45

Which of these processes will occur when a thousand maize plants are planted on a 1m plot?

A. association

B. inter-specific competition

C. succession

D. intra-specific competition

4. PCE/98/46/TYPE: C

The competition for an available resource by different species in an area leads to

A. primary succession

B. intra-specific competition

C. secondary succession

D. inter-specific competition

5. UME/2004/36/TYPE: 4

Which of the following is an example of intra-specific competition?

A. yam and potato shoots growing out through the same window

B. a lizard and an ant-eater chasing an insect

C. a worker termite and a soldier in a limited space

D. a hawk and an eagle targeting the same chicken

6. UME/2007/13

Which of the following factors can bring about competition in a population?

A. dispersion

B. emigration

C. drought

D. mortality

7. UME/2008/45

A major adaptive feature of endoparasites is the

A. loss of the organ of movement

B. presence of claws

C. loss of the central nervous system

D. presence of piercing mouthparts

8. PCE/07/44

Competition may occur in a rice field if

A. the environmental temperature increase

B. the field become weedy

C. nematodes are absent

D. nematodes are present

9. PCE/07/45

Communities at early successional stages are often characterized by

A. high competitive ability

B. high biological diversity

C. low productivity

D. high productivity

25.1. STRUCTURAL ADAPTATION FOR:

25.1.1 OBTAINING FOOD

10. UME/91/46

Which of these is NOT an adaptive features for arboreal life?

A. possession of a long tail

B. possession of claws

C. possession of teeth

D. counter shading in coat colour

11. UME/95/46
The slender, long and slightly curved beak of the sun-bird is an adaptation for feeding on
A. nectar
B. small seeds
C. big seeds
D. insects

12. UME/98/44
The group of insects that have mouth parts adapted for both piercing and sucking is
A. cockroaches, aphids and mosquitoes
B. aphids, houseflies and moths
C. mosquitoes, tse-tse flies and aphids
D. aphids, beetles and grasshoppers.

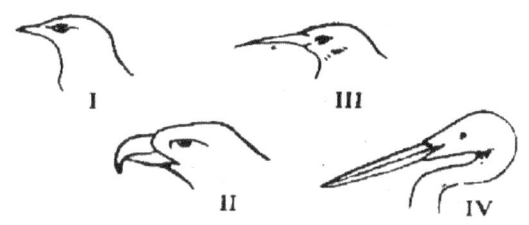

13. UME/98/45
The bird's adapted for fishing is labeled
A. I   B. II   C. III   D. IV

14. UME/98/46
Toes of the feet ending in a sharp curved hook suitable for holding and tearing are most likely to belong to the bird with the bill in
A. I   B. II   C. III   D. IV

15. UME/2002/27/TYPE: Y
Birds which are large with long straight pointed beaks, longnecks and legs are likely to be
A. insect eaters   B. fish catchers
C. nectar feeders   D. fruit eaters

16. UME/2003/48/TYPE: A
An insect with a mandibulate mouth part will obtain its food by
A. chewing   B. chewing and sucking
C. sucking   D. biting and chewing.

17. PCE/97/46
Floating water plants are adapted to their environment because they
A. have no cuticle
B. have plenty of air spaces in the roots and leaves
C. do not require dissolved air
D. have their root embedded in the mud

25.1.2. PROTECTION AND DEFENCE
18. UME/97/47
What combination of characteristics should a prey develop to survive in the environment of its predator?
A. Camouflage, well developed limbs and effective vision
B. Showy colours, big body and well developed limbs
C. Camouflage, big body and effective vision
D. Showy colours, well developed muscles and an acute sense of smell

19. UME/2000/34/TYPE: M
Which of the following are adaptations of animals to aquatic habitats?
A. Gills, streamlined bodies and lateral line
B. Lateral line, streamlined bodies and lungs
C. Gills, scaly skin and lungs
D. Gills, streamlined bodies and spiracles

20. UME/2000/35/TYPE: M
Which of the following is an adaptation of forest species?
A. few stomach   B. thick bark
C. buttress roots   D. reduced leave

21. UME/2003/46/TYPE: A
Spines and shells on animals are adaptations for
A. Physical defence   B. Camouflage
C. Chemical defence   D. Mimicry

22. UME/2009/36
The most important characteristic that makes reptiles to conquer terrestrial habitats is the possession of
A. long tail   B. scaly skin
C. sharp claws   D. amniotic egg

23. UME/2008/48
The ability of a chameleon to change its colour are adaptive feature for
A. Attraction   B. defence
C. Display   D. Attack

24. PCE/97/47
Most fishes that live near the surface of water are protected from predators because they are
A. dark on the ventral surface
B. silvery on the dorsal surface
C. silvery on the ventral surface
D. stripped and greenish on the sides

25.1.3.   SECURING MATES FOR REPRODUCTION
25. UME/94/48
Red coloration on the head of a male lizard helps it to
A. mark its territory
B. camouflage in the environment
C. secure its mate
D. defend itself

26. PCE/94/49
The structural adaptation for securing mates for reproduction is best exhibited by
A. snakes   B. newts
C. lizard   D. moths

27. PCE/03/1/TYPED: N
The act of fetching food by a cock to feed a hen is a display of
A. competition   B. mutualism
C. courtship   D. territoriality

25.1.4. REGULATING TEMPERATURE AND LOSS OF WATER
28. UME/92/47
An example of plant adaptation to a xerophytic environment is represented by the development of
A. fleshy tissues and reduced leaves
B. broad canopy and extensive surface
C. thick barks and broad leaves
D. rough leaves and shallow root system

29.   UME/95/47
Scales on reptiles are a feature for
A. Conserving water   B. Conserving food
C. Protecting the skin   D. Locomotion

30. UME/97/46
Water loss is regulated in plants and animals by both the
A. Scales and the skin
B. Thick leaves and the feathers
C. Thick leaves and the feathers
D. Leathery cuticle and the feathers

31.   UME/98/48
In the whistling pine, leaves are reduced to brown scales and young stems are green. This is an adaptation for
A. obtaining food   B. conserving materials
C. storing water   D. reducing transpiration

32. UME/2000/50TYPE: M
Which of the following structural features are adapted for uses other than water conservation?
A. Succulent stems   B. Scales in animals
C. Spines in plants   D. Feathers in birds

33. UME/2001/48/TYPE: T
The most important adaptation for xerophytes is the ability of the protoplasm to
A. Resist being damaged by loss of water
B. Store sugar and minerals in the vacuoles
C. Absorb water and swell
D. Shrink from the cell wall.

34. UME/2003/45/TYPE: A
The presence of sunken stomata and the folding of leaves are adaptations to
A. Prevent entry of pathogens
B. Prevent guttation
C. Remove excess water
D. Reduce water loss

35.   PCE/96/46
Which of the following set of structures is good at temperature regulation in vertebrates?
A. skins, scales and feathers
B. feathers, carapace and skin
C. hairs, skin and feathers.
D. scales, hairs and feathers.

36.   PCE/99/34/TYPE: E
One of the adaptations shown by hydrophytes is the
A. Lack of special absorbing and conducting tissues
B. Development of thick mechanical tissues
C. Possession of stomata on the under surface of the leaf
D. Development of thick cuticles

37. UME/2007/16
An example of a fish that aestivates is
A. shark   B. croaker
C. lung fish   D. cat fish

38. UME/2008/33
Stomata of some plant are sunken and protected by hairs. These are features of
A. mesophytes   B. epiphytes
C. hydrophytes   D.xerophytes

39. UME/2009/47
Which of the following animals is most adapted for water conservation?
A. earthworms   B. mammals
C. flatworms   D. insects

40. PCE/97/48
Thermal gapping is a common phenomenon in
A. lions   B. rabbits
C. birds   D. alligators

25.2. ADAPTATIVE COLOURATION AND ITS FUNCTIONS
41. UME/2003/49/TYPE:
An example of cryptic colouration is the
A. Bright marks on a poisonous tropical frog on variegated leaves
B. Bright colour of an insect pollinated flower
C. Mottled colours on moths that rest on lichens
D. Green colour of a plant.

42. UME/2001/49/TYPE: T
A green snake in green grass is able to escape notice from predators because of its.
A. disruptive colouration
B. counter shading
C. warning colouration
D. cryptic colouration

43. UME/2004/37/TYPE: 4
The spots and stripes of the leopard and tiger are examples of
A. warning colouration
B. counter shading
C. cryptic colouration
D. disruptive colouration

44.   UME/95/48
The colour of the ventral surface of a fish is lighter than that of the dorsal. This is mainly
A. an adaptation for movement
B. an adaptation for camouflage
C. for attracting mates
D. for regulating body temperature

45. UME/99/48/TYPE: D
The very bright colours in some types of Mushroom
A. are a warning that they may be poisonous
B. indicates that they are very tasty
C. attract potential transporters of their spores
D. perform the same function as bright colours in flower

46. PCE/2001/49/TYPE: T
The conspicuous colouration of flowers in plants is to aid
A. Fertilization   B. Seed formation
C. Pollination   D. Seed dispersal

47. UME/2009/48
The specialized pigment cells that are involved in colouration and colour change in animals are the
A. xanthophyll   B. chematophores
C. chlorophyll   D. melanin

25.3. BEHAVIOURAL ADAPTATION IN SOCIAL ANIMALS
48. PCE/2002/23/TYPE: 5
The chemical produced by ants for communication are the
A. Hormones   B. Enzymes
C. Vitamins   D. Pheromones

49. UME/98/47
In the honey bee colony, the drones are
A. sterile males with reduced mouth parts
B. fertile males with reduced mouth parts
C. sterile males with well-developed mouth parts
D. fertile males with well-developed mouth parts

50. UME/2001/46/TYPE: T
The role of male honey bee is to
A. clean the hive   B. ventilate the hive
C. mate with the queen   D. care for the young

51.UME/2002/24/TYPE: Y
A feature of the caste systems of bees and termites is that
A. the workers are sterile
B. the kings are bigger than the queens
C. only the workers perform duties
D. nuptial flight is performed by all members

52.   PCE/92/50
Which of the following should be killed in order to destroy a termite colony?
A. The king   B. The queen
C. The workers   D. The soldiers

53. UME/2004/48/TYPE: 4
The association between termites and the cellulose-digesting protozoan in their guts is an example of
A. Mutualism   B. Saprophytism
C. Commensalism   D. Parasitism

54. PCE/2000/48/TYPE: A
The caste of the termites that have protozoans living as sybionts in their guts are the
A. Soldiers   B. Queens
C. Reproductive   D. Workers

55. PCE/97/49
Workers in a bee hive perform a waggle dance to indicate that food is
A. less than 100m away   B. more than 100m away
C. Not available   D. Abundant

56. UME/97/47/TYPE: D
The loud cry made by a broading hen when a predator is around is meant to
A. alert the poultry attendants
B. attract cocks to come and fight
C. advertise the boundaries of its territory to intruders
D. warn chicks and other chickens of impending danger

57. UME/2000/49/TYPE: M
Complex social behaviour and organization are found mostly in
A. Insects   B. Birds
C. Reptiles   D. Mammals

58.UME/2003/47/TYPE: A
The inactive state exhibited by an animal during hot dry seasons is termed
A. Aestivation   B. Dormancy
C. Resting   D. Hibernation

59. UME/2009/49
During the dry season in the tropics, the body metabolism of same animals slows to a minimal level in a process referred to as
A. hibernation   B. aestivation
C. dormancy   D. senescence

60. UME/2010/34/TYPE: D
I. Adoption of appropriate nocturnal habits
II. Burrowing
III. Adjusting their internal body temperature
IV. Possession of many sweat pores
Which of the above are ways in which desert animals adapt to extreme heat of the environment?
A. II and III only   B. I and II only
C. I, II and III only   D. I and IV only

MISCELLANEOUS QUESTIONS (25)
61. Any feature which improves the chances of an organism surviving in its surrounding is called?
A. Speciation   B. Adaptation
C. Ecology   D. Predation   (PCE/94/48)

62. Among the vertebrates, mammals are generally the most successful because they
A. are capable of internal fertilization
B. have a well-developed brain
C. are warm-blooded
D. have a well-developed heart
(PCE/2005/49/TYPE: D)

63. When an animal has a dark-coloured dorsal surface and a light-coloured ventral surface this is an adaptation called?
A. Concealment colouration   B. Countershading
C. Colour blending   D. Disruptive colouration
(UME/2005/48/TYPE: D)

64. Long, sharp curved claws for catching and gripping prey are found in
A. Ducks   B. Domestic fowls
C. Hawks   D. Weaver-birds   (PCE/95/48)

65. An organism which uses 'shamming death' as a method of self preservation against its predator is the
A. Ladybird beetle   B. Praying mantis
C. Chameleon   D. Stick insect   (PCE/96/48)

66. The purpose of aestivation among animals is to?
A. Conserve water and food
B. Reproduce in large numbers
C. Survive unfavourable conditions
D. Hide from enemies   (PCE/98/48/TYPE: C)

67. One of the adaptations shown by hydrophytes is the
A. Lack of special absorbing and conducting tissues
B. Development of thick mechanical tissues
C. Possession of stomata on the under surface of the leaf
D. Development of thick cuticles
(PCE/99/34/TYPE: E)

68. Disruptive colouration is demonstrated in
A. Spots and stripes of a giraffe
B. Mixed colours on goats and sheep
C. Changing colours of chameleon
D. Counter-shading in fishes
(PCE/2000/74/TYPE: A)

69. The caste of termites that have protozoans living as symbionts in their guts are the
A. Soldiers   B. Queens
C. Reproductives   D. Workers
(PCE/2000/48/TYPE: A)

70. Intraspecfic competition is more severe than interspecific because of
A. Identical needs   B. Predation
C. Unidentical needs   D. Co-existence
(PCE/2002/24/TYPE: A)

71. Which of the following best illustrates behavioural adaptation?
A. Flocking   B. Camouflage
C. Defence   D. Countershading
(PCE/2004/49/TYPE: 9)

72. A dance by a worker bee indicating that food is close to the hive is called?
A. waggle dance   B. figure of eight dance
C. round dance   D. the straight run
(PCE/2001/48/TYPE: 4)

73. The long limbs of antelopes are an adaptation for
A. obtaining food   B. escape and defence
C. scaring other animals   D. structural adaptation
(PCE/03/2/TYPE: N)

Use the diagram below to answer questions 74 and 75

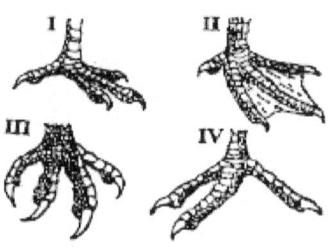

74. An animal with the type of foot represented II is
A. Pigeon   B. Hen
C. Duck   D. Vulture   (PCE 2001/46/Type 4)

75. An adaptation of the foot for catching prey is shown in
A. IV   B. III
C. II   D. I   (PCE 2001/47/ Type 4)

CHAPTER 26

EVOLUTION

Understand the following after reading this chapter:

Theories of evolution
Evidences of evolution
Patterns of evolution
Forms of evolution

Evolution (or more specifically **biological** or **organic evolution**)is the change over time in one or more inherited traits found in populations of individuals.

THEORIES OF EVOLUTION

Lamarck's Theory (Lamarckism)
Larmarck proposed that individuals lose characteris- tics that they do not use, and develop further on characteristics that they use a lot (**use and disuse**). And that these change would be inherited (**inheritance of acquired traits**). For example, an ath- lete would have stronger kids or that giraffesstret- ching for branches generation after generation would lead to them having longer necks.

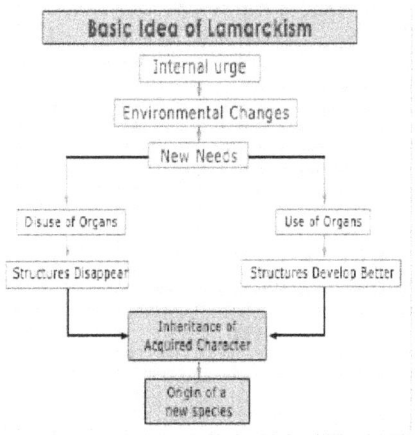

Darwin's Theory (Darwinism) or Theory of Natural Selection
In Darwinian evolution, genetically programmed traits which increase organisms' chances of surviving, do survive more, and spread through the population. Natural selection states that: 1. More organisms are produced than can survive because of limited resources 2. Struggle for existence   3. Variation in traits, some of which are heritable 4. Some traits are better adapted than others 5. Survival of the fittest.

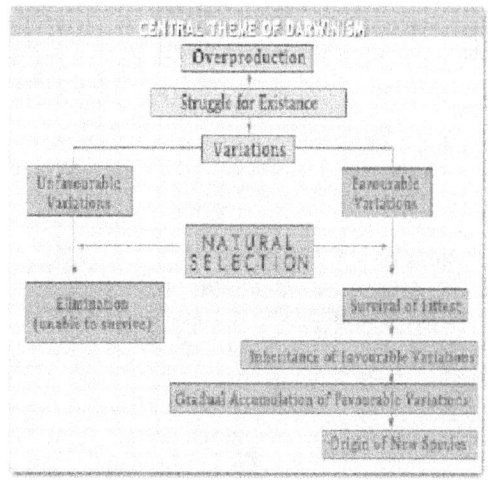

EVIDENCES OF EVOLUTION

Fossil records: Fossils are remains of organisms or their impressions preserved in earth's sedimentary rocks. Their ages can be determined by radioactive dating. The study of fossil is called palaeontology.

Biochemical similarities: the chemical make-up and biochemical process of all organisms are similar e.g. all organisms use energy in the form of ATP.

Comparative anatomy: body parts of organisms show structural similarity, progressive complexity and modifications e.g. all organisms have a cellular structure, all vertebrates show a basic body plan, there is a progressive complexity among vertebrates in the number of heart chambers. Vestigial parts like the appendix in man which functions as caecum in herbivores also show evolution.

Embryology: during early development, the embryo of an organism resembles that of its ancestors.

DIFFERENT PATTERNS OF EVOLUTION
Groups of species undergo various kinds of natural selection and, over time, may engage in several patterns of evolution: convergent evolution, divergent evolution, parallel evolution, and coevolution.

Convergent evolution
Convergent evolution is the process in which species that are not closely related to each other independently evolve similar kinds of traits. For example, dragonflies, hawks, and bats all have wings. None of these organisms owes its wings to genes inherited from any of the others. Each kind of wing evolved independently, suggesting that the trait of flight is a useful one for the purpose of survival and reproduction. These independently evolved wings are

called *analogous structures.*

Divergent evolution

Divergent evolution is the process in which a trait held by a common ancestor evolves into different variations over time. A common example of divergent evolution is the vertebrate limb. Whale flippers, frog forelimbs, and your own arms most likely evolved from the front flippers of an ancient jawless fish. Because they share a common evolutionary origin, these are examples of *homologous structures*.

An important consequence of divergent evolution is *speciation*, the divergence of one species into two or more descendant species.

Parallel evolution

Parallel evolution is sometimes difficult to distinguish from convergent evolution. Parallel evolution occurs when different species start with similar ancestral origins, and then evolve similar traits over time. This kind of thing happens because the two different species, though they do not necessarily share a common ancestor, experience similar kinds of environmental pressures and survive only by undergoing similar adaptations. A classic example of parallel evolution is found among plants, in which several similar but distinct forms of leaf evolved in parallel and are evident today.

Coevolution

Coevolution occurs when closely interacting species exert selective pressures on each other, so that they evolve together in a kind of conversation of adaptations. Examples of coevolution are common among predator-prey and host-parasite pairs. More picturesque examples of coevolution occur among hummingbirds and the flowers from which they seek nectar and unwittingly pollinate.

FORMS OF EVOLUTION

There are five main forms of evolution: natural selection, mutation, genetic drift, gene flow, and non-random mating.

Natural selection is the process by which genetic mutations that enhance reproduction become and remain, more common in successive generations of a population.

Mutations are changes in the DNA sequence of a cell's genome, which can be caused by radiation, viruses transposons and, mutagenic chemicals as well as errors that occur during meiosis or DNA replication. Mutation theory of evolution was proposed by Hugo de Vries in 1901.

Genetic drift refers to evolution occurring through random changes in allele frequency over time. It depends on chance alone.

Gene flow refers to the recombination and spread of genes in a population as they interbreed. Gene flow occurs in a non-random manner.

Nonrandom mating occurs when the probability that two individuals in a population will mate is not the same for all possible pairs of individuals. When the probability is the same, then individuals are just as likely to mate with distant relatives as with close relatives — this is random mating. Nonrandom mating can take two forms:

Inbreeding - individuals are more likely to mate with close relatives (e.g. their neighbors) than with distant relatives. This is common.

Out breeding - individuals are more likely to mate with distant relatives than with close relatives. This is less common.

Revision Questions

6.1. THEORIES OF EVOLUTION

1. UME/94/41

The differences and similarities among living things account for:
A. Diversity   B. Stability
C. Competition   D. Evolution

2. UME/92/49

The theory of natural selection was developed by
A. Lamarck and Darwin   B. Wallace and Mendel
C. Darwin and Wallace   D. Mendel and Lamarck

3. UME/2002/25/TYPE: Y

In his theory of evolution, Darwin implied that
A. The struggle for existence among living organisms is sporadic
B. The most successful organisms are those that best adapt to their environment
C. Organs of the body which are not regularly used by an organism will disappear
D. Any traits acquired by an organism during its lifetime can be passed on to its offspring.

4. UME/98/49

The best explanation for the theories of natural selection is that
A. All organisms have equal capacity for survival in their habitats.
B. Organisms have varying capacities for survival in their habitats.
C. Organisms compete for resources and better competitors survive and thrive.
D. Habitats allow only organisms that will not have to compete for survival.

5. UME/99/49/ TYPE: D

Which of the following is one of Lamarck's theories?
A. Some variations are more favourable to existence in a given environment than others.
B. All living organisms are constantly involved in a struggle for existence
C. The size of a given population remains fairly constant.
D. New species originate through the inheritance of acquired traits.

6. UME/2003/50/TYPE: A
An argument against Lamarck's theory of evolution is that
A. Acquired traits cannot be passed onto the offspring.
B. Disuse of body part cannot weaken the part.
C. Disused part is dropped off in the offspring
D. Traits cannot be acquired through constant use of body parts.

7. UME/97/50
Long neck in giraffe is used to illustrate the theory of
A. Use and disuses   B. Origin of species
C. Origin of life   D. Natural selection

8. PCE/97/50
The cutting of the tails of mice in successive generation by Wallace was used to disprove a theory postulated by
A. Darwin   B. De Vries
C. Mendel   D. Lamarck

9 .UME/2000/47/TYPE:M
The mutation theory of origin of evolution was propounded by
A. Gregory Mendel   B. Hugo de Vries
C. Jean Lamarck   D. Charles Darwin.

10. UME/98/50
The basic point of impact by changes which produces mutation is the
A. Gametes   B. Chromosomes
C. Phenotype   D. Zygote.

11. UME/2008/50
The natural process that produces adaptive evolutionary changes is
A. mutation   B. gene flow
C. genetic draft   D. natural selection

12. UME/2009/50
According to Darwin, the driving force behind evolutionary change is
A. natural selection   B. genetic drift
C. mutation   D. gene flow

13. PCE/06/38/TYPE: A
According to Charles Darwin, the main cause of natural selection is
A. gene flow   B. genetic drift
C. physiological pressure   D. environmental pressure

26.2. EVIDENCE OF EVOLUTION
14. UME/99/50/TYPE: D
From an evolutional standpoint, the older a fossil-bearing rock is the more likely to contain
A. Aves as opposed to amphibians
B. Invertebrates as opposed to vertebrates
C. Angiosperms as opposed to algae.
D. Vertebrates as opposed to invertebrates

15. UME/94/50
The anatomical evidence usually used in support of the evolutionary among whales, humans, birds, and dogs is the possession of
A. thick skin   B. pentadactyl limb
C. tail   D. epidermal structure

16. UME/93/50
Which of the following organisms has lost the pentadactyl limb structure?
A. Bat   B. Fish
C. Frog   D. Pigeon.

17. UME/95/49
The least evidence of the theory evolution is provided by the study of
A. anatomy   B. ecology
C. geology   D. embryology

18. UME/95/50
From which groups of animals are mammals generally believed to have most recently evolved?
A. Reptiles   B. Fishes
C. Amphibians   D. Birds.

19. UME/98/17
The heart of the adult frog consists of
A. two auricles and two ventricles
B. one auricle and one ventricle
C. two ventricles and one auricle
D. one ventricle and two auricles.

20. UME/2002/25/TYPE: Y
The structure that is common in the embryos of mammals, amphibians, birds, fishes and reptiles, and which is an evidence of their common ancestry is the
A. eye   B. chorion
C. allantois   D. gill slits

21. UME /2003/44/ TYPE: A
Which of the following requires the use of carbon-dating to prove that evolution has occurred?
A. Biochemical similarities
B. Molecular records
C. Fossil records   D. Comparative anatomy

22. UME/2004/39/TYPE: 4
An evidence of the relationship between living organisms and their extinct relatives can best be obtained from
A. paleontology   B. embryology
C. comparative anatomy   D. comparative physiology

23. PCE/93/50
Which of the following is a vestigial structure?
A. Appendix in man   B. Wings of birds
C. Forelimbs of toads   D. Wings of cockroaches

24. PCE/95/50
From which of the following is the most direct evidence of evolution?
A. Comparative anatomy   B. Embryology
C. Paleontology   D. Geographical

25. PCE/98/50/TYPE: C
The evidence of evolution that is based on the structural affinities of organisms is
A. embryology   B. biochemical similarities
C. comparative anatomy   D. fossil records

26. PCE/2000/50/TYPE: A
Evolutionary relationship are reflected in natural classification based on structure that are
A. homologous   B. heterozygous
C. homozygous   D. analogous.

27. PCE/2002/25/TYP: 5
The similarity that occurs between the wing of a bat and the human arm as a result of common evolutionary origin is called
A. embryology   B. paleontology
C. morphology   D. homology.

28. UME/2008/36
Which of the following is the most advanced evolutionary development in plants?
A. possession of unicellular structures
B. development of flowers
C. dispersal of spores
D. development of secondary thickening

29. PCE/06/36/TYPE: A
The description of how species are evolutionarily related to one another by modern systematic is based on
A. fossil records and comparative anatomy
B. fossil records and physiology
C. fossil records only
D. comparative anatomy only

MISCELLANEOUS QUESTIONS (26)
30. In evolution, Larmark's theory is known as theory of
A. mutation
B. inheritance of acquired characters
C. natural selection
D. survival of the fittest through competition
(PCE/94/50)

31. Which of the following belongs to the group that first attempted to colonize the terrestrial habitat?
A. Hawk
B. Lizard
C. Gorilla
D. Toad   (PCE/95/46)

32. Darwin's explanation of the way evolution occurs implies that
A. God determines which species should evolve
B. Organisms have evolved simultaneously
C. Those traits used most of ten, persist longer
D. Progressive adaptations enable perpetuation of species
(PCE/98/49/TYPE: C)

33. Which of the following represents the most ancestral in the evolutionary development among plants?
A. Development of fruits
B. Dispersal of spores
C. Unicellular plant body
D. Development of vascular system
(PCE/99/46/TYPE: E)

34. An example of a vestigial structure in the human body is the
A. Appendix   B. Heart
C. Diaphragm   D. Small Intestine
(PCE/2005/3/TYPE: P)

35. The most convincing evidence of evolution from embryology is the presence of
A. Head   B. Blood
C. Notochord   D. Tail (PCE/2005/5/TYPE: P)

36. The pentadactly limb arrangement of vertebrates is an evolutionary fact derived from
A. Morphological differences
B. Serology
C. Embryology
D. Comperative anatomy
(PCE/2004/50/TYPE: 9)

37. An evidence of a common ancestry for fishes, amphibians, reptiles, birds and mammals is the
A. Possession of wings by birds and bats
B. Cold-bloodedness of fishes, amphibians and reptiles
C. Possession of gill clefts in vertebrate's embryos
D. Possession of scales by fishes and reptiles
(UME/2005/50/TYPE: D)

38. An organism with a two-chambered heart is
A. Fish   B. Man
C. Toad   D. Bird
(PCE/2005/50/TYPE: P)

39. The evidence that supports the advancement of ferns over mosses is derived from
A. comparative anatomy
B. molecular records
C. biochemical similarities
D. physiological records (UME/2006/34)

40. Which of the following vertebrates has the most simple structured heart?
A. reptile   B. fish
C. mammal   D. amphibian
(UME/2006/35)

41. The theory which supports the view that the large muscles developed by an athlete will be passed on to the offspring was proposed by
A. Lamarck   B. Darwin
C. Mendel   D. Pasteur (UME/2007/14)

42. The flippers of a whale and the fins of a fish are examples of
A. coevolution
B. continuous variation
C. convergent evolution
D. divergent evolution (UME/2010/50/TYPE: D)

43. Gene flow indicates that
A. inerbreeding is possible between members of different species
B. iterbreeding is possible between members of the same species
C. some genes may spread more than others
D. random changes alter genetic information.
(PCE 2001/50/TYPE: 4)

44. Which of the following is **not** included in the theory of natural selection
A. Stuggle for existence
B. Offspring show variation
C. Inheritance of adaptive structures
D. Disappearance of functionless organs.

45. The order of evolutionary trend in plants is
A. Brophyta, Thallophyta, Pteridohyta and Spermatophyta
B. Thallophyta, Bryophyta, Pteridophyta and Spermatophyta
C. Spermatophyta, Pteridophyta, Bryophyta and Thallophyta
D. Bryophyta, Pteridophyta, Spermatophyta and Thallophyta.

46. A vestigial organ in human is
A. earlobe   B. toe bone
C. tail bone   C. spleen.

47. Natural selection is a consequence of
A. distribution of organism
B. adverse condition
C. variation in oranism
D. inbreeding.

MODELQUESTIONS SET I

1. A similarity between plant and animal cells is the possession of a
A. cell wall
B. nucleus
C. definite shape
D. chloroplast

2. Which of the following is true of the epithelial tissue in animals?
A. it binds other tissue together
B. it forms a protective covering over the body surface
C. it fills spaces in the body and provides support
D. it transmit impulses to various parts of the body.

3. Amoeba gets rid of excess water that enters the cells through the
A. pseudopodia
B. endoplasm
C. contractile vacuole
D. oral groove

4. The criteria used in classifying living organisms into the five kingdoms include complexity of the
A. organism's behaviour and its body size
B. organism's body and its habitat
C. organism's body and its mode of nutrition
D. cell structure and its body size.

5. In which of the following aquatic habitats is spirogyra found?
A. The lagoon
B. A pond
C. The ocean
D. A river

6. In cockroach, the organ which is sensitive to smell is the
A. pedipalp
B. mandible
C. maxilla
D. antenna

7. A typical life cycle of plants involves alternation of the
A. diploid sporophytes with the haploid gametophytes
B. haploid sporophytes with the diploid gametophytes
C. diploid sporophytes
D. haploid gametophytes

8. The main feature that distinguishes a bacterial cell from an eukaryotic cell is the
A. cell wall
B. cell membrane
C. nuclear material
D. cytoplasmic material

Use the diagram below to answer questions 9 and 10.

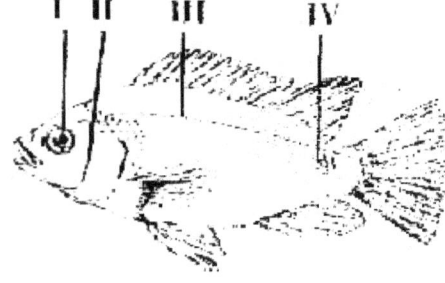

9. the structure used for detection of vibration in water is labeled
A. I
B. II
C. III
D. IV

10. The structure labeled II is the
A. neck
B. head
C. operculum
D. gill.

11. The body cavity of mammals is separated into two regions by the
A. thoracic cavity
B. abdominal cavity
C. placenta
D. diaphragm.

12. The widest uninterrupted zone in the transverse section of a herbaceous dicot stem is the
A. cortex
B. xylem
C. pith
D. phloem

13. Plants that have special devices for trapping and digesting insects are referred to as
A. parasite
B. saprophytic
C. carnivorous
D. symbiotic

14. The most suitable combination of food for a strict vegetarian includes
A. carrot, cheeses and cucumber
B. eggs, beans and yam
C. beans, rice and mushrooms
D. tomatoes, rice and fish.

15. Which of the following processes takes placed in the ileum of humans?
A. Absorption of water
B. Absorption of digested food
C. Secretion of hormone
D. Secretion of saliva.

16. The hardest material in the body of a mammal is the
A. dentine
B. pulp cavity
C. enamel
D. cement

17. A steroid that is found in the mammalian blood is
A. globulin
B. albumin
C. cholesterol
D. fibrinogen

18. The uptake of mineral ions into the root cells in plants is by
A. active transport
B. simple diffusion
C. osmosis
D. translocation

19. In the lungs, the air sacs are surrounded by a cluster of
A. bronchioles
B. capillaries
C. trachea
D. ribs

20. In mammals, the lungs serve as the major excretory organ for
A. urea
B. salts
C. carbon (IV)oxide
D. water

21. Muscles are attached to bones by the
A. ligament
B. cartilage
C. tendon
D. connective tissue.

22. If a germinating seed is attached horizontally on a revolving klinostat, the effect on the seedling after three days is that the
A. plumule will curve upwards while the radicle will curve downwards
B. plumule and radicle will not show any visible curvature
C. radicle will curve downwards while the plumule will not show any curvature
D. plumule will curve upwards while the radicle will not show any curvature.

23. In entomophilous flower, the function of the large petals is to
A. secrete large quantity of nectar
B. protect the stigma
C. attract insects
D. attach the flower to the shoot.

24. The hormone in the mammalian females that induces ovulation is?
A. testosterone
B. oestrogen
C. progesterone
D. leutenizing hormone

Use the diagram below to answer questions 25 and 26.

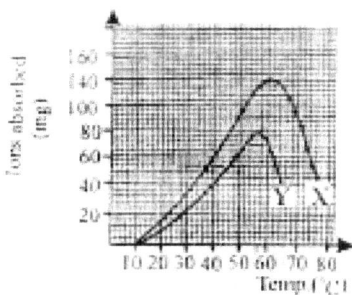

25. The optimum temperature for the absorption of ions by X and Y is between
A. 10 - 80°C
B. 40 - 50°C
C. 50 - 60°C
D. 50 - 80°C

26. What is the title for the graph?
A. relationship between temperature and growth in X and Y.
B. Effect of temperature on the rate of ion uptake in X and Y
C. rate of growth of X and Y in a medium
D. Interaction of temperature and ions absorbed in a medium.

27. The cells that transmit messages to the effectors are the
A. sensory receptors
B. relay neurons
C. motor neurons
D. sensory neurons.

28. The ecological condition that favours breeding of blackfly is
A. slow flowing streams
B. water in ponds and swamps
C. water in small containers
D. fast flowing streams.

29. The pair of abiotic factors specific to the distribution of organisms in aquatic habitats is
A. salinity and humidity
B. salinity and turbidity
C. turbidity and temperature
D. turbidity and rainfall.

30. The protozans living symbiotically in the stomach of termites is responsible for the
A. digestion of cellulose taken in by the termites
B. digestion of protein taken in by the termites.
C. supply of mineral salts to the termites
D. supply of vitamins to the termites.

31. Carbon in the form of carbon (IV)oxide is removed from the atmosphere during
A. evaporation
B. transpiration
C. photosynthesis
D. respiration

32. Breathing roots is an adaptive feature for plants that grow in the
A. mangrove swamp
B. Sahel savanna
C. rain forest
D. Guinea savanna.

33. Which of the following is the most common arboreal dweller?
A. Moths
B. Mosquitoes
C. Butterfiles
D. Blackflies

34. The Obudu cattle ranch in Nigeria is located in the
A. tropical rainforest
B. Guinea savanna
C. highland of montane forest
D. Sudan savanna

35. Which of the following is most important in the adoption of biological control for pests?
A. The knowledge of agricultural practices by the farmer.
B. The relative abundance of plants and animals in the farm.
C. The knowledge of the breeding habits of the predator.
D. The predator-prey relationship in the ecological community.

36. Excess growth of algae in a pound will result in the depletion of
A. nutrients   B. water
C. nitrogen   D. oxygen

37. The soil that has the lowest water retention capacity is
A. clay   B. sand
C. loam   D. laterite.

38. The loss of soil fertility through leaching is caused by
A. deforestation through agriculture
B. overgrazing by animals
C. draining of the nutrient contents
D. rill and gulley erosions.

39. Non- renewable resources can be conserved by
A. afforestation   B. deforestation
C. recycling   D. sanitation.

40. Which of the following parasites is transmitted by houseflies?
A. *Plasmodium sp.*   B. *Vibrio sp.*
C. *Trypanosoma sp.*   D. *Schistosoma sp.*

41. Which of the following can lead to variation in living organisms?
A. Budding   B. Fertilization
C. Fragmentation   D. Sporulation

42. The knowledge of blood group genotype is very important in counseling before marriage so as to prevent.
A. genetic diseases   B. communicable diseases
C. agglutination reactions D. infertility in couples

43. One of the major criticisms of the Mendelian laws arises from the fact that
A. one trait is often controlled by many pairs of genes.
B. a single factor inheritance is never a reality
C. complete dominance is always possible
D. complete dominance is never a reality.

44. In a cross involving a heterozygous black goat (Bb) and a white goat (bb), what is the probability that the offspring will be Bb?
A. ¼   B. ½   C. ¾   D. 1

45. The exchange of genes between homologous chromosomes is referred to as
A. mutation   B. crossing –over
C. back-cross   D. test-cross.

Use the diagram below to answer questions 46 and 47.

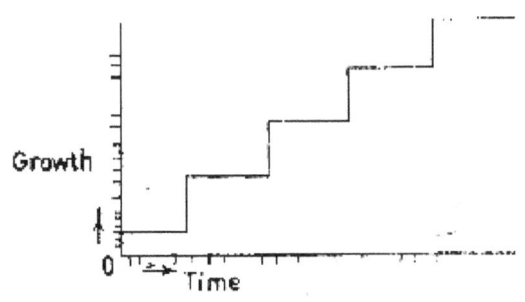

46. An adaptation for obtaining air in an aquatic habitat is shown in
A I   B. II   C. III   D IV

47. Which of the following is represented in II?
A. Stilt roots of mangrove
B. Breathing roots of mangrove
C. Prop roots of maize
D. Aerial roots of an epiphyte

48. Counter-shading in fishes, toads and snakes is an important adaptation for
A. regulating body temperature
B. warning and attacking predators
C. securing mates for reproduction
D. concealing the animals from predators

49. Which of the following is an adaptation for conserving water in plants?
A. broad leaves with hairs.
B. leaves with sunken stomata and thin cuticle.
C. leaves with sunken and thick cuticle.
D. reduced leaves with numerous stomata.

50. Which of the following castes is absent in a termite colony?
A. Queen   B. Drone.
C. Worker.   D King.

MODEL QUESTIONS SET II

Use the diagram below to answer questions 1 and 2.

1. The growth pattern illustrated above is an
A. isometric growth
B. Automatic growth
C. Intermittent growth
D. Hyperbolic growth

2. Which of the following groups of animals shows this pattern of growth?
A. cestoda
B. reptilia
C. amphibia
D. insecta

Use the diagram below to answer questions 3 and 4.

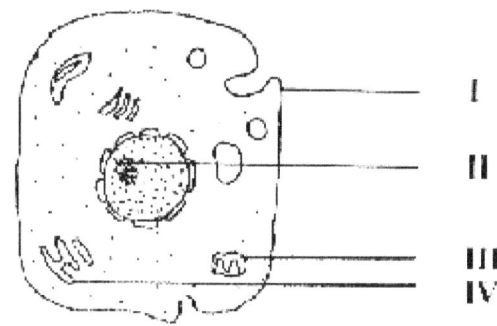

3. The organelle responsible for heredity is labeled
A. I
B. II
C. III
D. IV

4. The part labeled IV is the
A. mitochondrion
B. cell wall
C. endoplasmic reticulum
D. nucleus.

5. Which of the following is the lowest category of classification?
A. class
B. species
C. family
D. genus.

6. Plants that show secondary growth are usually found among the
A. thallophytes
B. pteridophytes
C. monocotyledons
D. dicotyledons

7. The fungi are a distinct group of eukaryotes mainly because they have
A. spores
B. no chlorophyll
C. many fruiting bodies
D. sexual and asexual reproduction.

8. An arthropod that is destructive at the early stage of its life cycle is
A. butterfly
B. mosquito
C. bee
D. millipede.

9. An animal body that can be cut along its axis in any plane to give two identical parts is said to be
A. radically symmetrical
B. bilaterally symmetrical
C. asymmetrical
D. symmetrical

10. Which of the following possesses mammary gland?
A. Dogfish
B. Whale
C. Shark
D. Catfish

11. The feature that links birds to reptiles in evolution is the possession of
A. feathers
B. beak
C. skeleton
D. scales.

12. Countershading is an adaptive feature that enables animals to
A. fight enemies
B. remain undetected
C. warn enemies
D attract mates.

13. Which of the following plant structures lacks a waterproof cuticle?
A. leaf
B. stem
C. root
D. shoot

14. In the mammalian male reproductive system, the part that serves as a passage for both urine and semen is the
A. urethra
B. ureter
C. bladder
D. seminal vesicle.

15. In plants, which of the following is required in minute quantities for growth?
A. copper.
B. potassium
C. phosporus
D. sodium.

16. Which of the following organisms is both parasitic and autotrophic?
A. Sundew
B. *Loranthus*
C. *Rhizopus*
D. Tapeworm

17. A function of the hydrochloric acid produced in the human stomach during digestion is to
A. neutralise the effect of bile
B. coagulate milk protein and emulsify fats
C. stop the action of ptyalin
D. break up food into smaller particles.

18. Which of the following is a polysaccharide?
A. Glucose
B. Sucrose.
C. Maltose.
D. Cellulose.

Use the diagram below to answer questions 19 and 20.

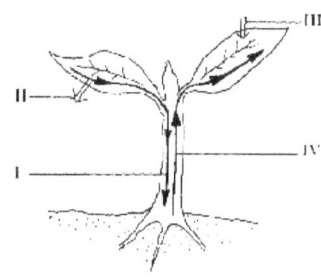

19. Transportation in the xylem is represented by
A. I   B. II   C. III.   D. IV

20. The arrow labeled II represents the
A. release of oxygen
B. intake of carbon (IV) oxide
C. movement of photosynthates
D. movement of nutrients

21. In the kidney of mammals, the site of ultrafiltration is the
A. uriniferous tubule   B. Bowman's capsule
C. loop of Henle   D. renal tubule.

22. Which of the following is involved in secondary thickening in plants?
A. Collenchyma and xylem cells.
B. Vascular cambium and collenchyma cells.
C. Vascular cambium and cork cambium.
D. Cork cambium and sclerenchyma.

23. An example of a fruit that develops from a single carpel is
A. okro   B. tomato
C. bean   D. orange

Use the diagram below to answer questions 24 and 25.

24. The developing embryo is usually contained in the part labeled
A. IV   B. III   C. II   D. I

25. The function of the part labeled III is to
A. produce egg cells
B. protect sperms during fertilization
C. secrete hormones during coitus
D. protect the developing embryo.

26. Plant growth can be artificially stimulated by the addition of
A. gibberellin   B. kinin
C. abscisic acid   D. ethylene

27. The autonomic nervous system consists of neurons that control the
A. voluntary muscles   B. heat beat
C. tongue   D. hands

28. Plants of temperate origin can be grown in tropical areas in the vegetation zones of the
A. rain forest   B. Guinea savanna
C. Sudan savanna   D. montane forest.

29. The water cycle is maintained mainly by
A. evaporation of water in the environment
B. evaporation and condensation of water in the environment
C. condensation of water in the environment
D. transpiration and respiration in plants.

30. Organisms living in an estuarine habitat are adapted to
A. withstand wide fluctuation in temperature
B. survive only in water with low salinity
C. withstand wide fluctuations in salinity
D. feed only on phytoplankton and dead organic matter.

31. The presence of stilt roots, pneumatophores, sunken stomata and salt glands are adaptive features of plants found in the
A. tropical rainforest   B. mangrove swamps
C. grassland   D. montane forest.

32. Which of the following animals can exist solely on the water they get from food and metabolic reaction?
A. Forest arboreal dwellers   B. Desert dweller.
C. Forest ground dwellers   D. rainforest dwellers

33. The most likely first colonizers of bare rock are
A. mosses   B. ferns
C. lichen   D. fungi

34. The carrying capacity of habitat is reached when the population growth begins to
A. increase slowly   B. increase exponentially
C. slow down   D. remain steady.

35. The abiotic factors that control human population include
A. disease and famine   B. space and rainfall
C. flooding and earthquake   D. temperature and disease

36. An indigenous method of renewing and maintaining soil fertility is by
A. clearing farms by burning
B. planting one crop type
C. adding inorganic fertilizers yearly
D. crop rotation and shifting cultivation.

37. The diseases caused by water-borne pathogens include
A. gonorrhea and poliomyelitis
B. typhoid and syphilis
C. tuberculosis and cholera
D. typhoid and cholera.

Use the diagram below to answer questions 38 and 39.

Frequency

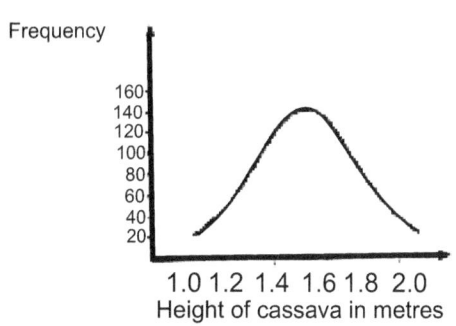

1.0 1.2 1.4 1.6 1.8 2.0
Height of cassava in metres

38. The graph illustrates
A. the highest frequency for height of 2 metres
B. a discontinuously varying character
C. a continuously varying character
D. total yield in a cassava farm.

39. the largest number of cassava plants has an approximate height of
A. 1.4m   B. 1.6m   C. 1.8m
D. 2.0m

40. Which of the following is true in blood transfusion?
A. A person of blood group AB can donate blood only to another person of blood group AB.
B. Persons of blood groups A and B can donate or receive blood from each other.
C. A person of blood group AB can receive blood only from person of blood group A or B.
D. A person of blood group O can donate only to a person of blood group O.

41. A yellow maize is planted and all the fruits obtained are of yellow seeds. When they are cross-bred, yellow seeds and white seeds are obtained in a ratio 3:1. The yellow seed is said to be
A. non-heritable   B. sex-linked
C. a recessive trait   D. a dominant trait.

42. When a colour-blind man marries a carrier woman. What is the probability of their offspring being colour blind?
A. 25%   B. 50%   C. 75%   D. 100%

43. The correct base pairing for DNA is
A. adenine à thrmine and guanine à cytosine
B. adenine à guanine and thymine à cytosine
C. adenine à cytosine and guanine à thymine
D. adenine à adenine and cytosine à cytosine

Use the diagram below to answer questions 3 and 4.

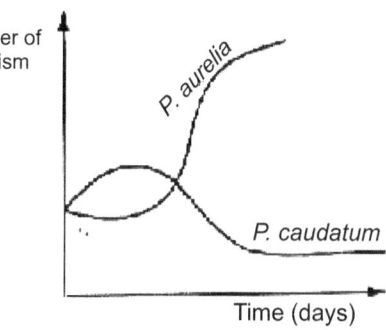

Number of organism

P. aurelia

P. caudatum

Time (days)

44. The type of interaction shown is referred to as
A. interspecific competition
B. intraspecific competition
C. mutualism
D. cooperation.

45. Which of the following statements is true of the interaction?
A. *P. aurelia* is better adapted for obtaining food than P. caudatum
B. *P. caudatum* is better adapted for obtaining food than P. aurelia
C. both organisms cannot coexist.
D. both organisms cannot reproduce.

46. The short thick beak in birds is an adaptation for
A. crushing seeds   B. sucking nectar
C. tearing flesh   D. straining mud.

47. The basking of *Agama* lizards in the sun is to
A. change the colour of their body
B. raise their body temperature to become active
C. fight to defend their territories
D. attract the female for courtship.

48. The significance of a very large number of termites involved in nuptial swarming is to
A. provide birds with plenty of food
B. ensure their perpetuation despite predatory pressure.
C. search for a favourable place to breed
D. ensure that every individual gets a mate.

49. the use and disuse of body parts and the inheritance of acquired traits were used to explain
A. Darwin's theory   B. Lamarck's theory
C. genetic drift   D. gene flow.

50. From his study of Galapagos finches, Darwin derived his theory of evolution from
A. comparative anatomy   B. comparative physiology
C. fossil remains   D. comparative embryology.

217

ANSWERS TO MODEL QUESTIONS SET I

1. B = Plant and animal cell have the nucleus in common.
2. B = Epithelial tissue in animals form a protective convering over the body surfaces.
3. C = Amoeba removes excess water (osmoregulation) via the contractile vacuole
4. D = Complexity of cell structure (prokaryotic or eukaryotic) and size are the criteria used in classifying living organisms into the five kingdoms namely: Monera, Protista, Fungi, Plantae, Animalia.
5. B = Spirogyra floats in masses in ponds, lakes, and slow moving streams. They are commonly called pond scum or water silk.
6. D = Cockroach antenna, also known as feelers, are responsible for their sense of smell.
7. A = The sporophyte is diploid (2n) while the gametophyte is haploid (n).
8. C = The bacterial cell is a prokaryotic cell, their nuclear material is scattered in the cytoplasm.
9. C = I- Eye, II - operculum (gill cover), III- lateral line IV- scales. The lateral line is used to detect vibration in water.
10. C
11. D = The diaphragm separate the thoracic cavity from the abdominal cavity.
12. A= Xylem is interrupted from phloem by the cambium. The cortex is uninterrupted. The cortex is wider than the pith.
13. C = Carnivorous plants (or insectivorous plants)
14. C = A vegetarian takes diets obtained strictly from plants. The body needs carbohydrate, proteins and lipids basically but also needs vitamins. Beans (proteins), rice (carbohydrate) mushroom (carbohydrate, fat, protein and vitamins)
15. B = Absorption of digested food takes place in the ileum.
16. C = The tooth enamel is the hardest and most highly mineralized substance in the human body. Enamel's primary mineral is hydroxyapatite, which is a crystalline calcium phosphate. The large amount of enamel accounts not only for its strength but also for its brittleness. Dentine is less mineralized and less brittle.
17. C = Cholesterol is a steroid in mammalian blood plasma. Globulin, albumin and fibrinogen are proteins in the blood plasma.
18. A = Active transport is the movement of a substance against its concentration gradient (from low to high concentration). It requires energy in the form of ATP. Example include the uptake of glucose in the human intestines and uptake of mineral ions in the root hair cells in plant.

19. C = The air sacs are the alveoli, which are surrounded by the blood capillaries for effective diffusion of gases. Capillaries have thin walls.
20. C = The percentages of gases in inhaled and exhaled air respectively is Nitrogen (78% & 78%), Oxygen (21% & 17%), Carbondioxide (0.04% & 4%) and water vapour (0.86% & 3%)
21. C = Tendon –muscle to bone. Ligament – bone to bone.
22. A = The plumule will curve upward (shoots of plants are negatively geotropic) while the radicle will curve downward (roots of plants are positively geotropic).
23. C = Entomophilous flowers (flowers that need pollinators) commonly have larger showy petals to attract insects or birds or large white petals to attract bats. Anemophilous plants do not need pollinators, they are pollinated by wind.
24. D = Leutenizing hormone (LH) induces ovulation, that is why there is LH surge at ovulation.
25. C = There was optimum absorption of by X and Y between 50-60ᵒC, 80mg of ions was absorbed by X (optimum) and at 60ᵒC, 140mg of ions was absorbed by Y (optimum).
26. B = The best title for the graph is the effect of temperature on the rate of ion uptake (absorption) in X and Y.
27. C = Motor neurons – CNS to effectors. Sensory neurons – receptors to CNS.
28. D= Black fly, *Simulium damnosum* breeds in fast flowing streams.
29. B = Salinity and turbidity are unique to aquatic habit.
30. A = The protozoan digest the cellulose taken in by the termite while the termite provide food and shelter for the protozoan.
31. C= Photosynthesis removes CO_2.
32. A = Pneumatophores (breathing roots) of the white mangrove allow intake of atmospheric air.
33. C
34. A = Obudu cattle ranch is situated deep in the tropical and mountainous rainforest of Northern Cross River State.
35. D = The predator – prey relationship in the ecological community e.g. cat to prey on rat
36. D = Oxygen is depleted by alga overgrowth in water bodies
37. B = Sand has the lowest water retention capacity. Clay has the highest.
38. C = During leaching, flood water (from heavy rainfall) permeates the soil and carries along with it, dissolved nutrients into the depth of the soil where roots cannot reach.
39. C = Non-renewable resources can be conserved by recycling.
90. B = Houseflies are vectors of *Vibro sp.,* the bacteria that causes cholera.
41. B = Sexual reproduction (fertilization) will result in variation because of gene combination.

218

42. A = Genetic diseases such as sickle cell anaemia.
43. A = A triat is often controlled by many pairs of genes
44. B = parents: Bb vs bb

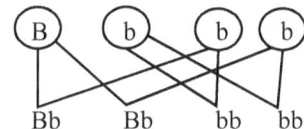

Gemetes: Bb   Bb   bb   bb

Probability of Bb is 2/4 i.e. ½
45. B = Chromosomal crossover (or crossing over) is an exchange of genetic material between homologous chromosomes.
46. A =   I – Breathing root of mangrove,
            II – Aerial roots of an epiphyte,
            III –Prop roots of maize,
      IV – Dicot leaves.
47. D
48. D = Countershading helps to conceal animals from predators making them remain unnoticed.
49. C = Leaves with sunken stomata and thick curticle with reduce loss of water by transpiration, helping to conserve water .
50. B = The drone is found in the bee colony.

ANSWERS TO MODEL QUESTIONS SET II

1. C = Growth patterns include:
   (i) Isometric growth: when parts of an organism grow at the same rate as the whole organism e.g. plant leaves.
   (ii) Allometric growth: When parts or organs of an organism grow at different rates from the whole organisms e.g. the head and brain of man during the early stage of life.
   (iii) Intermittent growth: This is a step-like growth; it occurs in arthropods due to moulting.
2. D.
3. B = The gene is the unit of inheritance, it is the constituent of the chromosome, located in the nucleus.
4. C =   I – Cell membrane,
           II – Nucleolus,
           III – Mitochondria,
           VI – Endoplasmic reticulum
5. B = The highest category of classification is kingdom and the lowest is species.
6. D = Secondary growths is found in the dicotyledons because of the presence of cambium
7. C = Fungi are distinct among other eukaryotes because of many fruiting bodies – sporocarps, a multicellular structure of which spore – producing structures, such as basidia or asci, are borne. The presence of spores, sexual and asexual reproduction; and of lack of chlorophyll is not unique to the fungi.

8. A = Butterfly (adult form) helps in pollination of flower while its larva form – caterpillar, is destructive.
9. A = Radial symmetry will give a mirror image when cut along its axis in any plane. Bilateral symmetry will only give a mirror image, when cut along its axis in one plane.
10. B = Whale is a mammal. Mammals are distinct for their pinna (external ear), mammary gland, hair and sebaceous gland).
11. D = Birds and reptiles have scales.
12. B = Countershading helps to conceal animals from predators making them remain unnoticed.
13. C = The epidermis of roots lacks the waxy cuticle found in the parts of the plant. This is so to allow effective absorpion of water and nutrients by the hair cells.
14. A = The urethra is a conduit for both urine (excretory) and semen (reproductive).
15. A = Copper is a trace element needed in small quantity. Major elements are: nitrogen, potassium, phosphorus calcium, magnesium and sulphur.
    Mnemonic: Nigerian Police Pick Criminals Most Saturdays.
16. B = Loranthus is a partial parasite (obligate), it has a haustoria with which it absorb organic food and mineral nutrients from host such as mango. It also has chloroplast in its leaves for photosynthesis – autotrophic nutrition.
17. C = Hydrochloric acid (HCl) in the stomach is to halt the action of ptyalin from the mouth, which can only function in an alkaline medium. It also kills bacteria in the food.
18. D= Monosaccharides – glucose, fructose, ribose. Disaccharides – sucrose, maltose, lactose.
    Polysaccharides – starch, cellulose.
19. D =   I and II – Translocation of nutrients,
           III – Transpiration of water vapor,
           IV – Transportation of nutrients in the xylem.
20. C = Movement of photosynthate (translocation).
21. B = Ultra filtration takes place in the Bowman's capsule.
22. B = Secondary thickening is as a result of the activity to two lateral meristems, the cork cambium and vascular cambium. The cause the lateral growth of the stem and root.
23. C = Bean is a legume. A legume is formed from one carpel and split along two sides.
24. C =   I – Oviduct (fallopian tube) – passage of eggs, site of fertilization.
           II – Uterus – where fertilized egg develop.
           III – ovary – produces eggs
           IV – virgina – where sperm is deposited.
25. A
26. A = Gibberellins promotes growth.
27. B = The autonomic nervous system consists of neurons that control involuntary (reflex) actions e.g. heart beat.
28. A = plants of temperate origin can be grown in tropical

areas in the vegetation zones of the rainforest. There are two types of rain forest, temperate and tropical.

29. B = Water leaves the earth crust via evaporation and transpiration (in plants), respiration (in plants and animals) and returns as rain via condensation and precipitation.

30. C = An estuary is a ecological zone where river and sea water meet, thus to establish brackish conditions. Brackish water has a salinity which fluctuates with the tides and the wet and dry seasons.

31. B = Plants of mangrove plants have stilt root (red mangrove) pneumatrophores (white mangrove), sunken stomata and salt glands.

32. B= Desert dwellers strictly conserve water.

33. C = The first colonizers of a bare rock surface are the lichen. They breakdown the rocks to provide soil for subsequent plants (mosses).

34. D = Carrying capacity is the number of individuals an environment can support without significant negative impacts to the given organism and its environment. Below carrying capacity, populations typically increase while above, they typically decrease. Population remain fairly steady at carrying capacity.

35. C = Flooding and earthquake are purely abiotic factors. Disease is a biotic factor because it involve living organisms e.g. vectors.

36. D = Indigenous methods refers to natural methods and shifting cultivation

37. D = Typhoid is caused by *Salmonella sp*. And cholera is caused by *Vibrio sp.,* both of which are carried in water. Gonorrhoea and syphilis are sexually transmitted. Tuberculosis is air–borne while poliomyelitis can also be water–borne.

38. C = The graph illustrates continous variation in height.

39. B = 1.6m

40. A = Blood group AB is a universal recipient but can only donate to AB blood group.

41. D = Yellow is a dominant trait.

42. B = Colour blindness is a sex-linked traits.

Parents: Colour blind man X^cY vs carrier women X^CX^c

Gametes:

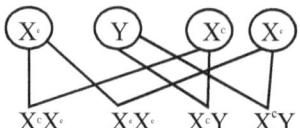

Offspring: X^CX^c X^CX^c X^CY X^cY

Carries female, colour blind female, normal male, colour blind male respectively.

43. A = AT; GC i.e. adenine thymine & guanine cytosine

44. A = *P.aurelia* and *P.caudatum* are two different species of *Paramecium*. Hence, it is interspecific competition.

45. A = From the graph, it is obvious that *P.aurelia* is better adapted for obtaining food than *P.caudatum*.

46. A = A short thick beak in birds in an adaptation for crushing nuts (seeds).

47. B= *Agama* lizards basks in the sun to raise their body temperature to become active. They are poikilothermic.

48. B = Large numbers of termite participate in the nuptial swarming in other to ensure perpetuation despite predatory pressures.

49. B = Lamarck propose the theory of use and disuse; and the inheritance of acquired traits

50. A = Charles Darwin made a comparative anatomical studies of the Galapagos finches to illustrate evolution.

UNIVERSITIES POST – UTME

UNIVERSITY OF BENIN POST UME 2005

1. When two genes for the same character are contain in the same individual, the character that shows is known as
A. important character   B. dominant character
C. superior character   D. controlling character

2. In Nigeria, a tropical rainforest can be found in
A. Sokoto   B. Kaduna
C. Abuja   D. Abia

3. The sum total of the biotic and abiotic factors that affect living things is referred to as
A. environment   B. lithosphere
C. hydrosphere   D. atmosphere

4. Hygrometer is an ecological instrument that measures
A. rainfall   B. humidity
C. temperature D. light

5. *Candida vaginalis* is a
A. bacterium   B. fungus
C. virus   D. Protozoan

6. Which one of the following is air borne?   A .
Malaria   B. Yellow fever
C. Cholera   D. Tuberculosis

7. All of these are vertebrates except
A. lizard   B. rat
C. star fish   D. tilapia

8. A box was left in the lawn for two days when the box was removed, the grass under had turned yellow due to lack of
A. carbon dioxide   B. light
C. oxygen   D. water

9. Which of the following deaminates excess amino acids?
A. Duodenum   B. Ileum
C. Liver   D. Kidney

10. In which of the following floral parts does meiosis occur?
A. Anther   B. Petal
C. Receptacle   D. Style

11. When an organism moves its whole body towards a stimulus, the organism is said to exhibit
A. tropic movement   B. trophic movement
C. tactic movement   D. nastic movement

12. A stable self sustaining environment produced by an interaction between the biotic and abiotic components its best described as
A. a niche   B. a community
C. a habitat   D. an ecosystem

13. Which of the following is not present in the nucleus of a cell
A. Chromosomes B. Nucleolus
C. Genes   D. Mitochondria

14. The release of useful substances from the cells of an organism is called
A. excretion   B. evacuation
C. metabolism   D. secretion

15. The head of the femur articulates with the pelvis and the
A. glenoid cavity   B. olecranon process
C. acetabulum   D. carnoid process

16. The mammalian cervical vertebrate invariably numbers
A. 4   B. 7
C. 12   D. 15

17. What mechanism is responsible for the rise of water to the top of tall trees?
A. Transpiration pulls   B. Root pressure
C. Osmosis   D. Capillarity

18. Antibodies in mammalian blood are formed by
A. platelets   B. white blood
C. red blood cells D. livers

19. Which of the following liquids supplies cells in the tissues of a mammal with oxygen and nutrients?
A. Blood   B. Plasma
C. Serum   D. Lymph

20. The cockroach and grasshoppers have mouth parts adapted for
A. sucking   B. piercing and sucking
C. biting and chewing   D. biting and lapping

21. The Sponging and lapping mouthparts occurs in
A. housefly   B. butterfly
C. cockcroach   D. mosquito

22. Which of the following is not an example of sex – linked character?
A. Baldness   B. Colour blindness
C. Haemophilia D. Height

23. Which of these is a vestigial structure?
A. Appendix   B. Caecum
C. Pancreas   D. Rotundus

24. Which of the following air pollutants causes acid rain?
A. sulphur dioxide   B. lead oxide
C. hydrogen dioxide   D. carbon dioxide

25. Which of the following does not have a well-developed tissue?
A. Moss
B. Fern
C. Whispering tree
D. Maize

UNIVERSITY OF BENIN POST UME 2006

26. Which is the largest single cell in the body?
A. a neuron
B. the ovum
C. muscle
D. Cell

27. Blue – green algae belongs to the phylum
A. cyanophyta
B. schizophyta
C. chlorophyta
D. chrystophyta

28. The concept of antibiotics started with the concept of
A. Gregor Mendel
B. Mary Slessor
C. Louis Pasteur
D. Alexander Fleming

29. The botanical name for yellow yam is
A. *Discorea cayenesis*
B. *Discorea ratundata*
C. *Discorea alata*
D. *Discorea dometorum*

30. Which of the following tissues is made up of dead cells?
A. Xylem vessels
B. Cambium
C. Mesophyll
D. Palisade

31. Which is not a function of the liver?
A. storage of iron
B. formation of bile
C. breakdown of excess amino acids
D. excretion of urea from the blood

32. Alkaline pyrogallol was used in an experiment. That experiment must have been connected with
A. respiration
B. photosynthesis
C. transpiration
D. excretion

33. The enzyme that breakdown sugar cane is
A. lipase
B. Ptyalin
C. invertase
D. peptidase

34. The largest single bone is the
A. scapula
B. humerus
C. femur
D. skull

35. Auxins are produced at the
A. root and stem apices
B. young leaves and nodes
C. flower buds
D. leaves apices

36. The hormone which tones up the muscle of a person in times of danger is from the
A. thyroid gland
B. pancreas
C. adrenal gland
D. sebaceous gland

37. A diet with a high concentration of iodine will probably be needed by a patient suffering from a malfunction of the
A. thyroid gland
B. adrenal gland
C. nervous system
D. Circulatory

38. Which of the following group of factors is completely abiotic?
A. soil, water, bacteria, salinity
B. salinity, tide, plankton, turbidity
C. wind, altitude, humidity, light
D. conifers, wind, pH, rainfall

39. Nitrogen fixing bacteria and cowpea demonstrate an ecological association known as
A. predation
B. parasitism
C. mutualism
D. Commensalisms.

40. Plants that live in salty water areas are called
A. hydrophytes
B. xerophytes
C. halophytes
D. salinophytes

41. A sample of soil was put into a measuring cylinders and water was added to it. After the mixture was shaken, the cylinder was left undisturbed for one hour. This experiment was probably performed to
A. compare the capillarity of different samples
B. find out the relative densities of different soil samples
C. find out the water retaining capacity of the soil
D. demonstrate the presence of air in the samples

42. A farmer X working in a swamp did not eat any food not drink any water. Which of these diseases can he not contract
A. bilharzia
B. malaria
C. cholera
D. sleeping sickness

43. Which of the following elements may be added to drinking water to lessen dental decay?
A. calcium
B. phosphorus
C. fluorine
D. chlorine

44. A man with normal haemoglobin marries a woman who has sickle cell haemoglobin. They have a child who had sickle cell trait. Which of the following genotype could be associated with the child's haemoglobin?
A. SS
B. AS
C. AO
D. AA

UNIVERSITY OF BENIN POST UME 2007

45. Hydra is able to perform all the following except
A. feeding
B. photosynthesis
C. moving
D. egestion

46. Spherical bacterial cells arranged in chains are called
A. staphylococci
B. streptococci
C. cocci
D. bacilli

47. One of the following factors that must be considered for safe blood transfusion is
A. social class of the donor
B. rhesus factors of the donor and the recipient
C. age of the recipient
D. nationality of the donor

48. The theory of use and disuse of organs was pro pounded by
A. Charles Darwin   B. Jean Lamark
C. Louis Pasteur   D. Robert Hooke

49. Which of these is not a function of the kidney?
A. production of urine   B. production of bile
C. osmoregulation   D. removal of urea

50. Auxins are produced in the
A. petioles of leaves
B. epidermis of roots and shoots
C. parenchyma of roots and shoots
D. apical regions of roots and shoots

51. Which of the following mammalian features acts as a shock absorber to developing embryo?
A. Umbilical cord   B. Amniotic fluid
C. Placenta   D. Amnion

52. The maintenance of a constant internal environment is called
A. environmental regulation   B. balance in nature
C. peristalsis   D. homeostasis

53. People in the blood group which can receive blood from another blood group fall into
A. AB blood group   B. O blood group
C. A blood group   D. B blood group

54. Which of the following crop disease is caused by a fungus?
A. Groundnut Rosette   B. Black pod disease of cocoa
C. Yam Mosaic   D. Cassava Blight

OBAFEMI AWOLOWO UNIVERSITY
POST UME 2008

55. One of the function of xylem
A. Strengthening the stem
B. Manufacturing food
C. Conducting manufactured food
D. None of above

56. People suffering from myopia
A. can see near objects clearly
B. can see far away objects clearly
C. cannot see any object clearly
D. are colour – blind

57. The cilia in paramecium are used for
A. Respiration   B. Locomotion
C. Protection   D. Excretion

58. Which of these types of skeleton is most appropriate to the cockroach
A. Hydrostatic skeleton   B. Exoskeleton
C. Endoskeleton   D. Certilaginous skeleton

59. When proteins are broken down they produce
A. Oxygen   B. Carbohydrate
C. Energy   D. Amino acids

60. The function of lenticel is
A. To receive excess water in the plant
B. To absorb water from the atmosphere
C. For gaseous exchange
D. To absorb light

61. Which of the following is characteristic of the animal cell
A. presence of large vacuoles
B. possession of a cellulose cell wall
C. absence of large vacuoles
D. presence of large vacuoles

62. In the life history of Schistosoma (Bilharzia) one of the following is the intermediate host
A. Man   B. Snail
C. Mosquito larva   D. Fish

63. The hormone which tones up the muscles of a person in the time of danger is from the
A. thyroid gland   B. pancreas
C. adrenal gland   D. spleen

64. The study of the organisms and the environment of abadoned farmland is the Ecology of
A. A community   B. A population
C. A species   D. An ecosystem

65. At fertilization
A. One chromosome from the male joins another from the female
B. One gene from the male combines with the another from the female
C. The male nucleus fuses with the female nucleus
D. One set of chromosome combines with another set from the female

66. The neck region of the tapeworm (Taenia spp) is responsible for the
A. The production of eggs
B. The storage of eggs
C. The formation of new segments
D. The development of the suckers

67. The movement of molecules from a region of higher concentration to one of lower concentration is
A. Diffusion   B. Transpiration
C. Osmosis   D. Plasmolysis

68. The region of cell division in a root is
A. Root cap   B. Endodermis
C. Xylem   D. Meristem

69. Which of the following is not an excretory organ?
A. Lungs   B. Kidney
C. Leaf   D. Large instestine

70. The part of the mammalian brain responsible for maintaining balance is
A. Medulla oblongata B. Cerebellum
C. Optic lobe   D. Cerebrum

71. A sugar solution was boiled with Fehling's solutitons A and B and the colour remain blue. The sugar tested was
A. Glucose   B. Maltose
C. Fructose   D. Sucrose

72. The blood vessel which carries digested foom from the small intestine to the liver is the
A. Renal vein
B. Renal artery
C. Hepatic artery
D. Hepatic portal vein

73. The maize grain is regarded as a fruit and not a seed because
A. It is covered by a sheath of leaves
B. The testa and fruit wall fuse after fertilization
C. It has both endosperm and cotyledon
D. The pericarp and seed coat are separate

74. Identical twins are produced under one of the following conditions
A. Two ova fertilized at the same time by the sperm
B. One ovum fertilized, divides to give two embryos
C. Two ova fertilized by one sperm
D. One ovum fertilized by sperms

75. Which of the following hormones is produced during fright or when agitated?
A. Insulin   B. Adrenalin
C. Tryroxine   D. Pituitary

76. Grasses à Grasshopper à Lizards à Snakes àHawks.
In the food chain, the organisms which are the least in number are
A. Grasses   B. Hawks
C. Lizards   D. Snakes

77. One significant difference between roots and stem is that
A. Branch root originate in the pericycle while branch stems do not
B. Stems are always below the ground while roots are always under the ground
C. Stems are positively geotropic while roots are negatively geotropic
D. Stems are sometimes used for storage while roots are never so use

78. The arrangements below are steps in protein digestion. Which is the correct sequences
a- polypeptides b-protein c- amino acids d- peptones
A. a → b → c→ d   B. c → d→ c → b
C. b → d → a → d   D. b → d→ a → c

79. Partially digested food ready to leave the stomach is called
A. chyme   B. curd
C. glycogen   D. paste

80. The particles found in the blood which play an important role in blood clotting are called
A. platelets   B. red blood cells
C. monocytes   D. granulocytes

81. In a cross between a normal male and a female carrier for haemophilic disease, the percentage of their sons expected to suffer the disease as hemophiliac is
A. 25%   B. 50%
C. 70%   D. 45%

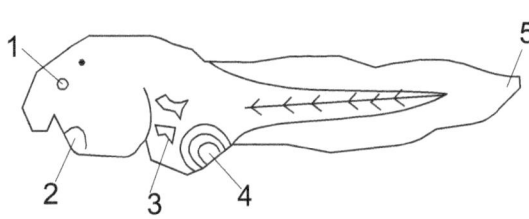

82. The structure labeled 4 is for
A. feeding   B. attachment
C. excretion   D. respiration

83. As the tadpole develops the structure labeled 5
A. grows longer   B. becomes shorter
C. becomes the hind legs D. remains unchanged

84. The gill rakers of fishers take part in
A. feeding   B. respiration
C. swimming   D. diffusion

85. Where is energy produced in a cell
A. Nucleus
B. Lysosomes
C. Mitochondria
D. Nucleolus

86. Which of the following organisms does not exist are a single free – living cells
A. Amoeba
B. Euglena
C. Chlamydomonas
D. Volvox

87. Euglena is an autotrophic organism because it
A. has flagella
B. has plant and animal features
C. can manufacture its food
D. moves fast

88. In which of the following organisms does a single cell perform all function of active movement, nutrition, growth excretion and photosynthesis?
A. Paramecium
B. Amoeba
C. Euglena
D. Hydra

89. What is the function of contractile vacuole in Paramecium?
A. produces enzymes
B. gets rid of excreta
C. stores and digests food
D. gets rid of excess water

90. The ability of organisms to maintain a constant internal environment is known as
A. diuresis
B. endosmosis
C. plasmolysis
D. homeostasis

91. Which of the following is the medium of transportation of nutrients within unicellular organism?
A. Lymph
B. Plasma
C. Protoplasm
D. Serum

92. In aerobic respiration, oxidative phosphorylation takes place in
A. cytoplasm
B. lysosome
C. mitochondrion
D. ribosomes

93. Bryophytes are different from flowering plants because they
A. are simple small plant
B. carry out alternation of generation
C. possess small
D. possess no vascular tissue

94. In lower plants like mosses, the structure which performs the functions of roots of higher plants is called
A. roots hairs
B. rhizoids
C. hyphae
D. roots

95. Which of the following components of an ecosystem has the greatest biomass?
A. primary producers
B. primary consumers
C. secondary consumers
D. tertiary consumers

96. The name of a bacterium which derives its energy from oxidizing nitrites into nitrates is
A. nitrosomonas
B. azotobacter
C. nitrobacter
D. Escheritchia coli

97. Potometer is used to measure
A. rate of osmosis
B. rate of diffusion
C. rate of transpiration
D. rate of photosynthesis

98. Meiotic cell division ensures that
A. many similar cells are produced
B. chromosome number of cells is halved
C. cells produced are doubled
D. cells produced posses the same chromosome number

99. The stem of young herbaceous plants keep upright mainly by
A. Osmotic pressure
B. Turgidity
C. Transpiration pull
D. Root pressure

100. Which of the following tissues is not found in the stem and root of monocotyledons?
A. xylem
B. cambium
C. pith
D. pericycle

101. Fruit enlargement can be induced by spraying young ovary with
A. gibberellins, ethylene and abscisic acid
B. auxin, cytokinin and gibberellin
C. auxin, cytokinin and abscissic acid
D. auxin, ethylene and gibberellin

102. A dry, indehiscent, winged fruit formed from one carpel is known as
A. schizocarp
B. caryopsis
C. samara
D. nut

103. A fruit which developes without fertilization is described as
A. simple
B. aggregate
C. multiple
D. pathernocarpic

104. A dwarf plant can be stimulated to grow to normal height by the application of
A. thyroxin
B. gibberelin
C. insulin
D. Kinin

105. The condition known as cretinism is caused by the deficiency of
A. Vitamin A
B. Insulin
C. Thyroxin
D. Vitamin C

106. The difference between viviparous and oviparous animals is
A. possession of yolked eggs
B. laying and brooding of eggs
C. possession of yolk less egg
D. laying of unfertilized egg

107. The following are features of the tropical rainforest except
A. loose and moist soil
B. short trees growing beneath tall trees
C. scanty trees with small leaves
D. presence of many animals

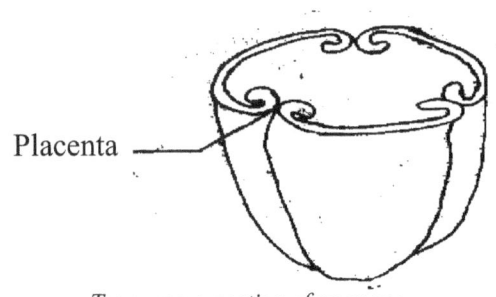

Transverse section of an ovary

108. The type of placentation shown in the figure above is
A. parietal
B. marginal
C. axile
D. free – central

OBAFEMI AWOLOWO UNIVERSITY
POST UME 2006

109. An amoeba moving towards a crumb of cake in a pond most likely exhibits
A. Phototropism
B. Chemotaxis
C. Thermotaxis
D. Nastic movement

110. Which of the following cells would probably contain the greatest number of Golgi bodies
A. muscle cell
B. secretory cell
C. nerve cell
D. white blood cell

111. A group of similar cells performing the same function is
A. an enzyme
B. an organ
C. a tissue
D. an organelle

Structures found in cells are listed below:
i. Cell wall   ii. Cell membrane
iii. Chloroplast   iv. Cytoplasm
v. Nucleus   vi Sap vacuole

112. Which of these structures are found in both animal cells and plant cells?
A. i, ii and v
B. i, ii and v
C. ii, iii and v
D. ii, iv and v

113. Which of the following is not present in the nucleus of a cell
A. Chromosomes
B. Nucleolus
C. Mitochondrion
D. Genes

114. A plant which grows on another plant without apparent harm to the host plant is called
A. A parasite
B. An epiphyte
C. A saprophyte
D. A predator

115. The petals of a flower are collectively called
A. Calyx
B. Capsule
C. Carpel
D. Corolla

116. Osmosis can be defined as diffusion of
A. water molecules from an area of high concentration to an area of low concentration
B. water molecules from dilute solution to a concentrated solution across a permeable membrane
C. Water molecules from a concentrated solution to a dilute solution through a semi-permiable membranes
D. Water molecules from a dilute solution to a concentrated solution through a semi-permeable membrane

$$6O_2 + 6H_2O \xrightarrow[\text{Chlorophyll}]{\text{Sunlight}} C_6H_{10}O_4 + 6O_2$$

117. The oxygen given off during the process in the above equation is derived from
A. Sunlight
B. Water
C. Carbondioxide
D. Atmosphere

118. When testing a leaf for starch why is it first placed in boiling water?
A. To extract the chlorophyll
B. To remove colour from the leaf
C. To dissolve the starch
D. To stop chemical reaction

119. Each of the following is an arthropod except the
A. crab
B. scorpion
C. spider
D. snail

120. The largest phylum in animal kingdom is
A. Cnidaria
B. Mollusca
C. Chordata
D. Arthropoda

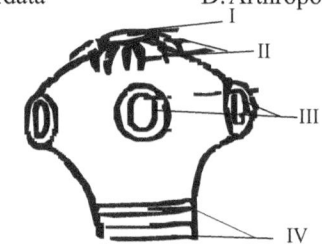

121. Which reference to the figure above, which of these are correct
A. I and II are proglottids and hooks
B. I and III are rostellum and suckers
C. III and IV are hooks and proglottids
D. II and IV are hooks and rostellum

122. Which of the following statement is not correct about the function of each group of mammalian vertebrae?
A. Caudal vertebrae support the tail and provide attachment for tail muscles
B. Thoracic vertebrae articulate with the ribs
C. Lumbar vertebrae provide attachment for abdominal muscles
D. Sacral vertebrae support the skull and allow nodding and rotating movement

123. In the adult toad, gaseous exchange takes place through
A. Buccal cavity, skin and spiracle
B. Buccal cavity, bladder and lungs
C. Buccal cavity, skin and lungs
D. Gills, skin and buccal cavity

124. The foot of the bird shows above is strong and has strong claws on its digits. This implies that the bird
A. is a scavenger
B. is a bird of prey
C. uses the foot to supplement wing action
D. uses the foot to scratch the soil

125. Which of the following is not a means of conservation
A. Replacing harvested nature timber trees with their seedlings
B. Prevention of poaching
C. Controlling excessive deforestation
D. Burning of vegetation before cropping

126. One o f the following statements is not true of viruses
A. They are micro – organisms
B. The are smaller than bacteria
C. They can be seen with an ordinary light microscope
D. They cause tobacco disease, polio and smallpox

127. The brain and the spinal cord constitute the
A. autonomic nervous system
B. sympathetic nervous system
C. somatic nervous system
D. central nervous system

128. Which of the following parts of the mammalian brain is involved in taking the decision to run rather than walk
A. Cerebellum   B. Medulla oblongata
C. Midbrain   D. Cerebrum

129. Which part of the ear is responsible for the maintenance of balance
A. Cochlea   B. Sympathetic nervous system
C. Eustachian tube   D. Semi – circular canals

130. Which of the following factors is a decomposer?
A. Spirogyra   B. Mushroom
C. Chlamydomonas   D. Paramecium

131. Which of the following is not a type of supporting tissue in plant
A. Cambium   B. Collenchyma
C. Paramedian   D. Xylem

132. The medium of transporation in higher plant is
A. cell sap   B. water
C. phloem   D. cytoplasmic fluid

133. The association between protozoa and termites is an example of
A. Symbiosis   B. Parasitism
C. Saprophytism   D. Predation

134. The organ that secrete both digestive enzymes and hormones is
A. Salivary gland   B. Pancreas
C. Spleen   D. Liver

135. Deficiency of calcium in a child diet causes
A. Rickets   B. Cretinism
C. Goitre   D. Scurvy

136. The part of the nephridium through which waste products are discharged in the
A. Muscular tube   B. Narrow tube
C. Nephridiopore   D. Nephrostome

137. Property of the enzymes excludes
A. Specificity   B. Clarity
C. Inexhaustibility   D. Sensitivity

138. Chlamydomonas is
A. a blue – green algae   B. a green algae
C. a clammy algae   D. a red algae

139. What substance make up the genetic material
A. deoxyribonucleic acid   B. Ribonucleic and
C. Adenosine   D. Proteins

140. The table shows three processes that contribute to transport across cell surface membranes which processes are active and which are passive?

	Diffusion	Endocytosis	Osmosis
A	a	a	a
B	p	a	a
C	p	a	p
D	p	p	p

Key: a = active process   p = passive process

141. Which statement is the best explanation of why sucrose rather than glucose is transported by phloem?
A. Sucrose can pass through plant cell surface membranes more easily
B. Sucrose is a disaccharide and a more easily converted to starch
C. Sucrose is a non-reducing sugar, so is less reactive
D. Sucrose synthesis requires less energy

142. What will break an ionic bond between amino acids?
A. Condensation   B. High temperature
C. Hydrolysis   D. pH change

143. From which organelle are nucleic acid absent?
A. Chloroplast   B. Golgi body
C. Mitochondria   D. Ribosome

144. The first in milk produced by the mother for a new-born baby contains antibodies. What do these antibodies provide?
A. Artificial active immunity
B. Artificial passive immunity
C. Natural active immunity
D. Natural passive immunity

145. Which protein has a fibrous structure?
A. Amylase   B. Collagen
C. Heamoglobin   D. Insulin

146. A girl has blood group A and her brother has blood group B. Which combination of genotypes cannot belong to their parents.

	Mother	Father
A.	$1^A 1^A$	$1^B 1^O$
B.	$1^A 1^B$	$1^A 1^B$
C.	$1^O 1^O$	$1^A 1^B$
D.	$1^B 1^O$	$1^A 1^O$

147. In which form is carbon dioxide mainly transported in blood?
A. as carbamino – hemoglobin   B. as carbonic acid
C. as hydrogen carbonate   D. in solution

148. Which cell component is absent from all prokaryotic cells?
A. cell surface membrane   B. cell wall
C. endoplasmic reticulum   D. flagellum

149. Which feature prevents xylem vessels from collapsing?
A. continuous lumen   B. lignified wall
C. narrow lumen   D. pitted wall

150. What causes an increase in the heart rate during exercise, excitement or stress?
A. accumulation of lactate in the muscle
B. decreased concentration of oxygen in the blood stream
C. hormonal action on receptors in the heart
D. increased concentration of glucose in the blood stream

151. What limits the number of tropic levels in a food chain?
A. blomass of the autotrophs
B. efficiency of energy conversion between levels
C. net productivity of the ecosystem
D. species diversity in the ecosystem

152. For what process is the large surface area of the cristae in the mitochondria important?
A. Energy radiation   B. Enzyme reaction
C. Gaseous exchange   D. Protein synthesis

153. Various steps are identified using the following procedure

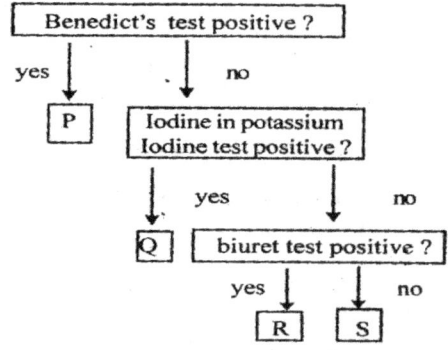

What could the fours substances be?

	P	Q	R	S
A	Glucose	Starch	Protein	Lipid
B	Glucose	Sucrose	Starch	Protein
C.	Sucrose	Protein	Lipid	Starch
D.	Sucrose	Starch	Lipid	protein

154. Which part of the lipid molecule contributes most to the thickness of a surface membrane
A. glycerol
B. hydrocarbon chain
C. hydrophilic head
D. phosphate group

155. Food tests are carried out on four solutions. Which solution contains only sucrose and protein

solution	Benedicts test	Acid Hydroysis Benedicts test	Iodine in KI Solution
A	-ve	-ve	+ve
B	-ve	-ve	-ve
C.	+ve	+ve	-ve
D.	-ve	+ve	+ve

156. Which of the following defines an ecological niche
A. the habitat in which an organism finds it's food supply
B. the habitat in which an organism finds the most suitable climate
C. the relationships between an organism and other species
D. the way in which the environment is exploited by an organism

157. Which region of the kidney nephron is the main site of glucose reabsorption
A. Bowman's capsule
B. Proximal convoluted tubule
C. Distal convoluted tubule
D. Glomerulus

Below are names of groups used when classifying organisms.
I. phylum II. kingdom III. class
IV. family V. species VI. genus
VII. order
Use it to answer question 158
158. The correct order when grouping organism is
A. I, II, III, IV, V, VI and VII
B. I, II, VI, V and VII
C. II, I, III, VII, IV, VI and V
D. II, IV, I, VI, VII, II and V

159. Elephantiasis is a disease caused by
A. filarial worms
B. hookworms
C. flukes
D. tapeworms

OBAFEMI AWOLOWO UNIVERSITY
POST UME 2009

160. One of these is not found in urine
A. Water
B. Sodium chloride
C. Nitrogenous compounds
D. Calcium chloride
E. Nitrogenous salts

161. An organism with a pair of indistinguishable genes is a
A. heterozygote
B. hybrid
C. allelemorph
D. homozygote
E. diploid

162. The fruit formed from a single flower having several free carpels is called
A. multiple fruit
B. simple fruit
C. aggregate fruit
D. dehiscent fruit
E.. indehiscent fruit

163. The function of ossicles (maleus, incus and stapes) in the mammalian ear to
A. transmit vibrations
B. regulate pressures
C. support of the inner ear
D. maintain balance during motion
E. secrete oil

164. "Jointed skeleton" is absent in the
A. Cockroach
B. Spider
C. Millipede
D. Snail
E. House fly

165. Mucor and Spirogyra can be put in a group because they
A. are unicellular
B. have spores that are dispersed by wind
C. can live independent lives
D. reproduce sexually
E. have bodies made up of thallus and filaments alternatively

166. The organ through which nourishment and oxygen diffuse into an embryo called
A. Amnion
B. Chorion
C. Umbilical cord
D. Oviduct
E. Placenta

167. A tapeworm fasten itself to the intenstine of its host with
A. Neck & sucker
B. Hooks & suckers
C. Rostellum & suckers
D. Proglottids & neck
E. Rostellum, hooks & suckers

168. Which of these is false about the piliferous layer of a root? It
A. has a thin cuticle
B. is the outermost layer of the cortex
C. may bear root hairs
D. breaks down with age
E. is replaced by cork in old roots

169. Anaerobic respiration in yeast produces
A. CO_2 & ethanol
B. CO_2 & H_2O
C. CO_2 & O_2
D. CO_2 & glucose
E. Ethanol & H_2O

170. Which of these set of factors is completely abiotic?
A. turbidity, tide, salinity, plankton
B. pressure, pH, soil, insect
C. water, soil, bacteria, salinity
D. pH, bamboos, wind rainfall,
E. light, altitude, wind, humidity

171. In which of these are flagella and cilia found?
A. flatworms
B. protozoa
C. coelenterates
D. annelids
E. nematodes

172. The three important organs that are situated close to the stomach are
A. liver, kidney & gall bladder
B. pancreas, liver & kidney
C. gall bladder, pancreas & spleen
D. liver, kidney & spleen
E. kidney, gall bladder, liver

173. The plantain reproduces asexually by
A. spores
B. buds
C. fragments
D. sucker
E. flowers

174. The major function of swim-bladder in fish is
A. Breathing
B. swimming
C. diving
D. repelling enemy
E. buoyancy

175. The part of the central nervous system concerned with answering an examination question is the
A. spinal cord
B. cerebellum
C. oryx
D. cerebrum
E. medulla oblongata

176. Wind pollinated flowers usually have
A. long styles
B. sticky stigmas
C. small and short stigmas
D. rough pollen grains
E. short styles and pollen

177. The radicle of a bean seedling grown most rapidly in the region
A. of the root tip
B. below the top soil
C. just around the root tip
D. just below the toot tip
E. just above the root tip

178. A key similarity between nervous and hormonal system is that both
A. involve chemical transmission
B. have widespread effect
C. shed chemicals into the blood stream
D. evoke rapid response
E. eliminate response

179. The bone of the neck on which the skull rest is
A. odontoid
B. axis
C. occipital
D. atlas
E. patella

180. A child with blood group genotype different from those of both parents and with a mother of genotype OO, can only have a father of genotype
A. A
B. B
C. OO
D. AB
E. AA

181. A true climax community
A. changes from year to year
B. persists until the environment changes
C. is the first stage in plant succession
D. consists of latest trees and small animals
E. is in a state of percubation

182. In a predator food chain involving secondary and tertiary consumers, the organisms become progressively
A. smaller
B. equal in number
C. larger and fewer along the food chain
D. parasitized along the food chain as consumers get bigger
E. sparse in distribution

183. Which of these is an adaptation to a xerophytic environment?
A. fleshy tissue with reduced leaves
B. extensive surface roots & broad leaves
C. thick barks and broad leaves
D. rough leaves and shallow root system
E. stunted growth and surface roots

184. When it is cold, the blood vessels of the skin
A. dilate to increase blood flow to the skin
B. constrict to reduce the amount of blood flowing to the skin
C. dilate to reduce the amount of blood flowing to the skin
D. constrict to increase the amount of blood flowing to the skin

185. The table shows three processes that contribute to transport across cell surface membranes which processes are active and which are passive?

	Diffusion	Endocytosis	Osmosis
A	a	a	a
B	p	a	a
C	p	a	p
D	p	p	p

Key: a = active process   p = passive process

186. Which statement is the best explanation of why sucrose rather than glucose is transported by phloem?
A. Sucrose can pass through plant cell surface membranes more easily
B. Sucrose is a disaccharide and is more easily converted to starch
C. Sucrose is a non-reducing sugar, so is less reactive
D. Sucrose synthesis requires less energy

187.   The diagram represent part of the nitrogen cycle. Which process is carried out by nitrifying bacteria?

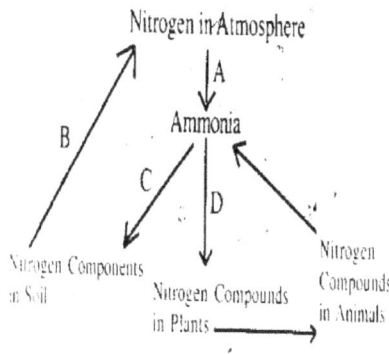

188. From which cell organelle are nucleic acid absent?
A. Chloroplast   B. Golgi body
C. Mitochondrion   D. Ribosome

189. What will break an ionic bond between amino acids?
A. Concussions   B. High temperature
C. Hydrolysis   D. pH change

190. The drawings show stages of the mitotic cell cycle

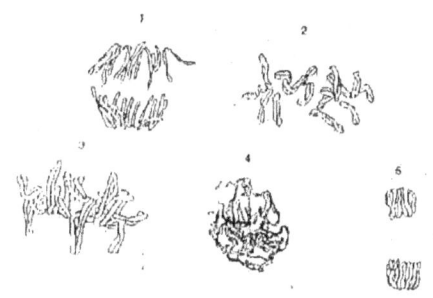

In which order do the stages occur?

	First →				last
A	2	1	3	5	4
B	2	4	1	5	3
C	4	2	1	3	3
D	4	2	3	1	5

191. The first in milk produced by the mother for a new-born baby contains antibodies. What do these antibodies provide?
A. Artificial active immunity
B. Artificial passive immunity
C. Natural active immunity
D. Natural passive immunity

192. Which protein has a fibrous structure?
A. Amylase   B. Collagen
C. Haemoglobin   D. Insulin

193. In which form is carbon dioxide mainly transported in blood?
A. As carbamino – haermoglobin   B. As carbonic acid
C. As hydrogen carbonate   D. In solution

Below are names of groups used when classifying organisms.
I. phylum   II. kingdom   III. class   IV. family
V. species   VI. genus   VII. order
Use it to answer question 194

194. The correct order when grouping organism is
A. I, II, III, IV, V, VI and VII
B. I, II, VI, V and VII
C. II, I, III, VII, IV, VI and V
D. II, IV, I, VI, VII, II and V

195. During sexual reporuduction, the maganucleus in paramecium
A. divide into 2 parts   B. divide into 4 parts
C. breaks into 4 parts   D. fuses with the micronucleus

196. Which of the following tissues represents supporting tissue?
A. columnar epithelium
B. cartilage
C. striated muscle
D. nervous tissue

197. The creamy semi-fluid of food in the stomach after some hours known as
A. chile
B. chyme
C. chyle
D. Choline

198. Which of the following structures is not in the male reproductive parts?
A. Uniferous tubule
B. Prostate gland
C. Epididymis
D. Seminal vesicles

199. The organelle in Amoeba which performs a similar function with that of the kidney in vertebrates is the
A. nucleus
B. contractile vacuole
C. mitochondrion
D. pseudopodium

200. Steady spread of the unpleasant odour from dead rat is a physical process called
A. cyclosis
B. diffusion
C. osmosis
D. plasmolysis

201. De-oxygenated blood from the head region first enters the heart through the
A. right auricle
B. right ventricle
C. left auricle
D. left ventricle

202. Which of the following is not a characteristic of enzymes?
A. they are proton which are activated by co-enzymes
B. they are specific and can act only on specific substrates
C. They are organic
D. Their reactions are irreversible

203. The best way of protecting natural resources is by
A. creating game and forest reserves
B. enforcing conservation laws
C. embarking on tree planting
D. prohibiting lumbering forest reserves

204. One basic characteristic feature of viruses is that they grow
A. as colonies
B. as jelly-like materials
C. inside living cells
D. as plant

205. Which feature is a characteristic of prokaryotic organism?
A. A cell
B. Circular DNA
C. Mitochodria
D. Rough endoplasmic reticulum

206. What is meant by resolution in light microscopy?
A. the product of the magnifications of the eyepiece and the objective lenses
B. the shorter distance between objects that can be seen as separate
C. the size of the smaller object that can be seen
D. twice the wavelength of the light use to illuminate the specimen

207. From which organelle are nucleic acid absent?
A. Chloroplast
B. Golgi body
C. Mitochondria
D. Ribosome

208. For what process is the large surface area of the cristae in the mitochondria important?
A. Energy radiation
B. Enzyme reaction
C. Gaseous exchange
D. Protein synthesis

209. Which level of protein structure maintains the globular shapes of enzymes?
A. primary
B. secondary
C. tertiary
D. quatenary

210. What bonds hold substrate to the active site of an enzyme?
A. disulphide
B. glycoside
C. hydrogen
D. peptide

211. Which of the following set of plants has a taproot system?
A. maize, bamboo, waterleaf
B. waterleaf, balsam, wheat
C. waterleaf, cowpea, balsam
D. wheat, rice and cowpea

212. Which adaptation would increase the efficiency of active transport of carbohydrates from a plant cell?
A. area where the cell wall is thin
B. increase permeability of the cell wall
C. large surface area of the cell surface membrane
D. large volume of the cell vacuole

LADOKE AKINTOLA UNIVERSITY OF TECHNOLOGY POST UME 2009

213. The nutritive layer of the eye in mammals is
A. choroid
B. conjunctiva
C. cornea
D. sclera

214. Ultrafiltration in the kidney takes place in the
A. Bowman's capsule
B. Pelvis
C. Loop of Henle
D. Proximal convoluted tubule

215. Which of the following bones is not a component of the forelimb?
A. olecranon   B. ulna
C. tibia   D. humerus

216. The condition in which the anther mature before the stigma is called
A. protandry   B. epigyny
C. hypogyny   D. protogyny

217. In most true ferns, sporangia are grouped into
A. indusium   B. sori
C. fronds   D. prothalli

218. The ratio of carriers to sicklers in the F, generation derived from a parental cross of two carriers of haemoglobin S gene is
A. 3:1   B. 1:3
C. 2:1   D. 1:2

219. In which part of a legumineous plant can bacteria like Azotobacter be found?
A. spongy mesophyll   B. root nodules
C. stem internodes   D. stem nodes

220. In a dicotyledonous stem, companion cells are found close to the
A. endodermal cell   B. sieve tubes
C. xylem vessels   D. pericyclic fibres

221. The position occupied by an organism in a food chain is referred to as
A. trophic level   B. niche level
C. energy level   D. feed level

222. The ventricles of the mammalian heart are more muscular than the auricles because the
A. auricles have smaller capacity
B. ventricules are larger in size
C. ventricles pump blood to distant organs
D. ventricles receive more blood

223. The young shoot of a plant is referred to as
A. Radicle
B. Plumule
C. Bud
D. branch

233

ANSWERS

It is my expectation that students refer to this answers when they have sincerely and satisfactorily answered the questions.

CHAPTER 1

1. B	2. C	3. C	4. D	5. D
6. D	7. C	8. D	9. B	10. B
11. C	12. C	13. B	14. B	15. C
16. C	17. C	18. B	19. C	20. B
21. A	22. A	23. A	24. C	25. C
26. C	27. C	28. B	29. B	30. B
31. C	32. B	33. A	34. B	35. A
36. D	37. A	38. D	39. B	40. A
41. B				

CHAPTER 2

1. A	2. A	3. C	4. B	5. B
6. B	7. D	8. B	9. C	10. E
11. D	12. B	13. D	14. D	15. C
16. D	17. D	18. C	19. C	20. B
21. C	22. E	23. B	24. B	25. C
26. D	27. D	28. C	29. E	30. D
31. B	32. C	33. C	34. C	35. A
36. B	37. B	38. D	39. B	40. B
41. C	42. A	43. D	44. D	45. D
46. B	47. B	48. B	49. B	50. D
51. B	52. B	53. D	54. B	55. E
56. A	57. E	58. A	59. A	60. C
61. A	62. C	63. A	64. C	65. C
66. E	67. D	68. C	69. C	70. C
71. D	72. D	73. D	74. C	75. B
76. B	77. D	78. C	79. A	80. C
81. D	82. A	83. C	84. C	85. D
86. C	87. B	88. C	89. D	90. B
91. B	92. B	93. A	94. B	95. D
96. D	97. A	98. B	99. B	100. D
101. B	102. D	103. B	104. B	105. C
106. B	107. C	108. C	109. A	110. C
111. C	112. B	113. C	114. B	115. B
116. B	117. C	118. A		

CHAPTER 3

1. D	2. C	3. E	4. B	5. B
6. D	7. A	8. E	9. A	10. D
11. C	12. C	13. B	14. C	15. C
16. B	17. B	18. A	19. B	20. B
21. C	22. C	23. C	24. D	25. A
26. C	27. A	28. A	29. D	30. D
31. D	32. B	33. B	34. C	35. D
36. D	37. B	38. B	39. B	40. B
41. C	42. B	43. B	44. E	45. C
46. D	47. A	48. D	49. D	50. C
51. D	52. B	53. D	54. D	55. C
56. C	57. C	58. D	59. B	60. B
61. A	62. C	63. E	64. B	65. E
66. C	67. B	68. C	69. D	70. E
71. A	72. A	73. A	74. C	75. C
76. B	77. B	78. E	79. C	80. C
81. B	82. C	83. E	84. E	85. C
86. D	87. B	88. A	89. D	90. C
91. C	92. B	93. C	94. D	95. E
96. C	97. D	98. B	99. A	100. D
101. A	102. D	103. A	104. C	105. B
106. C				

CHAPTER 4

1. B	2. D	3. C	4. C	5. A
6. C	7. D	8. E	9. B/C	10. B
11. D	12. C	13. D	14. B	15. B
16. D	17. C	18. C	19. E	20. D
21. C	22. C	23. C	24. D	25. B
26. C	27. A	28. D	29. B	30. E
31. D	32. A	33. B	34. B	35. C
36. B	37. A	38. D	39. A	40. B
41. A	42. D	43. A	44. C	45. A
46. A	47. A	48. C	49. A	50. A
51. D	52. C	53. C	54. A	55. B
56. D	57. C	58. A	59. D	60. A
61. C	62. C	63. B	64. B	65. B
66. B	67. D	68. A	69. E	70. E
71. E	72. A	73. C	74. D	75. C

CHAPTER 5

1. C	2. Nil	3. A	4. C	5. B
6. E	7. C	8. B	9. B	10. B
11. B	12. C	13. B	14. B	15. D
16. C	17. E	18. C	19. B	20. B
21. B	22. E	23. A	24. A	25. C
26. B	27. B	28. B	29. B	30. E
31. C	32. B	33. E	34. E	35. C
36. D	37. C	38. A	39. C	40. A
41. B	42. B	43. E	44. B	45. B
46. E	47. D	48. B	49. A	50. A
51. A	52. C	53. E	54. A	55. C
56. C	57. E	58. E	59. A	60. B
61. D	62. A	63. C	64. E	65. D
66. D	67. A	68. A	69. A	70. B
71. A	72. C	73. D	74. B	75. C
76. A	77. B	78. B	79. D	80. B

CHAPTER 6

1. A	2. B	3. A	4. C	5. D
6. B	7. C	8. B	9. D	10. B
11. D	12. A	13. B	14. D	15. C
16. A/B	17. B	18. A	19. A	20. A
21. E	22. C	23. A	24. B	25. D
26. D	27. C	28. C	29. D	30. E
31. C	32. B	33. C	34. B	35. D
36. C	37. D	38. A	39. D	40. C
41. C	42. A	43. C	44. C	45. B
46. B	47. B	48. D	49. A	50. C
51. A	52. A	53. A	54. C	55. B

CHAPTER 7

1. B	2. D	3. D	4. B	5. D
6. B	7. C	8. A	9. C	10. B
11. D	12. D	13. C	14. E	15. A
16. D	17. C	18. A	19. A	20. C
21. C	22. A	23. B	24. A	25. C
26. C	27. A	28. B	29. C	30. D
31. B	32. A	33. D	34. B	35. A
36. B	37. B	38. C	39. C	40. B
41. E	42. C	43. D	44. B	45. B
46. C	47. B	48. B	49. B	50. A
51. D	52. D	53. B	54. C	55. E
56. D	57. D	58. E	59. E	60. C
61. A	62. B	63. D	64. C	65. B
66. E	67. A	68. B	69. A	70. B

71. D	72. A	73. A	74. E	75. C
76. E	77. A	78. A	79. A	80. D
81. D	82. C	83. B	84. D	85. B
86. B	87. D	88. D	89. C	90. D
91. A	92. E	93. A	94. B	95. C
96. C	97. C	98. A	99. A	100. A
101. C	102. D	103. A	104. D	105. C
106. D	107. A	108. A	109. B	110. D
111. B	112. D	113. A	114. C	115. B
116. B	117. A	118. D	119. C	120. C
121. C				

CHAPTER 8

1. B	2. B	3. A	4. D	5. A
6. B	7. E	8. D	9. B	10. C
11. C	12. C	13. D	14. A	15. C
16. A	17. B	18. B	19. C	20. D
21. C	22. E	23. B	24. B	25. B
26. A	27. A	28. B	29. B	30. E
31. D	32. C	33. D	34. C	35. C
36. D	37. D	38. D	39. A	40. C
41. A	42. D	43. C	44. A	45. B
46. C	47. B	48. C	49. D	50. B
51. D	52. A	53. C	54. A	55. A
56. C	57. A	58. D	59. D	60. D
61. E	62. A	63. A	64. B	65. C
66. D	67. B	68. B	69. C	70. D
71. C	72. C	73. B	74. E	75. E
76. A	77. D	78. C	79. E	80. E
81. A				

CHAPTER 9

1. C	2. D	3. C	4. A	5. A
6. E	7. E	8. B	9. A	10. D
11. A	12. C	13. B	14. A	15. C
16. D	17. B	18. E	19. B	20. C
21. D	22. D	23. A	24. C	25. D
26. C	27. C	28. D	29. C	30. B
31. A	32. A	33. C	34. B	35. D
36. C	37. E	38. A	39. D	40. D
41. D	42. C	43. D	44. D	45. A
46. D	47. A	48. C	49. D	50. C
51. D	52. D	53. C	54. B	55. D
56. D	57. E	58. D	59. C	60. B
61. D	62. D	63. E	64. C	65. B
66. B	67. B	68. B	69. D	70. E
71. B	72. E	73. A		

CHAPTER 10

1. C	2. B	3. D	4. B	5. B
6. A	7. A	8. C	9. C	10. C
11. A	12. D	13. B	14. C	15. D
16. B	17. A	18. B	19. A	20. D
21. A	22. D	23. D	24. C	25. A
26. C	27. C	28. C	29. D	30. B
31. C	32. D	33. B	34. D	35. B
36. E	37. C	38. D	39. D	40. C
41. E	42. A	43. E	44. A	45. C
46. E	47. B	48. A	49. E	50. C
51. D	52. E	53. D	54. E	55. B
56. D	57. A	58. B	59. E	

CHAPTER 11

1. D	2. A	3. D	4. A	5. B
6. C	7. C	8. A	9. D	10. D
11. A	12. B	13. D	14. A	15. A
16. D	17. E	18. B	19. D	20. D
21. A	22. C	23. D	24. A	25. C
26. C	27. E	28. C	29. A	30. C
31. A	32. C	33. C	34. E	35. C
36. B	37. C	38. D	39. C	40. D
41. A	42. C	43. A	44. B	45. D
46. B	47. E	48. A	49. B	50. B
51. C	52. B	53. C	54. A	55. D
56. B	57. C	58. C	59. B	60. C
61. B	62. A	63. B	64. B	65. A
66. C	67. B	68. D	69. D	

CHAPTER 12

1. B	2. A	3. A	4. B	5. Nil
6. A	7. A	8. D	9. B	10. E
11. A	12. D	13. B	14. D	15. E
16. D	17. E	18. D	19. A	20. E
21. D	22. D	23. B	24. D	25. B
26. C	27. A	28. A	29. D	30. D
31. E	32. C	33. A	34. D	35. A
36. B	37. B	38. D	39. C	40. D
41. A	42. A	43. E	44. D	45. C
46. A	47. A	48. B	49. B	50. A
51. B	52. B	53. B	54. A	55. C
56. A	57. D	58. A	59. C	60. C
61. D	62. B	63. D	64. B	65. D
66. B	67. E	68. D	69. B	70. C
71. B	72. C	73. C	74. A	75. A

76. A	77. B	78. C	79. D	80. C
81. C	82. A	83. B	84. C	85. B
86. B	87. E	88. B	89. C	90. E
91. D	92. D	93. B	94. B	95. C
96. D	97. D	98. C	99. A	100. D
101. C	102. C	103. D	104. A	105. B
106. A	107. B	108. C	109. A	110. C
111. B	112. D	113. D	114. D	115. A
116. C	117. A	118. D	119. E	120. C
121. E	122. B	123. E	124. B	125. E
126. D	127. C	128. D		

CHAPTER 13

1. C	2. A	3. D	4. B	5. D
6. B	7. B	8. A	9. C	10. C
11. B	12. B	13. D	14. B	15. D
16. C	17. B	18. B	19. C	20. D
21. B	22. B	23. B	24. C	25. A
26. C	27. A	28. B	29. D	30. A
31. B	32. D	33. C	34. C	35. B
36. C	37. E	38. C	39. A	40. E
41. E				

CHAPTER 14

1. E	2. C	3. A	4. D	5. D
6. A	7. C	8. B	9. C	10. A
11. B	12. C	13. C	14. D	15. E
16. B	17. E	18. C	19. Nil	20. B
21. B	22. E	23. A	24. D	25. A
26. D	27. C	28. D	29. A	30. B
31. D	32. C	33. C	34. A	35. E
36. A	37. C	38. D	39. B	40. A
41. A	42. D	43. C	44. B	45. B
46. A	47. A	48. C	49. D	50. C
51. B	52. A	53. D	54. D	55. D
56. B	57. A	58. D	59. A	60. A
61. D	62. B	63. B	64. B	65. C
66. C	67. D	68. B	69. B	70. B
71. A	72. B	73. B	74. D	75. C
76. A	77. B	78. C	79. C	80. C
81. D	82. D	83. B/C	84. A	85. A
86. D	87. C	88. A	89. D	90. B
91. D	92. D	93. D	94. D	95. C
96. A	97. C	98. B	99. A	100. C
101. C	102. C	103. C	104. A	105. B
106. C	107. E	108. B	109. E	110. C
111. D	112. D	113. D	114. C	115. E

116. A   117. B   118. C   119. B   120. E
121. C   122. A   123. D   124. D   125. C

CHAPTER 15

1. E	2. B	3. C	4. B	5. C
6. C	7. B	8. A	9. C	10. B
11. B	12. A	13. C	14. A	15. A
16. D	17. A	18. A	19. C	20. B
21. B	22. C	23. B	24. D	25. B
26. C	27. D	28. B	29. A	30. B
31. C	32. A	33. C	34. B	35. D
36. C	37. B	38. C	39. B	40.
41. D	42. A			

CHAPTER 16

1. D	2. C	3. D	4. B	5. A
6. B	7. A	8. A	9. C	10. A
11. D	12. D	13. B	14. E	15. A
16. B	17. B	18. C	19. A	20. C
21. B	22. B	23. B	24. C	25. C
26. B	27. D	28. C	29. D	30. B
31. D	32. D	33. B	34. C	35. A
36. B	37. A	38. C	39. A	40. C
41. D	42. C	43. A	44. D	45. D
46. A	47. A	48. B	49. B	50. A
51. E	52. C	53. E	54. C	55. D
56. D	57. D	58. D	59. D	60. A
61. C	62. A	63. D	64. A	

CHAPTER 17

1. B	2. B	3. D	4. C	5. A
6. A	7. D	8. A	9. B	10. D
11. C	12. B	13. B	14. C	15. C
16. A	17. A	18. B	19. C	20. C
21. D	22. A			

CHAPTER 18

1. C	2. A	3. B	4. C	5. C
6. D	7. C	8. C	9. B	10. D
11. A	12. A	13. A	14. C	15. A
16. A	17. C	18. A	19. B	20. D
21. C	22. B	23. C	24. A	25. D
26. C	27. D	28. A	29. C	30. A

CHAPTER 19

1. B	2. C	3. D	4. D	5. B
6. C	7. B	8. D	9. B	10. D
11. B	12. B	13. B	14. C	15. C
16. A	17. C	18. C	19. B	20. A
21. C	22. D	23. B	24. B	25. D
26. A	27. B	28. B		

CHAPTER 20

1. B	2. B	3. D	4. D	5. D
6. E	7. C	8. A	9. D	10. D
11. B	12. D	13. A	14. A	15. B
16. C	17. D	18. C	19. D	20. A
21. D	22. D	23. E	24. D	25. A
26. D	27. B	28. C	29. C	30. D
31. A	32. B	33. C	34. A	35. B
36. A	37. B	38. D	39. C	40. A
41. A	42. C	43. C	44. B	45. B
46. C	47. D	48. B		

CHAPTER 21

1. A	2. E	3. D	4. D	5. D
6. A	7. D	8. D	9. B	10. D
11. A	12. D	13. D	14. A	15. B
16. A	17. B	18. A	19. D	20. D
21. D	22. D	23. A	24. D	25. C
26. D	27. D	28. C	29. D	30. D
31. B	32. C	33. B	34. C	35. C
36. D	37. C	38. A	39. D	40. A
41. B	42. A	43. A	44. A	45. A
46. D	47. D	48. C	49. B	50. A
51. B	52. B	53. B	54. C	55. A
56. C	57. A	58. C	59. A	60. C
61. B	62. D	63. D	64. A	65. C
66. C	67. B	68. B	69. D	

CHAPTER 22

1. C	2. A	3. C	4. B	5. B
6. B	7. A	8. A	9. A	10. B
11. E	12. C	13. D	14. C	15. B
16. A	17. D	18. A	19. D	20. B
21. A	22. A	23. B	24. D	25. C
26. C	27. C	28. A	29. A	30. B

CHAPTER 23

1. E	2. D	3. C	4. B	5. D
6. B	7. A	8. A	9. B	10. C
11. E	12. B	13. D	14. B	15. C
16. B	17. B	18. B	19. B	20. B
21. B	22. D	23. D	24. B	25. D
26. E	27. A	28. C	29. C	30. D
31. B	32. E	33. C	34. C	35. C
36. D	37. C	38. C	39. D	40. C
41. D	42. B	43. A	44. D	45. A

46. D 47. A 48. A 49. C 50. B
51. B 52. B 53. C 54. B 55. A
56. C 57. A 58. D 59. C 60. C
61. D 62. A 63. D 64. B 65. A
66. A 67. A 68. C 69. B 70. A
71. C 72. A 73. C 74. A 75. A
76. C 77. A

CHAPTER 24
1. A 2. E 3. A 4. A 5. C
6. C 7. B 8. C 9. A 10. D
11. D 12. A 13. A 14. A 15. A
16. A 17. C 18. A 19. D 20. A
21. A 22. D 23. B 24. D 25. C
26. A

CHAPTER 25
1. D 2. C 3. D 4. D 5. C
6. C 7. A 8. B 9. C 10. C
11. A 12. C 13. D 14. B 15. B
16. D 17. B 18. A 19. A 20. C
21. A 22. B 23. B 24. C 25. C
26. C 27. C 28. A 29. A 30. D
31. D 32. D 33. A 34. D 35. A
36. A 37. C 38. D 39. D 40. C
41. C 42. D 43. D 44. B 45. A
46. C 47. B 48. D 49. D 50. C
51. A 52. B 53. A 54. D 55. B
56. D 57. D 58. A 59. B 60. B
61. B 62. B 63. B 64. C 65. A
66. C 67. A 68. A 69. D 70. A
71. A 72. C 73. B 74. C 75. B

CHAPTER 26
1. D 2. C 3. B 4. C 5. D
6. A 7. A 8. D 9. B 10. B
11. D 12. A 13. D 14. B 15. B
16. B 17. B 18. D 19. D 20. D
21. C 22. A 23. A 24. C 25. C
26. A 27. D 28. D 29. A 30. B
31. D 32. D 33. C 34. A 35. C
36. D 37. C 38. A 39. A 40. B
41. A 42. D 43. B 44. D 45. B
46. C 47. B

ANSWERS TO POST UME QUESTIONS
1. B 2. D 3. A 4. B 5. B
6. D 7. C 8. B 9. C 10. D
11. C 12. D 13. D 14. D 15. C
16. B 17. A 18. B 19. B 20. C
21. A 22. D 23. A 24. A 25. A
26. B 27. B 28. D 29. A 30. A
31. D 32. A 33. C 34. C 35. A
36. C 37. A 38. C 39. C 40. C
41. C 42. C 43. C 44. B 45. B
46. B 47. B 48. B 49. B 50. D
51. B 52. D 53. A 54. B 55. A
56. A 57. B 58. B 59. D 60. C
61. C 62. B 63. C 64. A 65. A
66. C 67. A 68. D 69. C 70. B
71. D 72. D 73. B 74. B 75. B
76. B 77. A 78. D 79. A 80. A
81. A 82. B 83. B 84. A 85. C
86. D 87. C 88. C 89. D 90. D
91. C 92. C 93. D 94. B 95. A
96. C 97. C 98. B 99. B 100. B
101. B 102. C 103. D 104. B 105. C
106. B 107. C 108. A 109. B 110. B
111. C 112. D 113. C 114. B 115. D
116. D 117. B 118. C 119. D 120. D
121. B 122. D 123. C 124. D 125. D
126. C 127. D 128. D 129. D 130. D
131. C 132. A 133. A 134. B 135. A
136. C 137. C 138. B 139. A 140. D
141. C 142. C 143. B 144. D 145. B
146. A 147. C 148. C 149. B 150. C
151. B 152. B 153. A 154. D 155. D
156. A 157. B 158. C 159. A 160. E
161. D 162. C 163. A 164. D 165. D
166. E 167. E 168. A 169. A 170. E
171. B 172. C 173. D 174. E 175. D
176. A 177. A 178. A 179. D 180. D
181. B 182. C 183. A 184. B 185. D
186. C 187. C 188. B 189. C 190. D
191. D 192. B 193. C 194. C 195. A
196. B 197. B 198. A 199. B 200. B
201. A 202. A 203. B 204. C 205. B
206. B 207. B 208. B 209. D 210. C
211. C 212. C 213. A 214. A 215. C
216. A 217. B 218. C 219. B 220. B
221. A 222. C 223. B

CHAPTER ONE

1. B = All living organism respire to liberate energy. Movement in flowering plants nearly always involves growth and so unique to them. Feeding is used loosely to refer to nutrition in animals. Only plants photosynthesize and transpire.

2. C = Viruses are link between living and non-living things.

3. C = Spirogyra is an algae.

4. D = Examples of non-living things are: water, sun, bread, air etc.

5. D = Excretion, the ability to remove unwanted substances, is a characteristic exhibited by all living organisms.

6. D = A non-living object like crystals can also 'grow' in size. However, in non-living things, growth is external because it takes place due to the deposition (addition) of more particles on the outer surface of the crystal. Growth in living organism is internal as result of the addition of new protoplasm.

7. C = Onion bulb (plant cell) posses cellulose cell wall which is absent in cheek cell (an animal cell).

8. D = Golgi complex (or Golgi apparatus) modifies proteins and also transport organic materials in and out of the cell.

9. D = The nucleus contains thread-like called chromosomes. They carry the genes. The genes are made up of chromatis thread which joins to form chromatids. A chromosome consists of two strands of chromatids.

10. B = A matured plant cell has a cellulose cell wall, hence has rigid shape.

11. C = The reservoir of energy for the cell is carried by the mitochondria not the nucleus.

12. C = Centriole is peculiar to animal cells while middle lamella, starch grains and cellulose cell wall are peculiar to plant cells.

13. B = Endoplasmic reticulum transport substances within the cytoplasm and the nucleus

14. B = 1– cell membrane; 2 – rough endoplasmic reticulum; 3 – nucleolus, 4 – mitochondria; 5 – Golgi complex; 6 – nuclear membrane.

15. C = The nucleolus is a dark area in the nucleus where ribosomal RNA is being made.

16. C = The mitochondria is also known as the power house of the cell because they generate most of the cell's supply of ATP use as a source of chemical energy.

17. C = The chromosome contains genes, which are the units of inheritance.

18. B = Cell vacuole in plant cell is surrounded by the tonoplast.

19. C = Cell wall and chloroplast are peculiar to plant cells.

20. B = Ribosome is responsible for protein synthesis.

21. A = see 16

22. A = Cytoplasm in unicellular organism e.g. Amoeba is differentiated into ectoplasm and endoplasm just as the protoplasm is differentiated into nucleus and cytoplasm.

23. A = The role of lysosomes enables it to detoxify the cell of harmful substance. They are also called the 'suicide bag' of the cell.

24. C = The cell membrane is made up of proteins and lipids in the ratio of 1:50. It is also called the plasma membrane or plasmalemma.

25. C = Cell → Tissue → Organ → System → Multicellular organism.

26. C = Aggregates of cells form tissue; aggregates of tissues form organ; aggregates of organs form system. Well coordinated systems and organs form an organism.

27. C = An amoeba is a single cell. An unlaid chicken egg is also a single cell.

28. B = Rods and cones are cells present in the retina – a tissue in the eye.

29. B = Onion bulb is an organ – aggregates of epithelial tissues.

30. B = Tissue are aggregates of cells performing one or more functions.

31. C = Kingdom protista contains protists (Protozoa e.g. Amoeba and protophyta e.g. Chlamydomonas) are eukaryotic unicellular organisms. They lack specialized tissues and organs.

32. B = A virus possesses strand of nucleic acid which may be DNA or RNA.

33. A = Presence of DNA or RNA in viruses are responsible for the transmission of characters.

34. B = Mitochondria and chloroplast are responsible for cellular respiration and photosynthesis respectively.

35. A = The nucleus is a central organelle because it contains genetic materials used in controlling other activities in organism

36. D. = Water is a medium of all metabolic reactions in the body of living organisms.

37. A = Mitochondria will dominates in cells that requires more energy for its activities. Sperm cells contains many mitochondria to provide energy for the it propulsive movement.

38. D = Ribosomes are attached to the walls of the rough endoplasmic reticulum.

39. B = All cells have plasmalemma (cell membrane)

40. A = A bacterial cell (a prokaryotic cell) has no nucleus or any membrane – bound organelle hence its circular protein i.e. DNA lies in the cytoplasm.

41. B = Non-living things have relatively low energy content compared to living things.

CHAPTER TWO

1. A = Prokaryotic cells lack a defined nucleus; they lack membrane – bounded organelles but they possess a complex cytoplasm.

2. A = The pioneer organism in a plant succession on bare rock surfaces are some species of crutose lichen, followed by foliose lichen and then mosses.

3. D = Kingdom→Phylum→Class→Order→Genus→Species. As we go down the levels, the variety of organisms decreases and similarity of organism increases.

4. B = Kingdom monera have prokaryotic cell e.g. bacterial and blue green algae. Their nuclear materials e.g. chromosomes are found scattered in the cytoplasm as they lack a nuclear membrane.

5. B

6. B = Blue – green algae are prokaryote – kingdom monera, protophytes are plant–like protists e.g. Chlamydomonas– kingdom protista. All protists are eukaryotes and not prokaryotes.

7. D = Bacteria are prokaryotes. All prokaryotes are non-nucleated.

8. B = Bacteria will not survive in salt water.

9. C = Eukaryotic cells walls are usually made of cellulose or chitin (fungi). In contrast, most bacterial cell walls contain peptidoglycan a network of modified sugar polymers.

10. E = All of the above

11. D = Malaria is caused by a plasmodium, a protozoan. Pneumonia can be caused by different types of microorganisms, including bacteria, viruses and fungi.

12. B = Putrefying bacteria are usually anaerobic. They decompose animal proteins resulting in amines such a putrescine and cadaverine which have a putrid odour. They cause food poisoning.

13. D = Bacteria toxin are the most powerful human poisons known and retain high activity at very high dilutions.

14. D = Immunity is the state of having sufficient biological defenses to avoid infection, disease or other unwanted biological invasion. The case in the question gives an acquired immunity.

15. C = Pellicle is an animal–like feature in euglena. Although the possession of flagellum is also animal-like, it is present in some plants like Chlamydomonas.

16. D = The contractitle vacuole is for osmo-regulation.

17. D = Plant characteristics of Euglena include: (i) chloroplast, therefore, nutrition is holophytic (autotrophic); can be holozoic in the absence of light (ii) store food as starch–like carbohydrate, paramylon (iii) Possesses pyrenoid, an organelle found in many algae, which functions in the synthesis of paramylon.

18. C = Spirogyra is non-motile, Euglena moves by means by cilia and chlamydomonas moves by means of flagella.

19. C = The mushroom bears a microscopic "fruiting body" which consist of a foot (hyphae) which attaches it to the substrate, a stalk or stipe which bears an umbrella shaped portion at its end known as pileus or cap.

20. B = Lichen is formed by the combination of algae and fungi; the algal part is found in between two fungi parts; while the alga produces food by photosynthesis, the fungus protects the alga and also provides water for photosynthesis, hence, the two are in a symbiotic relationship.

21. C = Spore production and dispersion occur in all the classes of organisms mentioned.

22. E = Fungi lack chlorophyll; hence, they cannot photosynthesize. Conjugation occurs in algae and fungi only while mosses, ferns and some algae show alternation of generation. All of them reproduce by forming spores and not seeds.

23. B = Fungi are multinucleated and multicellular with exception of yeasts, which is unicellular.

24. B

25. C

26. D = Evolutionary trend is as shown: Schizophyta (Becteria)→ Thallophyta (Algae)→ Bryophyta (Mosses and Liverworts) → Pteridophyta (ferns) →Spermatophyta (seed plants). Spermatophyta is further divided into Gymnosperms (conifers) and Angiosperms (Flowering plants).
Memory Tip: Samuel The Biologist Planted Sugarcane.

27. D = Bryophytes and Pteridophytes inhabit moist terrestrial habitat (require moisture for fertilization). Spermatophytes (Gymnosperms and Angiosperms live exclusively on dry land.

28. C

29. E

30. D = Thallophytes (Algae) and Bryophyte (Mosses and Liverworts) are non – vascular.

31. B = The dominant phase in the life cycle of a bryophyte is the sporophyte.

32. C

33. C = Only Pteridophytes and Spermatophytes have their vegetative bodies differentiated into true roots, stems and leaves.

34. C

35. A

36. B = Sporophyte is the dominant phase in the life cycle of mosses and ferns, that show alternation of generation

37. B = Whistling pine is a gymnosperm and have cones, which are naked seeds.

38. D

39. B

40. B

41. C

42. A = Double fertilization is peculiar to flowering plants leading to the formation of fruit and seed.
43. D
44. D
45. D = This is because insects can live in almost all habitats and can survive unfavourable conditions and their mode of reproduction, undergoing metamorphosis makes them very difficult to eradicate.
46. B = An animal is said to be radially symmetrical when it can be divided into two equal halves along any plane e.g. hydra.
47. B = The bodies of coelenterates are composed of two clearly marked layers of cells, an outer ectoderm and a inner endoderm i.e. they are diploblastic animals.
48. B = Coelenterates have a cavity within their body, the entero/gastrocavity with a single opening to the outside (mouth) for ingestion and egestion.
49. B
50. D = Coelenterates reproduce asexually by budding
51. B
52. B = The cniodoblast is an ectodermal cell in which there is a large sac, the nematocysts, that are used for offense and defense.
53. D = Coelenterates are pseudecoelomates; they lack true coelom or body cavity called enteron.
54. B
55. E
56. A = The hypostome is the most apical structure, the domed portion of head where the mouth opens.
57. E = The testes are usually located below the tentacles i.e. above buds.
58. A = The ovary is usually located below the bud i.e. a little distance above the basal disc or foot.
59. A = ectodermal cells derived from the musculo-epithelial cells.
60. C = Interstitial cells form the reproductive organs and buds.
61. A
62. C = Excess water enters the enteron (body cavity) which is then remove via the mouth.
63. A = An example of symbiotic association.
64. C = There are 8 forms of movement in hydra namely: gliding, floating, swimming, walking, climbing, dragging, looping (forming 180° angle plane, most common type) and somersaulting.
65. C
66. E = Hydra is an hermaphrodite but prevent self – fertilization by protandry i.e. testis and sperms mature first before the ovary.
67. D
68. C = Coelenterates and platyhelminthes.
69. C = Annelids are metamerically segmented while nematodes are not.
70. C = As a result of their metameric segmentation.

71. D = Cocoon in earthworm is secreted by the clitellum. The clithelum is made up of about five or six glandular segments. Eggs are laid in the cocoon.
72. D = The soil is ground up in the gizzard.
73. D
74. C = Phylum Arthropoda is divided into four subphyla namely: Crustacea (e.g. crab), Insecta (e.g. cockroach), Arachnida (eg. spider) and Myriapoda (e.g. millipedes and centipedes) **Memory tip: CIAM**.
75. B
76. B = Snails have tentacles, earthworms have cheatea and millipedes have antennae.
77. D = Arthropods have a segmented body, exoskeleton made of chitin.
78. C = The thoracic section of the cephalothorax in crayfish has eight pairs of jointed appendages. The first three pairs are maxillipedes or "jaw-feet". They help to hold food material during feeding.
79. A = On the head of insects are a pair of antenneae (not two pairs) and a pair of compounds eyes.
80. C = The characteristics described are all true except in C. Each thoracic segment bears a pair of joints legs. The three NOT two.
81. D
82. A
83. C = Pedipalp, also known as feeler, is used for feeling.
NOTE: Spiders have spinnerettes on the abdomen from which the silk used in web-making comes. Not all spiders make webs, some hunt their prey.
84. C = Arachnids have no antennae.
85. D = A centipede has poison claws which it use to hunt its prey while millipedes feed on dead and decaying organic matter. A centipede possesses a pair of legs at each abdominal segment while millipede possesses two pairs of legs at each abdominal segment.
86. C = Insects have two compound eyes while millipedes have two simple eyes (ocelli).
87. B = Snail body is covered with hard calcareous (not chitinous) shell secreted by the mantle into which the head and food retracts.
88. C = Molluscs exist as free living organisms. Although some serve as intermediates host of parasitic worms.
89. D
90. B = respiration in molluscs is through gills or a lung which is within the mantle cavity. The upper surface of the mantle cavity is vascularized in land snail, which used muscular contractions to pump air in and out of the small respiratory pore at the anterior edge of the mantle cavity.
91. B
92. B
93. A = Viviparity is when fertilized eggs are produced, developed and nourished inside the body of the mother through placenta. Examples are all placenta mammals.

- Oviparity is the laying of eggs in which the embryos hatch outside the body of the parents. Examples are amphibians, most fish (e.g. tilapia) and birds (e.g. hen, pigen duck).
- Ovoviviparity is when fertilized eggs are produced, developed and nourished inside the body of the mother without placenta and hatch before or soon after eggs are laid. Examples are certain fishes and reptiles.
94. B = The first terrestrial vertebrates are reptiles; they evolved from amphibians (inhabit land and water).
95. D
96. D
97. A
98. B
99. B = Volvox exists as a colony.
100. D = Hyphal wall of fungi are made of the chitin.
101. B
102. D
103. B
104. B
105. C
106. B = Aves (birds) are completely oviparous.
107. C
108. C
109. A
110. C
111. C
112. B = Obelia is a coelenterate.
113. C = Metameric segment.
114. B
115. B
116. B
117. C
118. A

CHAPTER THREE

1. D = Spirogyra is a filamentous algae, the cells are joined end to end. Each cell can carry out nutrition, respiration, growth and reproduction independently.
2. C = Spirogyra have unbranched filaments (straight filament)
3. E = Pyrenoids are associated with the formation of starch.
4. B = Gametes are dissimilar usually "male" and "female" gametes.
5. B = Some species of spirogyra have rhizoids which they use to attach to surfaces.
6. D= Spirogyra cells is coated with a mucilage layer and gives a slippery feeling.
7. A
8. E = Autotrophic (holophytic) because of the presence chlorophyll.

9. A
10. D
11. C
12. C
13. B = The vegetative body of fungi is called a mycelium.
14. C = The rhizoid is the root-like structure of the mucor and rhizopus. It is formed from the hyphae.
15. C = The cell wall of fungi is not made of the normal cellulose found in other plants but chitin and a special type of cellulose called fungal cellulose.
16. B = Fusion takes place between two gametangia, which are multinucleated cells formed from branched hyphae of each.
17. B = Structurally, there is no difference. Sporangiophore is a vertically growing reproductive hypha, which grows away from the substrate.
18. A = Fungi e.g. Mucor, *Rhizopus* and yeast are non-green. They lack chlorophyll.
19. B = In some fungi, the filaments are divided into cells by cross walls (septate) e.g. *Penicillium*, but in others e.g. *Rhizopus* the cross walls are absent (non-septate) or coenocytic.
20. B
21. C
22. C = Carbohydrate is absorbed as glucose as in other heterotrophic organisms.
23. C = In fungi, carbohydrate food reserve is stored in the form of glycogen and not starch.
24. D = Rhizoid perform similar functions as roots of plants absorption.
25. A = In bryophytes, the male and female sex organs, the antheridia and archegonia respectively are produced in the gametophyte.
26. C
27. A = Asexual stage is the sporophyte generation while the sexual stage is the gametophyte generation.
28. A = Male gametangium (antheridium); female gametangium (archegonium). This is also true for fern.
29. D = Gametophyte stage: prothallus antheridium (spermatozoid) and archegonium (egg cell).
    Sporophyte stage: Fern plant Sporangium spore
30. D = An inflorescence is a reproductive structure found in flowering plants (angiosperms) consisting of many flowers while strobilus is the reproductive structure of gymnosperms consisting of a number of sporophylls.
31. D = The sporophyte on ferns is the visible fern plant; it is diploid and independent of the gametophyte.
32. B = I –Archegonium (female sex organ); II –Antheridium (male sex organ) III – Rhizoids.
33. B = The structure contains both male and female gamete. It is therefore equivalent to the gametophyte generation of moss.
34. C = The sporophyte is the dominant plant. Other descriptions are true of the prothalus.
35. D = Sporophyte = diploid (asexual reproduction).

Gametophyte = haploid (sexual reproduction).

36. D = Antherozoid + archegonium = Zygote (inside haploid archegonium).

37. B = Each spore germinates into a bisexual prothallus – gametophyte generation.

38. B = Dryopteris is a fern.

39. B = Many sporangium are organized into a sorus with or without a cover or indusium.

40. B = Spore dispersal is effected by means of the annulus and stomium under the control of changes in humidity. The hygroscopic annulus absorbs water from the air when the humidily is high.

41. C

42. B = Each sporangium is made up of a head and a stalk. The head is ovoid with a wall made up of thick – walled cells called annulus. A few cells of the wall however remain thin-walled forming the stomium.

43. B   44. E

45. C   46. D

47. A = Duration of sunlight determines the flowering period of a plant.

48. D = Axil is the angle between the leaf III and the stem in which the axillary bud II borne. Node is where leaves are attached. Internodes is the part between two successive nodes.

49. D = Photosynthesis occurs in the mesophyll cells of the leaf chlorophyllous parts in the shoot system. Photosynthesis is least likely to occur in the root (IV).

50. C = The characteristics combined in C are true for a monocot. They also have fibrous root system.

51. D = Root hairs develop in the region of differentiation (region of maturation), hence the region or zone of differentiation is also called the root hair region.

52. B

53. D = Cassava has tuberous root where food is stored.

54. D = White mangrove have breathing roots (pneumatophores) which allow the intake of atmospheric air.

55. C = Valamen is a spongy, multiple epidermis that covers the roots of some epiphytes. It is many cell layers thick and capable of absorbing atmospheric moisture and nutrients.

56. C = No secondary thickening in monocot because of the absence of cambium.

57. C = Medullary rays lie between two adjacent vascular bundles

58. D = Collenchyma cell walls are thickened at the corner while parenchyma cells consists of loosely packed, thin-walled, rounded cells with air spaces between them.

59. B

60. B = Periderm cells are cork cells in the stems of dicots. The periderm is produced by the phellogen (cork cambium) toward the outside of the stem.

61. A = Meristematic tissue continues to rapidly divide producing undifferentiated cells.

62. C = See page 36.

63. E = The leaves of onion are modified into a storage organ known as bulb.

64. B = Bulbs like garlic and onion store food in the form of sugar rather than as starch.

65. E = *Bryphyllum pinnatum* can be easily propagated by leaves because their leaves can develop buds and adventitious roots at the nodes along the margin. The buds develop into new plants.

66. C

67. B = Trichocyst in paramecium is for defence.

68. C = The contractile vacuole is present in both Amoeba and Paramecium.

69. D = Contractile vacuole perform a similar function to kidney, osmoregulation.

70. E = The cell membrane is also known as plasmalemma and provides a layer of protection to cell. It forms the outermost part of membrane in amoeba. Pellicle is the membrane covering that protects the paramecium like skin. It is responsible for the definite shape in paramecium.

71. A = Amoeba excrete ammonia through cell membrane (body surface) in dissolved form.

72. A = Excretion of ammonia– body surface; Osmoregulation– contractile vacuole.

73. A = Amoeba feeds predominantly on bacteria and diatom.

74. C = Contractile vacuoles are osmoregulatory in function as they are involved in the removal of excess water from the body.

75. C

76. B = Cysts formation is a means of survival under adverse conditions.

77. B

78. E = Amoebic dysentery (amoebiasis) is an infection of the intestine (gut) caused by an amoeba called *Entamoeba histolytica*, that among other things, can cause severe diarrhoea with blood.

79. C

80. C.

81. B = The surface of the paramecium is covered with cilia and contain bottle-shaped organelles, the trichocysts.

82. C

83. E = Cilia in the gullet create a swirling current of water which drives food mostly bacteria into the mouth–pore (cytostome). Food is egested through the anal–pore (cytopyge).

84. E = As in Amoeba, Paramecium removes metabolic waste by diffusion through the body surface. Osmoregulation is by contractile vacuole.

85. C = The contractile vacuole primarily expels excess water absorbed by osmosis. However, little nitrogenous wastes (ammonia) accompany the water.

86. D = Man is usually the primary host of tapeworm. Pig and cow are the secondary host.

87. B = Two species of tapeworm that commonly infect man are *Taenia solium* in Pig (with hook) and *Taenia saginata* in Cow (without hook).
88. A = The excretory ducts and the nerve cords pass down the entire length of the body.
89. D = I – hook, II – sucker, III – head or scolex, IV – neck, V – young proglottids.
90. C = Proglottids, which are flat segments budded off from the neck (IV) behind the scolex or head.
91. C = Tapeworm feeds by absorbing partly digested food in the host's intestine.
92. B
93. C = Have microtriches on their body surface for food absorption.
94. D
95. E
96. C = Eggs in the mature proglottid of the tapeworm are stored in the uterus.
97. D
98. B = It bores into the blood stream through the walls of the alimentary canal.
99. A
100. D
101. A
102. D
103. A
104. C
105. B
106. C = **Samuel The Biologist Planted Sugarcane**.

CHAPTER FOUR

1. B = The structure labeled I is the proboscis.
2. D = Proboscis is modified for piercing and sucking.
3. C = The larval stage of a mosquito is called wriggle because is moves by rapid wriggling.
4. C = Oil prevent the larval and pupae from breathing, hence they will die of suffocation.
5. A = The female *Culex* mosquito transmits *Wuchereria bancrofti*, a roundworm, causing filariasis (elephantiasis).
6. C = Saliva injected during mosquito bite is principally to prevent the host's blood from clotting.
7. D = Female *Anopheles* – malaria; female *Aedes* – yellow fever and dengue fever; female *Culex* – elephantiasis (filiariasis).
8. A = *Anopheles* mosquito have air–floats on eggs, which are laid singly. *Culex* mosquito lay eggs in batches which float easily.
9. B and C = *Anopheles* larva breath through the Spiracle, and anal gills on the 8th and 9th abdominal segment, *Aedes* larval breath through the siphon (breathing tube) connected to the tracheal system located on the 8th abdominal segment.
10. B = Movement is by the tail paddle (fins).
11. D = Female *Anopheles* mosquito transmits plasmodium, a protozoan parasite causing malaria.
12. C = mosquito undergoes complete metamorphosis. Cockroach undergoes incomplete metamorphosis.
13. D
14. B = Elytra is the fore-wing borne on the mesothorax (2nd segment of the thorax).
15. B = The correct sequence that describes the insect segmented leg is: Coxa (i), Tronchanter (iii), Femur (ii), Tibia (iv), Tarsus and Claws (v).
16. D = The claw (v) of insects is used for gripping or holding.
17. C = The mandibles or jaws are highly sclerotised, used for chewing.
18. C = The style is what differentiates the male cockroach from the female. Present in male and absent in female.
19. E = Metamorphosis is the stages which certain animals like insects, toads and frogs undergo before they attain adulthood.
20. D = After mating, female cockroach lays its egg in a honey egg – case called ootheca formed in the body.
21. C = The new emergent is called nymph.
22. C = The nymph are wingless, small in size and sexually immature compared to the adult.
23. C = The cerci has mechanoreceptive hairs called setae that are sensitive enough to detect the change in air pressure cause by fast approaching object, like a predator, hence an organ of defence.
24. D = The cockroach antenna, also known as feelers, are responsible for the sense of smell, without which the cockroach will not be able to locate its food.
25. B = I – compound eye, II – antenna, III – proboscis or sucking tube.
26. C
27. A = The mouthpart of housefly is modified for lapping and sponging. The proboscis is formed from a modified labium; it is a sponge – like organ composed of two halves called labella.
28. D = The head of a maggot (housefly larva) hears a hook with which the larva tears up food and drags itself along.
29. B = The hing wing of housefly is modified as halters also called balancers.
30. E
31. D = Malaria is transmitted by female Anopheles mosquito. Housefly could spread diseases such as typhoid fever, diarrhea, dysentery, cholera and poliomyelitis.
32. A = Termite can digest cellulose with the help of protozoan that live in worker termite's gut. This is symbiotic relationship.
33. B = In termites, new king and queen are produced from the adult reproductives from the same colony.
34. B = The lack of pigment in the workers suggest the common, though misleading name of "white ant" for these insects.
35. C

36. B = The termite colony will no longer exist without a queen.
37. A
38. D = Termites live in termitarium built with clay.
39. A = The queen is distinguished from other caste by the prominent and extended abdomen.
40. B
41. A = Adult butterfly helps in pollination whereas the larva (caterpillar) feeds on crops thereby destroying them and reducing their yield.
42. D = II – silk thread III – Spiracle IV – Silk pad.
43. A
44. C = I-antenna, II–compound eyes, III – mouth part, IV–coiled proboscis.
45. A = Mouthpart of butterfly.
46. A = Butterflies have thread-like antenna with a club-like knob at the end.
47. A
48. C
49. A = Mosquito larva are filter feeders.
50. A
51. D
52. C = Cockroaches and grasshoppers undergo incomplete metamorphosis.
53. C
54. A = Grasshopper larva (nymph) destroy plant leaves. Grasshopper adult also destroy plant leaves.
55. B
56. D = Moths are insects closely related to the butterfly. They are destructive in their larva (caterpillar) stage.
57. C = Ecdysis or moulting is the shedding of the exoskeleton
58. A
59. D = Except in the head, each segment of insect is covered by a dorsal tergum and a ventral sternum.
60. A
61. C = Spiders have no antennae. Antennae in cockroach and butterfly are long.
62. C = Nocturnal insects are active at night e.g. cockcroach.
63. B.
64. B = The exoskeleton (cuticle) is secreted by the epidermis and is composed partly of chitin and partly of protein.
65. B = The tracheal openings in insect is the spiracle found on the abdominal wall.
66. B = Elytra, the first forewing is borne on the mesothorax (2nd segment of the thorax).
67. D = The abdomen in insect is devoid of appendages except for a pair of jointed anal cerci which project laterally, one on either side, at the posterior end.
68. A
69. E = trumpet (respiratory trumpet) is found in the pupa of mosquito not all insect.
70. E = No stated differences in the options.
71. E = Adult insect reproduce to perpetuate the generation of the species.
72. A = Both have biting mouth parts.
73. C = Moulting in insects is the periodic shedding or replacement of outer cuticle (older cuticle).
74. D
75. C = Insect activities gradually increases with temperature.

CHAPTER FIVE

1. C = Fishes are cold-blooded (poikilothermic) animals whose body temperature varies according to changes in their external surroundings, others include amphibians and reptiles.
2. None of the options is correct = cartilaginous fishes have 5 to 7 pairs of gills without operculum (gill cover). Bony fishes have gills covered by an opercula flap. II and V = cartilaginous fishes IV = bonefishes.
3. A = Stream-lined shape with a neck minimizes friction during passage in water.
4. C = The swim bladder contains gas whose volume is adjustable with changing pressure; it keeps the fish buoyant. Only found in bony fishes.
5. B = The dorsal, ventral and anal fins help to stabilize the fish so that it does not roll in water.
6. E = The lateral line contains sensory cells that allow the fish to detect vibrations in water and to distinguish between differences in water.
7. C = Paired fins in bony fishes are equivalent to the limbs of higher vertebrates. They are the pelvic and pectoral fins.
8. B = Gill rakers prevent food and big particles from passing through the gills slits.
9. B = The pectoral and pelvic fins are used for steering, balancing and slowing movement.
10. B = Lung fish breath with the gills in water and also has a swim bladder that stores air for respiration
11. B = Pectoral fins lie just behind the gill covers.
12. C = The urea raises the body fluid osmotic pressure in other to prevent dehydrating.
13. B = Fertilization in cartilaginous fish is internal and only a few large yolky eggs are produced, each protected in horny egg cases. Development is direct. It has cartilaginous vertebrae.
14. B = Operculum is only found in bony fishes.
15. D
16. C = A fish moves forward by a combination of undulatory contractions of the body muscles and beating of the tail from side to side.
17. E = The swim bladder do not store excretory material. In some fishes, the swim bladder lies against the ear and act as an amplifier to enhance sound detection.
18. C = Olfactory lobe is more developed in the cartilaginous fishes because they depend more on their sense of smell. However, it is the cerebellum and optic

lobes in the bony fishes.

19. B = Cerebellum controls muscular activities.
20. B = A, C, and D are true
21. B. Development of fore limb is the last thing in the metamorphosis of toads.
22. E = The eggs laid in strings of jelly are without shell.
23. A = The long and coiled intestine is present in a young tadpole to enable it feed on weeds (herbivorous diet).
24. A = Fertilization is external in toad. During reproduction the male mounts on the female and as she extrudes her eggs into the water, the male discharges sperm over them.
25. C = The jelly spaces of the eggs is for aeration and also protects them from injury, predators and water uptake.
26. B = Poison glands are located on the dorsal surface of the skin to ward off predators.
27. B = Nictitating membrane covers the eyes and protect them from water–borne dirt.
28. B = Amphibians do not have scales, feathers or hairs.
29. B = The permeable skin allowed loss of water via osmosis.
30. E = see 27.
31. C = There are at least two types of skin secretion in toads, namely (1) granular, also called parotid gland, which secrete toxic, milk-white fluid and (2) mucus gland, which secrete non toxic, clear liquid. The granular gland facilitate defence mechanism while the mucus gland facilitate thermoregulation.
32. B = The web of the feet is part of the skin. They breath through: skin, mouth and lungs.
33. E = The forelimbs in male toad possess swollen copulatory (nuptial) pad on the first inner finger. In breeding season the copulatory and becomes dark vary thick and sticky. It is not called breeding pad.
34. E = It does not provide the eggs with mineral salt.
35. C = Toads do not have ciliated skin at any point of development.
36. D = Egg to adult.
37. C = The animal with the mentioned features is a lizard (reptile).
38. A = pre-anal pad, (anal pores) are present in male rainbow lizard (*Agama agama*) but absent in the female.
39. C = Lizards have dry scaly skin, which protect their body from desiccation.
40. A = Lizards eggs, unlike birds eggs, have a thin, pliable leathery shell.
41. B = The gular fold in lizard is lowered during courtship to attract mate or when the animal is frightened.
42. B = Homodont dentition, i.e. teeth all of same type.
43. E = Internal fertilization by means of copulatory organs.
44. B = Amphibians have oval stream-lined bodies with no neck.

45. B = Feathers which cover the body or birds also maintain their body temperature. They are homoeothermic, having uniform body temperature.
46. E = Webbed feet is an adaptation for swimming.
47. D = I – Double shell membrane, II – Germinal disc, III – Chalaza IV–Germinal disc. Chalaza suspend the yolk in the abdomen.
48. B
49. A = Down feather gives insulation.
50. A = Wing bears feathers.
51. A = Duck beaks is adapted for scooping (from mud, pond and water) and sieving food.
52. C = Birds show parental care e.g. building of nest for eggs, feeding the newly hatched, protecting the young from predators.
53. E = All true
54. A = Skin of bird is waterproof, non-gland (except preen glands) at the base of the tails
55. C = The spaces on the bird's body without feather tracts are referred to as apteria.
56. C
57. E. Feathers perform all the stated roles.
58. E = The uropygial gland (oil gland, preen gland or rump gland). It performs so many functions in birds including water proofing and keeping the skin, feathers and bill supple; the oils is said to have an antibacterial function.
59. A = The serrated edges helps to scoop and sieve food.
60. B
61. D = Mammals have their bodies covered with hairs.
62. A = The skin of mammals have the following glands: sebaceous, sweat and mammary glands.
63. C = Mouth of rodents have long curved incisors (Chisel – shaped) for gnawing food.
64. E = The organ is placenta, nourishment and oxygen diffuse to the developing embryo through the umbilical cord.
65. D = Evolutionary trend in vertebrate is
PARAM: Pisces (fish), Amphibians, Reptiles, Aves (birds), Mammals.
66. D = Ovoviviparity (oviviparity), is a mode of reproduction in animals in which the embryos develop inside eggs that are retained within the mother's body until they are ready to hatch. Examples are found in some species of fish (cartilaginous fishes and reptiles).
67. A = The dark hairy covering of the rat body aids camouflage (disruptive colouration) and escape from predators.
68. A = I – Whisker, II – Pinna of ear, III – Tail, IV – Hair. I (whisker) are photosensitive for feeling the way in the dark burrow.
69. A = Viviparous reproduction occurs only in mammals. It means development of the embryo inside the body of the mother, eventually leading to live birth, as opposed to laying eggs.
70. B

71. A = caudal fin is the tail fin.
72. C = The fore limbs of toad are adapted for absorbing shock on land while the hind limbs are adapted for hopping from the ground.
73. D
74. B = Fishes, Amphibians, and Reptiles are cold blooded (poikilothermic); Birds and mammals are warm blooded (homeothermic).
75. C.
76. A = birds and reptiles have bony structure of head.
77. B = Pectoral muscles assists the feathers for flight.
78. B = Bones of birds are light because they are thin and hollow for ease of flight.
79. D
80. B = pigeon are monogamous, one sexual partner.

CHAPTER SIX

1. A = Unlike the epidermis of the stem, the piliferous layer of the root has no cuticle.
2. B = A dicot stem and monocot root have the central pith in common.
3. A = In the root, the xylem and phloem strands are arranged on alternate radii while in the stem, they are arranged on the same radii.
4. C = **Memory Tip: Playing Xylophone Can Produce Emotional Pleasure Called Ecstasy**.
5. D = Cambium is present between the phloem and the xylem. Cambium cells are constantly dividing and can develop into cells of different types (secondary thickening).
6. B
7. C = Sieve tubes lie along with the companion cells. They, along with phloem parenchyma constitute the phloem tissue.
8. B = Collenchyma is living at maturity but thick-walled. Xylem vessels are dead elongated tubes. Sieve tubes are unlignified, elongated living cells. Sclerenchyma is dead at maturity but thick walled.
9. D = The guard cells control the opening and closing of the stomata through which air and water move in and out of the mesophyll layer of leaf.
10. B = Mesophyll parenchyma also called photosynthetic parenchyma is the principal site of photosynthesis. It contains two cells namely palisade cell (main site of photosynthesis) and spongy cells.
11. D = Secondary thickening is made possible by cambium.
12. A = The vascular bundles of the lead, which are continuous with those of the stem, form the veins.
13. B = I – Pericycle, II – Root hairs, III – Xylem IV – Phloem. Root hairs are used for absorption of water and salt.
14. D = Pericycle I, which is inside the endodermis. The endodermis is also referred to as starch sheath. It stores starch hence when stained with iodine solution, it turns blue – black.
15. C = See 2.
16. A and B = Monocot stems have no cambium, hence cannot undergo secondary thickening. Monocot stem also lack pith.
17. B = The pericycle lies between the endodermis and the vascular bundle.
18. A = In dicot stem, between the epidermis and pericycle is the cortex.
19. A = Xylem and phloem are conducting tissues in plants similar to artery and vein in animals.
20. A = Xylem comes before phloem from inside to outside.
21. E = The organ is placenta, nourishment and oxygen diffuse to the developing embryo through the umbilical cord.
22. C = Lungs and heart are in the thoracic cavity also called the anterior half.
23. A = The heart is enclosed in the pericardial cavity.
24. B = Left auricle receives oxygenated blood from the lung via the pulmonary vein. Right auricle receives deoxygenated blood from the body through the vena cava.
25. D = Hepatic portal vein starts and ends in a capillary network. It link the villi of the intestine to the liver.
26. D = The thoracic and abdominal cavities are separated by the muscle, diaphragm.
27. C = Mammals are unique for the glands in their skin namely sweat, sebaceous and mammary glands.
28. C = Heterodont, i.e. of different types.
29. D = Perineum seperates the anus from the scrotum in the male and the anus from the vulva (the labial opening to the vagina) in the female.
30. E = Mammalian ears have external pinna.
31. C = Hair provides insulation from extreme cold.
32. B = Bile duct helps the digestion of fat by emulsification of fat.
33. C = The pancreatic duct opens into the duodenum where it releases its secretions.
34. B = The caecum has a blind ending called appendix.
35. D = The pancreas is both a ductless gland and an accessory organ of digestion.
36. C = The liver is the biggest organ. The skin is the most extensive organ.
37. D = Urea formation takes place in the liver.
38. A = The heart pumps blood round the body.
39. D = HCl is secreted by the parietal cells concentrated in the middle part of the stomach. Cells in the pyloric region secretes gastrin and mucus.
40. C = Thyroid gland is located on the trachea in the neck.
41. C
42. A = The liver stores iron.
43. C = Absorption of water takes place in the large intestine colon, as ingested food passes through it.

44. C = Urine enters the urinary bladder through the ureter and leaves through the urethra.
45. B = I – Upper epidemis, II – Palisade mesophyll, III – Bundle sheath, IV – Spongy mesophyll. Palisade cells are positioned towards the upper surface of the leaf and contain the largest number of chloroplast per cell in plants. This makes them the primary site of photosynthesis in plant leaves.
46. B = Palisade cells have a very large surface area in order to absorb more light during photosynthesis.
47. B = I – Endodermis, II – Phloem tissue, III – Xylem tissue, IV – Epidermis.
48. D = Xylem tissue, III, transport water and mineral salts.
49. A = The stele consists of vascular bundle, it is bounded on the outer side by one-or-two cell thick layer of thin walled cells called pericycle.
50. C = Stomata is found between two guard cells to the lower epidermis.
51. A
52. A
53. A = Sclerenchyma cell are dead at maturity, collenchyma, parenchyma and endodermis are alive at maturity.
54. C
55. B

CHAPTER SEVEN

1. B = Insects are attracted to insectivorous plants, not to pollinate it, but to feed on the secretions.
2. D = Saprophytes digests their substrate outside their body before absorption. This is extracellular digestion.
3. D = Fungi lack chlorophyll hence they are heterotrophic, they are either saprophytes or parasites.
4. B = Nitrobacter is a nitrifying bacteria – chemosynthetic, *Ascaris* is an obligate parasite in man – parasitic, *Homo sapien* (Humans) ingest food and digest within the body – Holozoic.
5. D = Hydra exhibit both intra-and extra-cellular digestion. Rhizopus, a fungi, digestion is extracellular. Housefly is saprozoic, digest food extracellular by secreting saliva on food.
6. B = Pitcher plant is a carnivorous (insectivorous) plant which feed on insects. This is a heterotrophic mode of nutrition. Other examples are Sundew, Bladderwort.
7. C = Tapeworm is an obligate parasite in the intestine of man.
8. A = A haustoria or sucker is a developed structure with which plant parasites like Cassytha absorbs nutrient from its host's conducting tissues.
9. C = see 1.
10. B

11. D = Insectivorous plants such as sundew and bladderworts are primarily autotrophs because they have chloroplasts, hence they are both autrophic and heterotrophic.
12. D = In chemosynthesis, energy from the oxidation of inorganic molecules is used to generate organic molecules as food.
13. C
14. E = The oxygen given off during photosynthesis is derived from water.
15. A = The light stage of photosynthesis involve four steps namely: (1) Activation of chlorophyll (2) Photolysis of water (3) Transfer of NADP (Nicotinamide Adenine Dinucleotide Phosphate) (4) Formation of ATP (Adenosine triphosphate).
16. D = Formation of glucose using energy from NADPH (Nicotinamide Adenine Dinucleotide Phosphate Hydrogen) is the dark stage of photosynthesis.
17. C = $NaHCO_3$ enriches the water with carbondioxide.
18. A = Glucose – direct product, starch – first visible product.
19. A = In bright light because photosynthesis occurs faster than respiration hence, in dim light there may be equal rates.
20. C = Light energy and chlorophyll are the principal requirement for photosynthesis.
21. C = Photosynthesis occurs in the part of plants (shoot system only) that contain chlorophyll.
22. A = In the dark stage, carbondioxide, absorbed by leaves, combines with a 5-carbon compound, ribulose phosphate, to form an unstable 6-carbon sugar.
23. B = The dark stage of photosynthesis takes place in the stroma of the chlorophyll.
24. A = Photosynthetic pigments include: Chlorophylls (a and b) – in green plants, carotenoids (carotenes and xanthophylls) – in higher plants and also present in the green algae and brown algae, phycobilins -- only in red algae and cyonabacteria.
25. C = The pigments of chloroplasts absorb blue and red light most effectively, and transmit or reflect green light, which is why leaves appear green.
26. C = Chlorophyll contains metallic magnesium.
27. A = Carotenoids are accessory pigment of photosynthesis, there are two classes: carotene (yellow) and xanthophylls (orange).
28. B = CO_2 in gaseous form as CO_2 molecule.
29. C = Nitrogen is required for formation of protein, enzymes and nucleic acid.
30. D = Magnesium and iron are required for the formation of chlorophyll.
31. B = Yellowing of leaves chlorosis, could be due to the deficiency of nitrogen, magnesium, iron, sulphur.
32. A: In addition to carbon, Hydrogen and oxygen. Major elements required by plants are: Nitrogen, Phosphorus, Potassium, Calcium, Magnesium, Sulphurs. **Memory tip: Nigerian Police, Pick Criminals, Most Saturdays.**

33. D
34. B = Lack of phosphorus will cause the symptoms mentioned.
35. A = Trace or micro-elements are needed in small amount for the formation of pigments and enzymes.
36. B = Poor growth and yellowing of leaves are the deficiency symptoms of iron. Little growth and yellowing of leaves are the deficiency symptoms of magnesium.
37. B = see 34
38. C = Nitrogen is needed for protein synthesis.
39. C = Food is the material requirement of organisms.
40. B = Gastrin is a hormone secreted in the stomach.
41. E = Some vitamins are catalysts or enzymes in important metabolic processes. They yield no energy.
42. C = Deficiency of iodine causes goitre.
43. D = Large deficiency of protein in growing children results in kwashiorkor, a disease conditioned characterized by distended abdomen and lean body.
44. B = Carbohydrates are made of sugar units. They contain carbon (C), hydrogen (H) and oxygen (O) with H and O in the ratio 2:1, formular is $Cx(H_2O)y$.
45. B = Calcium is needed for strong bones, teeth and also contraction of muscles. Deficiency results in retarded growth.
46. C = Fats, carbohydrates and proteins are the foundational organic substances in human body.
47. A = Vitamin A is required for normal vision and keep skin healthy.
48. B = Vitamin K is required for the synthesis of blood clotting protein
49. B = Carbohydrates (main energy) source yields 17 kJ/g. Proteins (energy during starvation) yields 17kJ/g. Lipids (high energy) yield 39kJ/g.
50. A
51. D
52. D
53. B = Deficiency of vitamin E causes sterility.
54. C = Osmotic pressure of the body fluids is regulated by mineral salts. Also needed for proper functioning of the body.
55. E = Lactose is a disaccharide sugar that is found most notably in milk and is formed from galactose and glucose.
56. D = Blood sugar.
57. D = Sugar solution boiled with Fehling solution A and B, colour turns from blue to green and finally to orange (red) as a precipitate. Except for sucrose, unless HCl is added to it (hydrolyses) before the experiment.
58. E = Sample reducing sugar (e.g. glucose) can be tested by adding Fehling's reagent in neutral or alkaline medium and boil. Red precipitate is formed.
59. E = Lipids (fats and oils) shows a red colour with sudden (III) solution

60. C = Biuret's test for protein.
61. A = Biuret's test for protein.
62. B = Test for starch.
63. D = Benedict's test for reducing sugar e.g. glucose
64. C = Dental formula for man is C, rabbit is A, sheep and cow is B, dog is D.
65. B = Incisors – cutting, canines – tearing; premolars and molars – grinding and crushing.
66. E = The total number of molar teeth in man is 12, check the dental formula, multiply the upper and lower teeth by 2 and add up.
67. A = In herbivores, canines are inconspicuous or absent the toothless gap is called a diastema.
    Canines are used for tearing food.
68. B = Pulp cavity contains connective tissues, nerves and blood vessels.
69. A = 1, 2, 3, and 4 are molar, premolar, canine and incisor respectively.
70. B
71. D = Enamel (outermost) → dentine → pulp cavity (innermost).
72. A = Last upper premolar and first lower molar form the carnessial teeth in advanced carnivores.
73. A = Absorption of food takes place in the ileum. Food is not absorb in the stomach except alcohol.
74. E = Cardiac sphincter, upper part of the stomach. Food leaves through the pyloric sphincter.
75. C = Rectum, which terminates as anus.
76. E = Excretion of urea is by the kidneys.
77. A = Oral cavity, Oesophagus, Stomach, Duodenum, Jejunum, Ileum, Large intestine (Colon), Caecum, Rectum.
78. A
79. A = Bile from gallbladder emulsify lipids to increase the surface area for digestion.
80. D = Kidneys are not close to the stomach.
81. D
82. C = Liver and pancreas.
83. B = The entrance and exit of the stomach is guarded by cardiac and pyloric sphincters respectively.
84. D = The stomach secretes pepsin, which converts proteins to peptide.
85. B = The pancreas secretes hormones (insulin and glucagon) and enzymes (amylase, trypsin and lipase).
86. B = Colon is the large intestine.
87. D = The pH of the entire alimentary canal is alkaline, except the stomach, which is acidic.
88. D = Epiglottis prevent the passage of food into the larynx of the trachea.
89. C = Renin coagulates milk protein (caesin) before it is acted upon by pepsin.
90. D = Protein à Peptones (peptides) à Polypeptides à Amino acids.
91. A = The food particles acted upon by enzymes is referred to as the substrate.

92. E = Enzymes are inactivated by extreme temperature and pH.
93. A = Chyme and chyle are partially digested food in the stomach and intestine respectively.
94. B = Carbohydrate digest fastest.
95. C = The pancreatic juice contains amylase (amylopepsin), trypsinogen and lipase (steapsin) for the digestion of carhohydrate, protein and lipids respectively.
96. C
97. C = Deamination leads to the formation of urea.
98. A
99. A
100. A
101. C = Absorption of digested food materials takes place in the ileum via the villi while absorption of water takes place in the colon (large intestine).
102. D
103. A = Carbon, Hydrogen and Oxygen are common to carbohydrates, proteins and lipids.
104. D
105. C = Hepatic portal vein carries blood from the alimentary canal to the liver. It starts and ends in a capillary network.
106. D
107. A
108. A = Osmic acid test is used to test for lipids.
109. B = Photosynthesis and chemosynthesis are autotrophic mode of nutrition.
110. D = With millon's reagent on heating, proteins gives a deep red colour or precipitate.
111. B
112. D = I – Enamel, II – Dentine, III – Gum, IV – Pulp cavity. Dentine is extremely sensitive to heat and cold.
113. A
114. C
115. B = Lipase is an enzyme that digest lipids.
116. B
117. A
118. D
119. C = Osteomalacia is the softenining of the bones caused by the defective bone mineralisation. In children it is known as rickets. It could be caused by deficiency of calcium, vitamin D or phosphate.
120. C = Polished rice lacks vitamin B1 (thiamin). The deficiency of vitamin B1 causes beri – beri.
121. C

CHAPTER EIGHT

1. B = Small (unicellular) organisms have large surface area (S.A) to volume ratio (V.R) and do not need a transport system because diffusion is efficient, large (multicellualar) organisms have low surface area to volume ratio, so require a transport system.
2. B = Organism IV has the lowest S.A/V.R (1:5=0.2).

Therefore, it will have the most complex transport system.
3. A = Pulmonary artery carries deoxygenated blood.
4. D = The sequence of blood flow is: Heart→ arteries → capillaries→venules→veins→Heart.
5. A
6. B = The lymph is carried in lymphatic vessels to the subclavian veins in the neck, where it is passed back into the blood.
7. E
8. D = Blood in the right ventricle will leave the heart to the lungs for oxygenation via the pulmonary artery. Oxygenated blood will return to the left auricle via the pulmonary vein and will finally leave the heart via the aorta, to the toe without passing through the liver.
9. B = Hepatic portal vein begins and ends with capillaries.
10. C
11. C = Superior (anterior) vena cava returns blood from structures above the heart back to the heart.
12. C = The wall of the left ventricle is further thickened to enable it to pump blood with sufficient pressure round the body.
13. D
14. A
15. C
16. A= Systemic circulation starts from the left ventricle to reach all parts of the body except the lungs and back to the right auricle. Pulmonary circulation starts from the right ventricle to the lungs through the pulmonary artery and back to the left auricle through the pulmonary vein.
17. B
18. B = The sieve tubes of the phloem tissue translocate manufactured food and mineral ions from leaves or storage organs to other parts of the plant.
19. C
20. D= The xylem tracheids and vessels transport water and mineral salts from the soil to the leaves by osmosis.
21. C = The xylem vessels will appear red because the ink travels only in the xylem.
22. E = Red blood cells (erythrocytes) are non-nucleated at maturity.
23. B = Blood clotting begins with the exposure of blood to air from a cut and ends with the trapping of red blood cells and formatting of clot.
24. B = Lymph nodes produce certain white blood cells, lymphocytes, and filter out bacteria and other particles.
25. B = The lymph is much like the blood plasma but it lacks plasma proteins and red blood cells. It contains white blood cells.
26. A = The main function of blood is to transport excretory materials from tissues and supply of oxygen to tissues.

27. A = The exchange of nutrients and metabolic products occurs between the lymph and body cells.
28. B = Oxygen is carried by the erythrocytes.
29. B = The cell sap is the medium of transpiration in higher plants.
30. E = A potometer is an instrument for measuring the rate of transpiration of a plant shoot at a very short interval and under different environmental conditions. The limitation of a potometer is that it only measures the water absorption and not the transpiration rate directly.
31. D = Without the phloem tissue, translocation is not possible.
32. C = Increase in temperature, light intensity and air movement increase the rate of transpiration. Increase humidity lowers transpiration rate.
33. D = A transparent polythene is not semi permeable hence it will not enhance osmosis.
34. C = Water from conducting tissues in the veins moves into the mesophyll cells, from there to the substomatal cavity and not to the atmosphere through the stomata pores.
35. C = Mineral salts are absorbed into the plant roots against their concentration gradient by active transport, a process that requires energy. As they became concentrated in the root cell, they tend to diffuse to other parts of the plant.
36. D = Movement of mineral salts is hardly affected because mineral salts are not transported by the phloem. The plant dies after several weeks.
37. D = Guttation is the loss of water in a liquid form through the margins of leaves of plants growing in a warm and moist soil in a humid atmosphere.
38. D = The correct route of water movement in plants is the sequence shown in D.
39. A = Translocation is the transport of manufactured food substances from the leaves to all parts of the plants.
40. C = X will shrink as a result of plasmolysis. Y will become turgid as a result of osmosis.
41. A
42. D
43. C = Blood serum is the fluid expressed from clotted plasma; it can be defined as plasma deprived of fibrin but which contains other plasma contents.
44. A = The function is to reduce friction.
45. B = Water vapour – transpiration, water droplets – guttation.
46. C= The Pigment, haemoglobin, imparts dark red colour to blood.
47. B = Double circulation takes place in mammals. It involves pulmonary and systemic circulation.
48. C
49. D = Osmosis regulates the opening and closing of stomata. The stomata open when the guard cells are turgid and close when the guard cells are flaccid.

50. B = Inferior (posterior) vena cava returns blood from structure below the heart back to the heart.
51. D = Leucocytes (white blood cells) defend the body against diseases.
52. A = High osmotic pressure of root cells is able to exert an influence on the absorption of water.
53. C = Heart to liver – hepatic artery.
54. A = Semi-lunar valves (pulmonic valves) one located at the opening of the aorta and the other at the opening of the pulmonary artery, they prevent blood from flowing back into the ventricles.
55. A = Plasma proteins are proteins found in blood plasma, including albumin, fibrinogen and globulin. Albumin, alongside others, aid in regulating the water balance between the blood and tissues.
56. C
57. A = The insect circulatory system is an open system, where blood flows freely inside the body cavity without the aid of blood vessels (arteries and veins)
58. D = Turgor pressure pushes the plasma membrane against the cell wall of plant bacteria and fungi cells as well as those protist cells which have cell walls.
59. D = Vapour form
60. D
61. E = The rate of water absorption was affected.
62. A = Phloem tissue becomes impaired.
63. A
64. B = Red blood cells do not fight bacteria.
65. C = Heat is produced more in the liver and the muscle than any other parts of the body. This heat is distributed round the body through the flow of blood.
66. D
67. B
68. B
69. C = The heart is enclosed in a membranous structure called pericardium.
70. D = The ventricles are thicker than auricles because they transport blood while the auricle receives blood. The left ventricle is further thicker than the right right ventricle.
71. C
72. C = Arteries take blood away from the heart. So the colour of the blood and its oxygen content are not the determining factors.
73. B = Capillaries possess very than walls to allow diffusion of gases.
74. E = Carotid artery carry blood to the head.
75. E = The celiac artery supplies oxygenated blood to the liver, stomach, abdominal esophagus, spleen and the superior half of both the duodenum and the pancreas.
76. A = A portal venous system occurs when a capillary bed pools into another bed through veins, without first going through the heart. Examples of such systems include the hepatic portal system and the hypophyseal portal system.

77. D = The left subclavian artery supplies blood to the left arm while the right subclavian artery supplies blood to the right arm.
78. C = Lymph node manufacture lymphocytes, a type of white blood cell.
79. E = Oxygen is not transported in the lymph.
80. E = Fatty acids and glycerol enter the blood stream through the subclavian vein via the lymphatic system.
81. A = The iliac artery supplies blood to the lower limbs and gonads.

CHAPTER NINE

1. C = Pyruvic acid produced during glycolysis, in the absence of oxygen, is converted to CO_2 and ethanol in plants and lactic acid in animals.
2. D = After about 6-12 hours, the rise in the coloured water level in the glass tube of the animal specimen will be higher than in the glass tube of the plant specimen. Thus, the experiment is a comparison of oxygen uptake (respiratory rates) in plants and animals.
3. C = Manometer, used to measure the pressure of liquids and gases.
4. A = Glycolysis in plants and animals yields pyruvic acid.
5. A = Respiration oxidize food to release energy.
6. E = Crustaceans (e.g. crab) respires by means of gills.
7. E = Insects (e.g. grasshopper) respires by means of tracheal tubes.
8. B = Epiglottis is a regulatory structure that leads to the larynx (voice box).
9. A = Respiratory surfaces must not be dry and freely permeable.
10. D = In mammals, air moves from mouth and nose →pharynx→ trachea→bronchi→bronchioles→ alveoli. Each alveolus is surrounded by network of capillaries and gas exchange takes place in it.
11. A = Spiders respires by means of lungbooks.
12. C = Gas exchange in Aves (birds) occurs in the airs sacs of the lungs.
13. B = Mammals respires by means of lungs and whales is a mammal.
14. A = In insects, gas exchange occurs in the tracheoles of the trachea (air tubes) which open out as spiracles.
15. C = Oxygenated water is drawn in through the mouth, the dissolved oxygen is then absorbed by the blood capillaries in the gills and deoxygenated water flows out through the operculum.
16. D = Toads and frogs respires by means of gills at tadpole stage; skin (cutaneous), mouth (buccal) and lungs (pulmonary) at adult stage.
17. B
18. E = During inhalation volume (size) of the thoracic cavity increase while the pressure in the lungs decrease.

19. B
20. C = Increase in air pressure of the lungs occur during exhalation, while the diaphragm relaxes.
21. D = Exhaled air is warm, has more CO_2, water vapour and less oxygen.
22. D = Diaphragm separate the thorax and abdomen.
23. A = Contraction of the intercostals muscles move rib cage upward and forward.
24. C = Oxygen will diffuse from lungs (higher concentration) to the blood (lower concentration).
25. D = The presence of intercellular air spaces in the cells of the stem and leaf allow the circulation of air.
26. C
27. C = Lenticels are found in woody stems, while stomata are found in leaves. They help exchange gases.
28. D = CO_2 releases by respiring plant cells will be used for photosynthesis.
29. C = Increase in solute concentration in the guard cell makes the guard cell become turgid and thus the stomatal pores open. Decrease in solute concentration in the guard cell makes the guard cell become flaccid and thus the stomatal pores close.
30. B = During Kreb's cycle, hydrogen is release. Hydrogen reduces oxygen into water vapor.
31. A = During Kreb's cycle, the oxidative part of respiration takes place in the mitochondria.
32. A = CO_2 the gas produced during tissue respiration can be identified by calcium hydroxide (lime water), which it turns milky.
33. C = The significance of tissue respiration is the oxidation of food and the release of useable energy.
34. B = Artificial respiration is an emergency procedure where breathing is maintained artificially.
35. D
36. C = The bubbles of CO_2 released is what make the dough to rise.
37. E = This is due to anaerobic respiration of an organism such as yeast.
38. A
39. D = 2ATP molecules of energy is produced during anaerobic respiration.
40. D
41. D = Cramp and muscle fatigue are caused by the accumulation of lactic acid and the resulting increase in acidity on the muscle due to insufficient oxygen during strenuous exercise.
42. C = 2ATP molecules of energy and alcohol.
43. D = Glycolytic pathway – cytoplasm. Kreb's cycle – mitochondria.
44. D
45. A = Each alveolus is surrounded by a dense network of capillaries for effective gas exchange.
46. D
47. A = Yeast carry out anaerobic respiration.
48. C

49. D
50. C = The trachea in insects open outsides as spiracles on the cuticle.
51. D = Gas exchange in earthworm is via the skin by diffusion.
52. D = Respiratory substrates are monosaccharide (simple sugar) primarily glucose.
53. C = Glycolysis – anaerobic. Krebs's cycle – aerobic.
54. B
55. D
56. D
57. E = In the dark, respiring plant release large quantity of CO_2 because photosynthesis is not taking place. CO_2 is used for photosynthesis.
58. D= Oxygen is soluble in water. Plant root hairs absorb the oxygen in solution with water.
59. C = Air enters the insects body through the spiracle.
60. B = Tracheoles are supplied with capillaries, non-cuticularized.
61. D = The gill filaments are richly supplied with blood vessels i.e. vascularised, to enhance the utilization of oxygen.
62. D = Gill cover or operculum.
63. E = When water enters the mouth of the fish (by opening of mouth and lowering the floor of the mouth and closing the operculum), the oral valves are closed, the floor of the mouth is raised and the opercula are open, water is forced through the pharynx over the vascularized gills where gaseous exchange occurs.
64. C
65. B = Air sac function as an oxygen reservoir.
66. B = The palate complex.
67. B
68. B = Mammals do not have air sacs.
69. D = The larynx is the 'voice box' having the vocal cords.
70. E = The ribs, intercostals muscles and diaphragm.
71. B = Diaphragm flattens during inspiration and become dome shape during expiration.
72. E = In the lungs, there is an exchange of gases by diffusion via the alveoli.
73. A

CHAPTER TEN

1. C = The leaf is not an excretory organ.
2. B
3. D = Flame cells are only for excretion in flatworms.
4. B = All insects carry out excretion with the malpighian tubule.
5. B = Earthworm is an annelid.
6. A = The Bowman's capsule and glomerulus of the nephron form the malpighian capsule.
7. A = The proximal convoluted tubule, loop of Henle and distal convoluted tubule form the uriniferous tubules.
8. C = Concentration of sodium chloride.
9. C = Malpighian corpuscle – cortex; loop of Henle and collecting duct-- Medulla.
10. C = I – Glomerulus, II – Bowman's capsule, III – Distal convoluted tubule, IV – Proximal convoluted tubule.
11. A = In the proximal convoluted tubule, Na^+ is reabsorbed primarily with $HCO3^-$ and organic solutes such as glucose and amino acid, 67% of Na^+ is reabsorbed. In the distal convoluted tubule, Na^+ is reabsorbed primarily with Cl^- but without organic solutes, 5% of Na^+ is reabsorbed.
12. D = The segment of the loop of Henle are responsible for counter current multiplication, which is essential for the concentration and dilution of urine, 25% of Na^+ is reabsorbed; 3% is reabsorbed from the collecting duct.
13. B = Glomerulus →Bowman's→capsule →Proximal convoluted tubule→Henle's loop→Distal convoluted tubule→Collecting tubule.
Memory tip: Great Britain Prosecuted Hitler's Deputy Commander.
14. C
15. D = The Glomerulus and Bowman's capsule form the malpighian body, which is located in the cortex.
16. B = Ureter links the kidney to the bladder, the bladder opens out through the urethra.
17. A = The sweat gland helps the skin to remove Na^+ as sweat.
18. B = I – Sebaceous gland, II – Nerve fibre, III – Blood vessel V – Sweat gland.
19. A
20. D = The malpighian layer is an undulatory row of meristematic cells. Cells contain melanin pigment
21. A = The sebaceous gland is a sac of glandular cells which opens into the hair follicle, secretes oil which keeps hair supple and water proofs the skin.
22. D = Sap in a medium of transportation in higher plants. Oxygen is a waste product from photosynthesis.
23. D = The main waste products formed in plant are water in form of vapour from transpiration and respiration, carbon dioxide from respiration and oxygen from photosynthesis.
24. C = Anthocyanin imparts colours to plant flowers on accumulation.
25. A = Gaseous excretory products are released through the stomata and lenticels e.g. CO_2 from respiration, oxygen from photosynthesis and water vapour from transpiration.
26. C = CO_2 released as waste product of respiration is used for photosynthesis.
27. C = Tannins are used for tanning leather. They are obtained from mangrove bark.

28. C = Faeces is an undigested remains of food. It is expelled via the anus.
29. D = Insects excrete uric acid as a means of conserving water.
30. B
31. C
32. D
33. B = Uric acid, sweat and urea are not excretory products in plants.
34. D
35. B = Allantois helps the embryo exchange gases and handle liquid wastes.
36. E = Excretion is the removal of waste product of metabolism. Not all waste product are poisonous although may become poisonous on accumulation.
37. C
38. D = The kidneys absorb less water on cold days.
39. D =
40. C = The processes of ultrafilteration, reabsorption and secretion in a nephron summarizes the excretion of urine.
41. E
42. A
43. E
44. A = The pyramids consists of loops of Henle and the collecting tubules of the nephron. The pyramids are found in the medulla.
45. C = Outer zone-- cortex. Inner zone – medulla.
46. E = All are parts of the kidney.
47. B
48. A = The lungs eliminates CO_2 and water vapour.
49. E = On a wet and cold day water reabsorption by the kidney is less.
50. C
51. D = The liver produce urea from deamination of amino acids.
52. E
53. D
54. E
55. B = Some insect excrete nitrogenous waste within their fat bodies as uric acid to conserve nitrogen.
56. D = Nitrogenous waste is not lost during metamorphosis.
57. A = Contractile vacuole remove excess water.
58. B = The gills remove CO_2 and water vapour during respiration
59. E

CHAPTER ELEVEN

1. D = Xylem conduct water and ions; and also support the plant.
2. A
3. D = Parenchyma cells support the plant by turgor.

4. A = Phloem translocate food from the leaves to other parts of the plants. The phloem parenchyma is a storage and packing tissue.
5. B = Sclerenchyma in plants is the functional equivalent of bone in animals, that is, sclerenchyma and bone provide elastic support (flexible and inflexible).
6. C = Lignin is largely responsible for the strength and rigidity of plants. Sclerenchyma is lignified to provide strength and rigidity..
7. C
8. A = Haptotropism or thigmotropsim is a response to touch e.g. on touching a support, tendrils of a climbing plant twine round it.
9. D = Roots of plants are positively hydrotropic and geotropic.
10. D = There is a greater concentration of auxin on the shaded side of the shoot, which in turn causes increased cell elongation and thus bending of the shoot.
11. A – Taxism involves the whole organism.
12. B
13. D
14. A
15. A
16. D = Nastic movements are to general (diffuse) stimuli such as touch.
17. E
18. B = Nastic movement is non-directional.
19. D = The breathing roots of mangrove trees rows against the direction of gravity (negatively geotropic).
20. D = Plants the grows horizontally are neutrally geotropic.
21. A = The klinostat is used to show that plant roots respond positively to gravity.
22. C = Illuminating a young seedling from the top will give a diffuse source of light (not unilateral). So, the plant will remain straight.
23. D = The differential changes in turgor in the lower and upper side of the pulvinus of the Mimosa pudica causes the dropping of the leaves. This is elicited by touch.
24. A = Negative geotropism helps the plant shoot to grow upright against the direction of gravity.
25. C = There is a greater elongation of cells on the shaded side as a result of accumulation of auxin of the shaded side.
26. C = Light is responsible for the formation of carbohydrate in the guard cells, which in turn increases the concentration of the guard cells and cause the turgidity of these cells by osmosis. The stomata open when the guard cells are turgid.
27. E = All the options are true.
28. C = Scales, carapace, and hoofs are external skeleton. But carpals are bones of the wrist.

29. A = Bones and cartilage constitute the internal skeleton of animals.

30. C = Earthworms have a fluid-based skeleton called hydrostatic skeleton

31. A = Bone to bone – ligament. Muscle to bone – tendon.

32. C = Exoskeleton is an outer skeleton made up of chitin; it is present in arthropods. Maggot and caterpillar are larvae of housefly and butterfly respectively. Earthworm has hydrostatic skeleton.

33. C = Axial skeleton is made up of skull, vertebra column, ribs and sternum.

34. E = Appendicular skeleton is made up of girdles and limbs.

35. C = Atlas before Axis.

36. B = Odontoid process of the axis fits into the neural canal of the atlas. The odontoid process makes the axis bone peculiar.

37. C = The pectoral and pelvic girdle are part of the appendicular skeleton. The pectoral girdle has scapular. Ischium is present in the pelvic girdle.

38. D = The pre-and-post zygapophysis articulate successive vertebrae to allow bending movement and prevent rotational movement in the vertebral column.

39. C = The spinal cord is enclosed in the neural canal.

40. D = Ball and socket joint allow movement in about 360^0 direction.

41. A

42. C = Transportation is not a function of skeleton.

43. A

44. B

45. D = The axis has two post-zygapophyses but not pre-zygapophyses. The atlas has no zygapophyses. The atlas shape allow the head to nod 'yes' and the axis shape allows the head to shake 'no'.

46. B = The neural canal encloses the spinal cord.

47. E = The chief peculiarity of the atlas is that it has no body but axis has a body. They both have vertebral canals for passage of arteries, neural spine, and transverse processes with varying degree of modifications.

48. A = Thoracic vertebra has long neural spine backwardly directed.

49. B = In between consecutive vertebrae are compressible cartilage pads known as intervertebral discs which allow the vertebrae to move slightly.

50. B = The lumbar vertebrae provide attachment to abdominal muscles (not back muscles) and bears considerable weight of the body.

51. C = Sacral vertebrae usually fused to form a mass of bone called the sacrum.

52. B = Ulna – longest bone of the lower arm. Humerus – longest bone of the upper arm.

53. C = The longest bone of the thigh and the body is the femur.

54. A

55. D = The synovial fluid secreted by the synovial membrane lubricates the joints.

56. B = Swimming to a warm region from a cold region by bacteria is a positive response to temperature. Positive thermotaxism.

57. C = Hydrostatic skeleton is found in earthworm (class: Oligochaeta, phylum: Annelida).

58. C = Thigmotropisin is a movement on which an organism (plant) moves or grows in response to touch or contact stimuli.

59. B

60. C

61. B = I – Cervical vertebrae, II – Lumbar vertebrae

62. A = Chitin is embedded in a protein matrix and strengthened by mineralization, usually in the form of calcium carbonate.

63. B = Vertebrates have endoskeleton.

64. B = The biceps contract and triceps relax when the fore arm is raised.

65. A

66. C = *Mimosa pudica*, Venus fly trap and Pitcher plant show nastic movement. It is non-directional.

67. B = Cork cells are bark cells. The cells are dead at maturity. Cork is also called phellem

68. D

69. D

CHAPTER TWELVE

1. B = Spirogyra reproduces asexually (vegetatively) by fragmentation. Fragment of a matured spirogyra breaks off to live an independent life.

2. A = Exponential increase in population is common in microorganisms that reproduce by binary fission.

3. A = The nucleus, during binary fission, divides by mitosis into two.

4. B = Budding is a form of asexual reproduction common to both hydra and yeast. A bud is an outgrowth of the parent organism which develops into a new organism.

5. NIL = *Mucor/Rhizopus* – asexual reproduction by spore formation. Amoeba/Paramecium – asexual reproduction by binary fission. Spirogyra -- asexual reproduction by fragmentation. *Penicillium* -- asexual spore (conidium). *Ascaris* – sexual reproduction. No option contain group that exclusively reproduce sexually, which is the implication of the question.

6. A = Banana, yam, pineapple and cassava are normally propagated asexually. Banana (suckers) pineapple (off sets); yam (stem tuber); cassava (stem cutting).

7. A

8. D = In asexual reproduction, there is no exchange of genetic material, hence no variation among the offspring.

9. B= Rhizomes are elongated horizontal underground stems swollen with food reserves e.g. cannalily, ginger.

10. E = Stem tubers are swollen terminal ends of axillary branches which goes down into the soil. They are enlarged and swollen with food e.g. Irish potato.

11. A = A corn is a short globular underground stem swollen with food and have lateral buds along the nodes. E.g. cocoyam.

12. D

13. B = Bulbs have condensed conical stems from which leaves arise at its dorsal side. The leaf bases are swollen with food e.g. onion, garlic, tulip.

14. D = Runners are horizontal stems which creep on the ground producing roots and aerial shoots wherever they touch the ground. When the internodes die out, the shoots become independent plants e.g. sweet potato.

15. E = Suckers are shoots produced from below the ground level by underground stems or roots running near the surface e.g. plaintain, banana.

16. D = Cassava is propagated by stem cutting.

17. E = When the leaf of Bryophyllum falls on moist soil, buds and adventitious roots develop at the notches along its margin. The buds develop into new plants.

18. D = The offspring are better adapted to their environment of origin but may not survive in changing environment conditions due to lack of variation.

19. A = A scion is a detached living portion of a plant (as a bud or shoot) joined to a stock in grafting. It needs the stock to survive.

20. E = Stock – old established plant; scion -- a desired plant.

21. D

22. D = Each micronucleus divides mitotically into two and later into four.

23. B = Spindle fibres are formed during metaphase. Each chromosome is seen attached to spindle fibres at the centromere.

24. D = Parthenogenesis is a form of asexual reproduction where growth and development of embryos occur without fertilization.

25. B

26. C = The protruding protoplast is active and may be called the male cell, while the other is passive (its contents stay in the same cell) and may be called the female cells.

27. A = The chiasmata is thought to be the point where two homologous non-sister chromatids exchange genetic material during chromosomal crossover during meiosis. It is visible in the diplotene stage of prophase I.

28. A = During anaphase II, centromeric division occurs and the sister chromatids are separated. They start to migrate towards the opposite poles.

29. D = Cleavage is a period of cell proliferation, which converts the unicellular zygote into a multicellular embryo. It is a form of mitotic division.

30. D = Pair of homologous chromosomes segregates during anaphase I of meiosis so that diploid (2n) chromosome can be halved to haploid (n) in other to restore the genetic content of zygote to that of the parents.

31. E = At metaphase, the chromosomes are individually arranged on the metaphase plane (metaphase plate) or the middle of the cell.

32. C = The pistil (gynoecium) is made up of an ovary, one or more styles and one or more stigmas.

33. A = Petals are highly coloured and so attract pollinating agents.

34. D = Protection and photosynthesis.

35. A = Monocarpous flowers are composed of one carpel (a simple pistil). The terms apocarpous or syncarpous refers to compound pistils (one or more carpel). Apocarpous flowers contain two or more distinct carpels. In syncarpous flowers, two or more carpels are fused together.

36. B

37. B = An inflorescence may be defined as cluster flowers, all flowering arising from the main stem axis or peduncle.

38. D = Actinomorphic (regular) are flowers that are radially symmetrical in that they are able to be bisected into similar halves in more than one vertical plane, forming mirror images. Zygomorphic (irregular) are flowers that are bilaterally symmetrical so that they are able to be bisected into similar halves in only one plane, forming mirror images.

39. C = Dichogamy is having pistils and stamens that matures at different times, thus promoting cross-pollination rather than self – pollination. Homogamy is the maturation of pistil and stamen of a flower at the same time, ensuring self – pollination.

40. D

41. A

42. A = Dichogamous flowers are either protandrous or protagynous, protandry is when the stamens mature before the pistil. Protogyny is when the pistil matures before the stamen.

43. E

44. D = Wind pollinated flowers have long styles so that the stigma can project from the flower catching pollen grains flowing in the air.

45. C = They do not have sticky grains or stamen.

46. A = Cross pollination allows variable characteristics.

47. A = Insect visit in other to feed but pollinate in the process of feeding.

48. B = Crotalaria is protandrous: when a bee alights on the wing petals, it's body weight pushes down the wing petals and the keel; the pollen in the keel is pushed out of it into the undersurface of the bee by the rounded anthers.

49. B = After fertilization, a flower dries, its petals drop, the top of the pistil falls off and the ovary swells as the

seeds form inside it. After its ovules are fertilized, the ovary becomes the plant's fruit.

50. A = Flowers with inferior ovaries are termed epigynous, the flowers are above. Flowers with superior ovaries are termed hypogynous, the flower are below.

51. B = Double fertilization ensures the formation of fertile embryo (diploid) and endosperm (triploid). Diploid (male gamete + egg cell). Triploid (male gamete + definitive nucleus).

52. B = See 24.

53. D = During fertilization, there are three nuclei in the pollen tube – a tube nucleus and two male nuclei from generative nucleus.

54. A = I – Integuments, II – Polar nuclei, III – Egg nucleus, IV – Two synergids, V – Micropyle. Micropyle V allows the passage of pollen tube and pollen nucleus.

55. C = Female gamete is the egg nucleus. One male nucleus fuses with the egg nucleus to form zygote (2n), the other migrates to the centre of the embryo sac and fuses with the two polar nuclei to form a triploid (3n) primary endosperm cell.

56. A = Simple fruits are produced from a single flower in which the pistil is made up of one carpel or several fused carpels.

57. D = Indehiscent fruits – non-splitting fruits. Dehiscent fruits – splitiing fruits.

58. A = Caryopsis – pericarp and seed coat have become fused together e.g. grains of cereals (maize, guinea corn, millet) and grasses.

59. C = Follicle – formed from one carpel, split along one side only e.g. cnestus, magnolia. Legume – formed from one carpel, split along two sides (dorsal and ventral) e.g. crotalaria. Capsule – formed from two or more fused carpels and splits by many longitudinal slits e.g. okra, cotton. Schizocarp – many seeded fruit white breaks up into one seeded parts e.g. Desmodium.

60. C = Drupe – Pericarp consists of 3 layers: epicarp--thin outer skin mesocarp -- middle fleshy or fibrous layer; and endocarp – hard inner layer enclosing the seed.

61. D = Multiple or compound fruits are produced from many flowers, the ovaries are fused together, e.g. pineapple and fig.

62. B

63. D – In parietal placentation, ovules are arranged along walls of a syncarpous ovary. In marginal placentation, ovules are arranged along the joined edges of a single curpel. In axile placentation, ovules are arranged at the centre of the ovary of a syncarpous ovary.

64. B

65. D

66. B = A maize grain is a fruit and not a seed because the testa and the fruit wall fuse after fertilization.

67. E = An aggregate fruits is formed from a single flower with several free carpels.

68. D = A schizocarp is a dry fruit formed from a syncarpous ovary (fused) in which the carpels break into several parts each containing one seed, i.e. mericarps.

69. B = Groundnut is a legume, the pericarp is not hard and tough. Nut have tough, hard and stony pericarp e.g. cashew nut.

70. C

71. B = A tree fruits is formed from fertilized ovary. Seeds are found from ovules.

72. C

73. C = Complete – possess all the floral parts; regular (actinomorphic) can be cut into two halves from any plane (radially symmetrical): hermaphroditic – possesses both stamen (androecium) and carpel (gynoecium); inferior ovary (perigynous) – outer flora part appears to be above the ovary.

74. A = Samara is an achene with pericarp extended to form wings e.g. combretum.

75. A

76. A = Castor oil seed is a dicotyledonous seed while maize is a monocotyledonous seed.

77. B = An aggregate achene or aggregate fruit.

78. C

79. D

80. C = 1 carpel = follicle and legume; 2 or more fused carpels = capsule.

81. C = Possession of floss (fine silky thread) around seeds e.g cotton and silk cotton enhance their dispersal by wind.

82. A = Matured pararubber explode, dispersing seed up to 15 metres. Mango and oil palm and dispersed by man. Tecoma is dispersed by wind. Triumfetta fruits stick to the fur of animals that disperse them.

83. B = I – Ureter (from the kidney), II – Prostate gland, III – Vas deferens IV – Urethra.

84. C = Vasectomy is achieved by severing the *vas deferens* (sperm duct).

85. B = Oviduct and uterus are part of female reproductive system clasper is found in some fishes.

86. B = The epididymis store spermatozoa which are released on ejaculation.

87. E = Seminal vesicle (vesicula seminalis) secrete a significant proportion of fluid that ultimately become semen.

88. B = *Vas deferens* transport sperm. Sperm is transferred from the vas deferens into the urethra, collecting secretions from the male accessory sex glands such as the seminal vesicles, prostate gland and the bulbourethral glands which form bulk of the semen.

89. C = Urine and spermatozoa pass through the urethra

90. E = The bladder store urine only.

91. D = see 89

92. D

93. B = I – Ovary, II – Uterus, III – Cervix, IV Vagina V– Vulva.
94. B = Fertilized egg is implanted and developed in the uterus, II.
95. C = The ovum matures in the graafian follicle, where it is released at ovulation into the uterus. It is transported to the uterus and unless fertilization occurs, and out through the vagina.
96. D = The placenta functions as a selective maternal – fetal barrier. The placenta is an organ that connects the developing foetus to the uterine wall to allow nutrient uptake, waste elimination and gas exchange via the mother's blood supply.
97. D = Fertilization typically takes place on the upper part of the oviduct (fallopian tube).
98. C = Sex determination occur at fertilization.
99. A = Identical twins – One ovum fertilized by one sperm. Non identical twins – two ova fertilized by two sperms.
100. D = An ectopic pregnancy is an abnormal pregnancy that occur outside the womb (uterus). The baby (foetus) cannot survive, and often does not develop at all in this type of pregnancy.
101. C = Parturition is the process of delivering the baby and placenta from the uterus to the vagina to the outside world. Also call labour and delivery.
102. C = Once superficial implantation occurs, the embryo begins receiving nourishment directly from the cells lining of the mother's uterus. Later when the yolk sac is formed it provides nutrients from the mother to the embryo before the placenta is ready to function.
103. D = The outer cells of the blastocyst form a covering of two membranes, an outer chorion and inner amnion containing amniotic fluid which protects the embryo from shock and desiccation.
104. A
105. B = Birth control pills used by females contain oestrogen, which depress follicle stimulating hormone (FSH); and progesterone, which thickens the cervical muscles so that entry of sperms into the uterus is impeded.
106. A = Cervix is a muscular ring closing the lower end of the uterus.
107. B = Amniotic fluid is found in pregnancy.
108. C =
109. A = The placenta is responsible for gaseous exchange (lungs), removal of waste products (kidneys) and nutrition (digestive system).
110. C = Some placenta hormones are human chorionic somatomamotropin (hCS), human chorionic gonadotropin (hCG), progesterone and oestrogen.
111. B = Chorion – outermost membrane; amnion – inner membrane.
112. D = Groundnut is a fruit, fruits are product of fertilized ovary (sexual reproduction).
113. D

114. D
115. A
116. C
117. A = Pineapple crown of leaves are called offsets. They can be used to propagate pineapples.
118. D = Stigma before anther--protogyny; anther before stigma -- protandry.
119. E = Yam is an underground. Underground stems are parts of plants which helps it to reproduce. It stores food for the plant and MUST be underground.
120. C = The petiole secures the leaf to the stem.
121. E = The leaves of the onion are modified into a storage organ known as bulb.
122. B = Food is stored in bulbs (e.g. onion and ginger) as sugar not as starch.
123. E = *Bryophyllum pinnatum* can be easily propagated because the leaf is capable of reproducing vegetatively by producing buds.
124. B
125. E = Binary fission is a method of reproduction in unicellular organisms by splitting into two.
126. D = Gonads are reproductive organs. Testes in male, ovary in female. They produce gametes – reproductive cells.
127. C = Lizards do not show parental care.
128. D = Airspaces are characteristics of seeds or fruits dispersed by water so that they can float easily without sinking.

CHAPTER THIRTEEN

1. C = Growth is an irreversible increase in the size and weight of an organism by the addition of new protoplasm.
2. A = Without the addition of new protoplasm, growth cannot be said to have occurred.
3. D = Changes in dry weight gives a reliable estimate of growth.
4. B = Growth occurs when anabolism (build up of complex molecules) exceeds catabolism (breakdown of complex molecules).
5. D = Optimum growth is attained when the organism can reproduce itself.
6. B
7. B = The radicle is a miniature root; a little above the root tip is the actively dividing meristematic cells, the outermost layer of which forms the root cap and the piliferous layer.
8. A = Etiolation occurs when plant grows in the dark. It results in lengthening of internodes, reduced and pale yellow leaves due to absence of chlorophyll.
9. C = In the curve, III represent the highest growth rate, I represent birth.
10. C = Environment resistant is greatest at the point with little growth IV. V represent negative growth or senescence (old age). The shape of the curve is

sigmoid.

11. B = Unilateral illumination of shoot causes auxin to move away from the light source. This gives a greater concentration of auxin on the shaded side of the shoot, which in turn causes increased cell elongation and thus bending of the shoot.

12. B = Growth in animals is equal in all parts of the body (intercalary) and limited to certain periods in life. Growth in plants is restricted to growing regions (apical meristem) and continuous throughout life.

13. D

14. B = During germination, the embryo develops into a seedling.

15. D = 1 – Fused pericarp and tasta (seed coat), 2 – Endosperm, 3 – Embryo, 4 – Scutellum, 5 – Point of attachment to cob.

16. C = The endosperm stores food which nourish the embryo and the growing parts.

17. B = A dormant seed is a viable but dry seed that can remain for a long period without germinating. Dormancy ensures germination during a favourable season. Seed dormancy may be reduced by growth inhibitors, impermeability of the testa, delayed development and concentration of seed.

18. B = In epigeal germination, the cotyledons are carried above the soil due to the elongation of the hypocotyl. The hypocotyls is the part of the embryo just below the cotyledon.

19. C

20. D = In hypogeal germination, the cotyledons remain in the soil due to the elongation of the epicotyls. The epicotyls is the part of the embryo just above the cotyledon.

21. B

22. B = The micropyle is a try hole through which water is absorbed during germination.

23. B = Water helps to activate enzymes; it provides a medium for metabolic reactions and it transports nutrients to the growing regions of the embryo.

24. C = Intrinsic factors such as growth hormones – Auxin (indole acetic acid), gibberellins; and enzymes are most critical for plant growth.

25. A = Condition necessary for germination are: sufficiency of water, adequate supply of oxygen, suitable temperature, food or nutrient and light.

26. C = Oxygen is needed for tissue respiration, leading to the production of energy.

27. A

28. B = Alkaline pyrogallol is used in the experiment to show that oxygen is necessary for germination, its function is to absorb oxygen.

29. D

30. A

31. B = Endospermous seed are swollen with stored food e.g. maize seed.

32. D

33. C = Auxin (indole acetic acid) a powerful growth stimulant is effective in extremely low concentrations.

34. C

35. B

36. C = Heat is produced as seeds respire.

37. E = The sequence of events in seed germination is as follows: (i) Seed absorbs water and swells (ii) Testa splits (iii) Radicle appears and grows downwards (iv) Plumule appears and grows upwards.

38. C

39. A

40. E = Transition zone

41. E. In epigeal germination e.g. bean and castor oil, the cotyledons, when exposed to light, become green and carry out photosynthetic function.

CHAPTER FOURTEEN

1. E

2. C

3. A = Receptor to CNS – sensory neuron, CNS to effector – motor neuron.

4. D

5. D

6. A

7. C

8. B

9. C

10. A = A neuron is made up of cell body, dendrites and axon.

11. B

12. C

13. C = Another classification of various receptors divides them into (i) teleceptors "distance receivers", which are concerned with events at a distance; (2) exteroceptors, which are concerned with external environment near at hand; (3) interoceptors, which are concerned with the internal environment; and (4) proprioceptors, which provide information about the position of the body in space at any given instance.

14. D = Blinking of the eye is an involuntary (reflex) action, controlled by the medulla oblongata.

15. E = The primary function of the peripheral nerve is to transmit signals from the spinal cord to the rest of the body or to transmit sensory information from the rest of the body to the spinal cord.

16. B = The olfactory lobes of the brain eventually interpret the sensations of smell.

17. E = Hypothalamus is responsible for temperature regulation.

18. C = The spinal cord is an extension of the medulla oblongata of the brainstem.

19. None of the above. The left cerebellar hemisphere. Coordinates the left arm and leg.

20. B = The brain and spinal cord is covered by meninges,

which have 3 layers, an outer pia matter, middle dura matter and inner arachnoid matter.

21. B = The name ventral fissure was given because it is wide.

22. E = The dorsal root ganglion (or spinal ganglion) is a nodule on a dorsal root that contains cell bodies of neurons in afferent spiral nerves.

23. A = The grey matter contains neural cell bodies, (nerve cells) while the white matter contains myelinated nerve fibres. It is white because of the myelin.

24. D

25. A

26. D = Motor neuron only exist in the CNS (brain and spinal cord).

27. C = I – node of Ranvier

28. D = Na^+ influx and K^+ efflux causes depolarization which leads to generation of action potential.

29. A = Transmission of impulse across a synapse is a chemical process. Acetylcholine is a neurochemical secreted by the synaptic knob which facilitate neuromuscular transmission (nerve to muscle).

30. B

31. D

32. C = Ganglion contains cell bodies of neurons.

33. C = I – sensory neuron, 2 – relay neuron, 3 – grey matter, 4 – motor neuron.
All the cell bodies of the neurons in the spinal cord are found in the grey matter (3) while the nerve fibres of the cell bodies are present in the white matter

34. A = In reflex action impulse flows as shown: Receptor in skin→sensory neuron (1)→ relay neuron (2) →motor neuron (4) →effector in muscle.

35. E = 3-4-1-5-2

36. A = I– Choroid, II – Fovea (yellow spot), III – Vitreous humour, IV – Optic nerve. The choroid nourishes the eye.

37. C

38. D = Outer sclera, middle choroid and inner retina.

39. B = In the retina are the cone-shaped and rod-shaped cells for bright and dim vision respectively.

40. A = During image formation, light rays are bent most at the air/cornea surface and the lens. NOTE: The aqueous and vitreous humour also bend light rays.

41. A = When the gaze is directed at a near object, the ciliary muscle contracts. This decreases the distance between the edges of the ciliary body and relaxes the lens ligaments, so that the lens springs into a more convex shape (bulges).

42. D

43. C = Ciliary muscles and the lens tend to lose their elasticity at old age.

44. B = In myopia (short-sightedness), image is formed in front of the retina.

45. B = Nocturnal animals (e.g. bat) have more rods than

cones. Diurnal animals (e.g. man) have more cones than rods.

46. A = In correcting myopia, concave lens is used to diverge incoming rays and bring them to focus on the retina. In correcting hypermetropia, convex lens is used to converge incoming rays to bring to focus on the retina.

47. A = That is long sightedness. Corrected with convex lens.

48. D = I – Auditory meatus, II – Ear ossicles, III – Oval window, IV – Cochlea, V – Eustachian tube. Hearing is brought together by I, II, IV.

49. D

50. C = Sound entering the hear follow this path: Auditory meatus, eardrum (tympanum), ear ossicles, oval window.

51. B = Outer ear – pinna and auditory canal; middle ear – tympanum, ear ossicles, and eustachian tube; inner ear – semicircular anal, ampulla, utriculus, sacculus, cochlea, and auditory nerve.

52. A = Vibrations are amplified and carried across the oval window by the ear ossicles.

53. D = Tip – sweat, sides of centre – salty; side – sour; back – bitter.

54. D = The taste buds are sensitive to the four primary taste.

55. D = Chemicals in the surrounding air dissolve in the mucus film and stimulate the chemico-receptors which send off sensory impulses through the olfactory nerves to the brain.

56. B = I – Sebaceous gland, II – Nerve fibre, III – Blood vessel, IV – Sweat gland.

57. A

58. D = Endocrine system secretes hormones. Hormones are responsible for chemical coordination.

59. A = Endocrine system releases hormones (a chemical); nervous system, for example, in neuromuscular transmission, releases acetylcholine. Therefore, they both involve chemical transmission.

60. A = Growth hormones are secreted in the anterior pituitary gland, which is in the brain.

61. D = Antidiuretic hormone (ADH) and oxytocin are released from the posterior pituitary gland.

62. B = Deficiency of thyroxine (from thyroid gland) causes cretinism –poor physical and mental development.

63. B = Thyroid gland secrete thyroxine, which regulate metabolic rate, growth; mental and physical development.

64. B = Adrenalin is released during emergency to increase heart beat, and elicit some sympathetic actions which increase efficiency in danger.

65. C = The islet of Langerhans of the pancreas secretes insulin and glucagon. They decrease and increase blood glucose respectively.

66. C
67. D = I – Pituitary gland, II – Thyroid and parathyroid gland, III – Adrenal gland; and IV – Gonad.
68. B = Parathyroid gland secrete parathormone which raise the level of calcium ions in the blood.
69. B = Growth hormone is also called somatotropic hormone.
70. B = Insulin is secreted by the beta cells of the islet of Langerhans in the pancreas.
71. A
72. B
73. B = Photopism is caused by the unequal distribution of auxin, auxin accumulate on the shaded side of the shoot.
74. D = Abscissic acid is a growth inhibitor. Options A, B, and C contain abscissic acid.
75. C = Auxins are produced in the shoot and root apices, hence control apical dominance.
76. A = Gibberellins do not induce the formation of adventitious root, unlike auxin. This is why auxin can be used in stem cutting.
77. B = Auxin have no effect on the growth of dwarf plants but if gibberellins was applied to them, they will grow to normal height.
78. C = Gibberellins have their effect on plant stem in the same way as auxins, by stimulating cell elongation in particular. Rapid cell division is by cytokinin.
79. C = Ethylene induces ripening of fruits. Unripe fruits are exposed low level of ethylene to ripen them for the supermarket shelves. The effect of ethylene on the plant is to raise the rate of respiration and this cause the ripening of fruits.
80. C = Auxin can affect the metabolism of the cells, making them to respire excessively. These effects cause the death of the plant. As a result, synthetic auxins can be used as very effective weed killers. They are absorbed much more effectively by broad – leaved (dicotyledonous) plants than by monocotyledonous plants and are therefore particularly useful for removing broad-leaved weeds from monocotyledonous cultures as in lawns and fields.
81. D = Both gibberellins and cytokins break seed dormancy.
82. D = Both auxins and gibberellin are produced in the meristems of apical buds and embryo; and are both responsible for cell elongation and apical dominance. NOTE: Cytokinin and auxin are both responsible for cell division.
83. B and C = Homeothermic animals are able to maintain a relatively constant body temperatures. When the environment is cold, the body needs to gain heat. Conduction, convection and radiation can cause both heat loss and heat gain to the body, evaporation is a mechanism of heat loss only, in which a liquid is converted to gas. Sweat is also a mechanism of heat loss only.
84. A = Osmoregulation is the process by which a constant balance of osmotic pressure is maintained in the body of an animal by control of salts, ions and water content.
85. A = Poikilothermic animals are also called cold blooded e.g. fish, amphibian and reptiles while homeothermic animals are also called warm blooded e.g. birds and mammals.
86. D
87. C = Homeostasis is a feedback reaction that ensures a fairly constant internal environment. It is a reflex action.
88. A = When the body is too hot, it decreases heat production and increase heat loss by peripheral dilation, the dilation of blood vessels in the skin. When these vessels dilate, large quantities of warmed blood from the core of the body are carried to the skin, where heat loss may occur via radiation, convection and conduction. Sweating also causes cooling.
89. D
90. B = Increase osmotic pressure of blood (increase Na$^+$) will cause the increase in reabsorpiton of water and decrease reabsorption of salts by the kidney nephron. Decrease osmotic pressure of blood (increase water) will cause the increase in reabsorption of salts and decrease in the reabsorption of water by the kidney nephron.
91. D
92. D
93. D
94. D = The semicircular canal helps the brain to analyze complex body movements and positions of the head in man to maintain balance.
95. C
96. A
97. C = The malpighian layer is a row of meristematic cells, they produce new epidermal cells by mitosis.
98. B = The cells of the malpighian layer also contain melanin pigment, which gives the skin its colour.
99. A = The removal of the pancreas would affect the metabolism of glucose due to insulin deficiency and digestion of carbohydrate, protein and lipids due to the absence of amylase, trypsin and lipase.
100. C
101. C
102. C = The neurons are the functional unit of the nervous system.
103. C = Synapses are the small gaps occurring between the nerve fibres.
104. A
105. B = The neurolemma (neurilemma or sheath of Schwann) is a nucleated cytoplasmic layer of Schwam cells that surrounds the myelin sheath of axion in the PNS.
106. C = The Nodes of Ranvier are small gaps between the myelin sheath along the length of a nerve fibre. There is no neurilemma in the CNS.

107. E = None of the above. Reflex actions are involuntary.
108. B
109. E = Cranial nerves are nerves from the brain while spinal nerves are nerves from the spinal cord.
110. C = The taste buds and olfactory organs respond to chemical stimuli.
111. D
112. D = The vitreous humour is more viscous (thicker) than the aqueous humour.
113. D
114. C
115. E = The rays of light that enters the eyes are refracted by the cornea, lens, aqueous humour and vitreous humours.
116. A
117. B = The anatomy of the inner ear is dominated by large fluid-filled spaces. The bone tube (sometimes called the bony labyrinth) is filled with fluid called perilymph.
118. C
119. B = The ossicles are made up of malleous, incus and stapes.
120. E = The eustachian tube connect the ear to the oropharynx. Vibration in the ear is lost down the Eustachian tube.
121. C
122. A = The function of the cochlea is to convert vibration by using cilia and fluid filled chambers to send nerve impulses to the brain in order to convert these vibrations to sound (hearing).
123. D
124. D
125. C

CHAPTER FIFTEEN

1. E = Ecosystem in a self supporting unit consisting of the biotic (living) and the abiotic (non-living) components that are in a continuous interaction.
2. B = Population is defined as a collection of similar organisms that are found in the same habitat while a specie is defined as a collection of similar organisms that are capable of interbreeding freely to produce a viable offspring.
3. C
4. B = A community consists of different population of organisms living together in a particular habitat.
5. C = Physical factor are abiotic factors.
6. C = Guinea grass = 15; *Ipomea spp* = 5
   Sida spp = 7, *Imperata spp* = 23   Total = 50
   % of *Imperata spp* = 23/50 x 100/1 = 46%
7. B = A caterpillar and an aphid living in different parts of the same plant are occupying different ecological niche – the smallest area within an habitat.
8. A
9. C = A biosphere is the part of the earth and its

atmosphere in which living organisms exist or that is capable of supporting life.
10. B
11. B
12. A
13. C = Abiotic factors are non-living factors, plankton, insect, conifers and bacteria are living things.
14. A = Biotic factors are living factors. Temperature, water and soil are non-living things.
15. A = Terrestrial and aquatic habitats will have these in common; rainfall, temperature, pressure, light, wind, and pH.
16. D = Soil factors are referred to as edaphic factors.
17. A = pH affects the distribution of animals in marine and fresh water habitats.
18. A = An environment is everything surrounding an organism that affects it. Physical environments are abiotic, this, with the biotic environment, affects an organism.
19. C = Intertidal zone is exposed to wave action and the organism are subjected to dessication.
20. B = Density dependent factors affect population growth by either increasing or decreasing as population density changes e.g. predation, food, competition, diseases. See page 123.
21. B = Competition is a biotic factor (living) which affect any habitat with limited resources.
22. C = A turbid pond will impair the penetration of water thereby limiting the intensity of light needed for photosynthesis.
23. A = The main problem of intertidal organisms is exposure and the resulting drying out (desiccation) when the tide ebbs; to overcome this, organisms in sandy and muddy shores burrow into the soft substratum; those with shells e.g. molluscs, withdraw into their protective shells.
24. D = Marshes and swamps are characterized by high salinity, tidal movement, water-logged soil and anaerobic soil resulting in high acidity (low pH).
25. B
26. C
27. B = Transect method is one of the methods used for estimating the number and types of plants in a particular area.
28. B
29. A = Microscope is used to magnify and view organisms or part of organisms.
30. B
31. C
32. A
33. C = Miroscope, potentiometer, chemical balance, telescope are not used to measure ecological factors.
34. B
35. D
36. C = Capture and recapture methods are used to monitor the census of bird, fish, and insect population.

A quadrant method is used to determine the frequency or population of inactive animals like snail, millipedes or aphids on a leaf.

37. C = Bernacles and limpets are found in the marine habitat. They compete for space in the intertidal zone.

38. A = Deserts have low, irregular rainfall, low temperature at night and high temperature by day, low relative humidity, strong-wind and high intensity of sunlight, very sparse vegetation. Temperature is the most important of these abiotic factors.

CHAPTER SIXTEEN

1. D
2. C = X is a saprophyte – feeds on dead decaying organic matter. Y is a substrate.
3. D = Symbionts are unrelated.
4. B = Epiphytes derives only support from host without causing any damage.
5. A = Saprophytism is between a saprophyte (living organism) and a substrate (non-living organism).
6. B = Lichen is a symbiotic association between algae and fungi.
7. A = Nitrogen fixing bacteria are found in the root nodules of leguminous plants.
8. A = *Mycorrhiza* is a symbiotic association between fungi and roots of higher plants. The fungi aid the root in transport of inorganic nutrients from the soil into the plant while the plant supplies the fungi with organic nutrients.
9. C = Mistletoe is a partial plant parasite. It is parasitic due to its possession of a modified root called haustoria to obtain inorganic nutrient when it pierces the host's stem. It is autotrophic because it can carry out photosynthesis.
10. A = *Hydra viridis* have *zoochlorellae* algae cells in their bodies. These algae photosynthesize; producing sugars and other things which are both used by the algae and the Hydra. Hydra also supplies the algae with its mineral requirements.
11. D = Bacteria in the caecum of ruminants help to produce cellulase for the digestion of cellulose diets of the ruminants and when digested, they produce vitamin B and K. The bacteria in turn, gain shelter, protection and ideal environment.
12. D = See 8.
13. B = All food chains and webs begins with a producer (an autothroph).
14. E = The number of individual organisms decreases progressively down the trophic level in the food chain i.e. from producers to consumers.
15. A = Only a fraction of a solar energy that reaches the earth's surface is absorbed by the plants not all the solar energy.

16. B
17. B = Decomposers help to recycle nutrients e.g. fungi and bacteria.
18. C = Mouse is an omnivore (consume both plant and animal materials).
19. A = In a pyramid of number, hawk will have the least population compared to the others.
20. C = Every trophic level begins with producers (autotrophs) and then the consumers (heterotrophs).
21. B = The concentration of non-biodegradable substance will increase as it goes higher up the food chain. E.g. DDT in aquatic body, fish (2nd trophic level) 0.001ppm, seagull (3rd trophic level) 0.01ppm.
22. B
23. B = As we move forward in a food chain, there is a decrease in the number of individuals.
24. C = The highest energy is at the level of producers, there is always a great loss of energy as food is transferred from one trophic level to the next.
25. C = Pyramid of numbers represents the number of individuals at each trophic level of food chain at a particular time.
26. B = Pyramid of energy represents the rate of flow of food energy through each trophic level of a food chain.
27. D = The most efficient is from the sun to the producers while the least efficient is from consumers to decomposers.
28. C = Organisms become larger and fewer along the food chain.
29. D
30. B = Algea (producer), tadpole (primary consumer), Tilapia (secondary consumer), shark (tertiary consumer).
31. D = Second law of thermodynamics.
32. D = Energy transfer decreases along the trophic level. This leads to a reduction in the population size of organisms at the higher trophic level. Pyramid of biomass represents the total wet or dry mass of an organism in each trophic level. It takes into account the size and number of individual organism.
33. B = Carbondioxide is released into the atmosphere during respiration, combustion, decay and volcanic eruption.
34. C = Carbondioxide will be fairly maintained by the balance between photosynthesis (depletes CO_2) and combustion (replenishes CO_2).
35. A = I – Evaporation, II – Transpiration, III – Precipitation, IV – Surface run-off.
36. B
37. A = Evaporation releases water to the atmosphere where it condense in the cloud. Precipitation returns water to the earth as rain.
38. C = Capillary water is held in capillary pore of the soil. Plant roots are able to absorb it, hence it is the available water for plants.
39. A = During thunderstorm, nitrogen in the atmosphere

combines with oxygen to form nitrogen (II) oxide; with the addition of more oxygen, nitrogen (IV) oxide is formed. This dissolves in water to form trioxonitrate (V) acid HNO_3 which reacts with bases in the soil to form nitrates.

40. C
41. D = Nitrification is the conversion of ammonia or ammonium salts into nitrites by *Nitrosomonas* and nitrite into nitrates by *Nitrobacter*.
42. C = Leguminous plants have symbiotic nitrogen fixing bacteria in their root nodules; the bacteria fix nitrogen thereby enriching the soil with organic nitrogen.
43. A = Denitrification is the process by which soil nitrates are reduced to atmospheric nitrogen through the action of denitrifying bacteria. All other processess would help to add nitrate to the soil.
44. D = See 41.
45. D = Dead plants and animals are acted upon by saprophytes to form ammonia. The ammonia is converted into nitrites by *Nitosomonas* and nitrite is converted into nitrates by *Nitrobacter*.
46. A
47. A = Most denitrifying bacteria respire anaerobically, e.g. *Thiobacillus denitrificans*, obtain its energy by anaerobically oxidizing sulphur compounds; nitrates is used as the oxidizing agent and in the process it is reduced to gaseous nitrogen.
48. B = Denitrification is the reduction of nitrates in the soil into nitrogen gas by denitrifying bacteria e.g. *Escherichia coli* and *Thiobacillus denitrificans*.
49. B
50. A = Soil nitrite *Nitrobacter*→Soil nitrate.
51. E = An example of symbiont living in the root nodules of leguminous plant is *Rhizobium*, not given in the option.
52. C = The *Rhizobium* obtain photosynthetically derived carbohydrates and an ideal environment to live.
53. E = Nitrates are made into proteins in plants.
54. C = Putrefaction is one of the seven stages in the decomposition of the body of a dead animal. Some of the end products of putrefaction are: ammonia and ammonium compounds, hydrogen sulphide and hydrogen phosphide.
55. D = organisms that are not dependent are capable of surviving on their own i.e. autotrophs. In parasitism, commensalism and symbiosis, organisms either depend on each other or one depending on the other while the other is either unaffected or harmed.
56. D   57.   D   58.   D
59. D   60.   A
61. C = Zooplankton form autotrophs in the aquatic habitat.
62. A = Mineralization refers to the process where an organic substance is converted ton an inorganic substance.

63. D = *Pseudomonas denitrificans* is a denitrifying bacteria.
64. A

CHAPTER SEVENTEEN

1. B = In plants with floating leaves, stomata may be found only on the upper epidermis; submerged leaves may lack stomata entirely.
2. B = Scorpion and ant-lion are not present in fresh water.
3. D = Brackish water is formed when freshwater and seawater mixes at estuaries. Brackish water has salinity which fluctuates with tides, wet and dry seasons. In the lagoon, for example, the salinity of the water increases during the dry seasons and decreases during the rainy season.
4. C = Halophytes – plants adapted to salty environment. Xerophytes – plants adapted to dry habitats. Hydrophytes – plants adapted to aquatic habitats. Mesophytes – plants that live in an environment with adequate water supply. Epiphytes – plants which depends on another plants for support only.
5. A = Aerenchymatous tissues are tissues of thin-walled cells with large, air-filled intercellular spaces found in the roots and stems of some aquatic and marsh plants for buoyancy.
6. A = Air cavities (or air-filled intercellular spaces) which enhances buoyancy to float on water.
7. D = Stilt roots of the red mangrove enable it to stand firm in strong ocean winds and keep the branches above high tide levels. Pneumatophores (breathing roots) of the white mangrove allow intake of atmospheric air.
8. A = In mangrove plants, seed germinate on the parent plant, i.e. are viviparous, and develop long radicle.
9. B = Pneumatophores (breathing roots) of the white mangrove.
10. D = There are more species (specie diversity) in the rainforest than other habitats.
11. C = Littoral zone is along the edge of the stream.
12. B = Tilapia is found in freshwater. Tortoise – land and fresh water. Achatina is the land snail.
13. B = Abiotic factors are physical or non-living factors.
14. C = Date-palm thrive best in almost rainless districts like in grassland and desert.
15. C = Water conservation is an important adaptation required by terrestrial organisms particularly in the Sahel savanna and desert. Food, predator and space affects all organisms regardless of the habitat.
16. A
17. A = Rainfall is the most important ecological factor that affects biotic community in a terresrial habitat.
18. B

19. C = Reduction in the number of stomata on the leaves reduces the loss of water vapour via transpiration.
20. C   21.   D   22.   A

264

CHAPTER EIGHTEEN

1. C = Guinea savanna can be found in Kaduna, Enugu, Benue, and Kwara.
2. A = Rivers state (Port-Harcourt) is an estuarine and Yobe state (Damaturu) is a Sahel. Moving from: estuarine→ rain forest→ Guinea savanna→Sahel.
3. B = Up north is Sahel, down south is estuarine.
4. C = Non-epiphytic ferns and fern allieds live in the shade created by the canopy formed by the tall broad-leaved trees.
5. C = Rain forest is rich in epiphytes and climbers. They use tall trees as support to reach for the light.
6. D = Savanna trees possess thick, fire-resistant barks. In the Sahel, plant bark posses thorns.
7. C = Southern Guinea savanna has more rainfall and tall grasees compared to the Northern Guinea savanna.
8. C = Savanna grasses have underground stems (*Rhizomes*) helping them to root the soil properly.
9. B = Guinea savanna is between the rain forest and Sudan savanna.
10. D
11. A = Cactus is a plant found in the desert and mammals also undergo aestivation for long periods in the desert.
12. A = Perennating organs are organs possessed by plants that enable them to survive from one growing season to the next e.g. bulb, runner, sucker, corm e.t.c.
13. A = The greatest species diversity occur in the tropical rainforest. Transect method is difficult to use in estuarine. It is also difficult to use in Guinea and Sahel savanna because of few plant species.
14. C = Northern Guinea savanna and Sahel savanna can be found in Sokoto state.
15. Trees are planted as shelter belts in the Sahel to reduce the harmattan wind.
16. A = Desert plants have devices to reduce transpiration e.g. reduced leaves and stomata
17. C = Desert has low rainfall and high temperature.
18. A Relative humidity is the amount of moisture in the atmosphere; it is high when there is high rainfall and low temperature, and low when there is low rainfall and high temperature.
19. B = Katsina state falls in the Sahel savanna, which is very susceptible to desert encroachment.
20. D = Countershading is more important in some aquatic organisms like fishes.
21. C = Between the littoral and benthic zone, light can only penetrate the littoral zone while benthic zone is dark and cold.
22. B = Grasses are predominant in Guinea savanna, which grow tall during the rainy season with sparse trees.
23. C
24. A
25. D = Cactus plant is structurally adapted to survive by possession of swollen stems that store food and water;

and the leaves are reduced into thorns.
26. C = Mambilla plateau is a high grassland plateau with an average elevation of about 1,524 metres found in Taraba state of Nigeria. It is the highest plateau in Nigeria.
27. D = Mangrove swamp in Nigeria is restricted to the tropical rainforest.
28. D = Montana vegetation is typical of Adamawa and Jos plateau.
29. C = Tropical deserts has a temperature ranges from 27^0C to 33^0C in summer; to 15^0C in winter. The absence of cloud in desert skies result in hot days and very cold nights, rainfall is low and fast blooming plants during rainy season.
30. A

CHAPTER NINETEEN

1. B = Species are group of similar organisms that can interbreed freely. Population deals with the total numbers of organisms of the same spice in a habitat.
2. C = Density dependent factors affect population growth by either increasing or decreasing as population density changes. Those that can limit a population include competition, disease, predation, parasitism, food shortage.
3. D = Population will increase when there is high birth rate, low mortality rate, high immigration rate and low emigration rate.
4. D = Seed dispersal provides means of preventing overcrowding; seeds are dispersed away from the parent plant.
5. B
6. C = Capture and recapture methods are used to monitor the census of bird, fish, and insect population. Quadrant method is used to determine the frequency of population of plants and inactive animals like snail, millipedes or aphids on leaf.
7. B = Overcrowding can be caused by immigration (movement into), natality (birth), high survival rate. Overcrowding can be reduced by emigration (movement out of), mortality, and low survival rate.
8. D = Turbidity is peculiar to aquatic habitat. It is the depth of visibility in water, which is a measure of penetration of light.
9. B = Exponential (logarithmic) growth doubles the size of population at a constant interval. It occurs in bacteria.
10. D = Population density is the average number of individuals of a species per unit area of habitat.
    Density = Population size / Area of habitat.
    Population frequency is how often a specie occur at different sites in its habitat.
11. B = Population equilibrium is attained when there is a balance in nature (Predator-prey) as a result of the

attainment of its carrying capacity.

12. B = Food scarcity increases mortality rate by increasing competition.

13. B = Population dynamics is the changes (increase or decrease) in population size over time.

14. C = The pioneering organism in a plant succession on bare rock surfaces are some species of crustose lichen, followed by foliose lichen and then mosses after the lichen have changed the rock to soil.

15. C = Primary succession starts on sites which have not previously borne vegetation.

16. A = Competition will lead to succession by the emergent organisms.

17. C = In the given trend, Spirogyra is the simplest plant and water lily is the most advanced.

18. C = Climax forms the final stage of succession. It is characterize by stable composition of plant and animal species.

19. B = Primary succession involves the formation of new soil. Secondary succession is faster than primary succession because soil is already present.

20. A = As succession approaches climax, there is an increasing stability.

21. C

22. D = Growing of vegetation on an abandoned farmland is an example of secondary succession.

23. B = Intra- specific competition arises when organisms of the same species compete for a limited resources.

24. B = In a normal succession, species diversity increases until it becomes fairly constant due to the attainment of a climax community.

25. D

26. A

27. B = Space by organism in an habitat
$$= \frac{\text{Area of habitat}}{\text{Number of organisms}}$$

28. B = Territorial behaviour helps organisms to guard against intruders thereby preventing overcrowding.

CHAPTER TWENTY

1. B = Clay contains the finest particles. Silt is finer than sand but still feels gritty.

2. B = Granite is a form of igneous rock and not a soil type it contains readily recognized mineral crystals such as muscovite, quartz, biotite and hornblende.

3. D = Mineral particles in the soil are derived from the chemical breakdown of rock particles or from decomposing organic matter in a less degree.

4. D = By sedimentation, heavier particles settle at the bottom leaving lighter ones to suspend.

5. D = Clay soil is the most difficult to plough in a wet season because of its sticky nature.

6. E = The experiment is to compare the capillarity of various soil samples; capillarity is high in clay, low in sand and average in loam.

7. C = The experiment is to determine the percentage of humus in soil sample. The presence of humus release the smoke.

8. A = After sedimentation, particles settles in this order (light to heavy): humus, clay, silt, sand, gravel (stones).

9. D = Clay has high water retaining capacity. Although humus can also retain water, it has air spaces for soil aeration. Sand, however, is very porous; hence a soil devoid of sand and low in humus with high clay content would retain the highest water.

10. D = Laterite is a red soil; it is common in the tropics. It is produced as a result of leaching ; the soil tends to be acidic because lime potash from the topsoil are leached away by rain action.

11. B = Water retention in soil is greatest in clay soil and poorest in sandy soil. It indicates the capillarity of the soil.

12. B = Nitrates are particularly prone to loss through leaching due to their high solubility in water.

13. A = Irrigated lands after some years become unfertile.

14. A = Continuous cropping reduces soil fertility.

15. B = Leaching is the removal of soluble material from a substance, such as soil or rock, through the percolation of water.

16. C = Gravel (>2mm); feel coarse, sand (2- 0.05 mm); feels gritty, silt (2-0.05mm); feels like flour but still gritty, clay (< 0.002 mm); feels sticky when wet.

17. D = Weight of crucible alone = 5 gm
Weight of crucible + fresh soil before being heated = 10 gm
Weight of fresh soil being heated is (10-5)gm = 5 gm
Weight of crucible + fresh soil after being heated = 8 gm
Weight of water in the soil is (10-8) gm = 2gm
Percentage % of water = $\frac{2}{5} \times \frac{100}{1}$ = 40%

18. C =
Volume of air = vol. of water + vol. of soil added - vol. of water + soil after stirring
= $(500 + 350)\,cm^3$ - $800\,cm^3$.
= $850\,cm^3$ - $800\,cm^3$
= $50\,cm^3$
% of air in soil sample = $\frac{\text{Volume of air}}{\text{Volume of soil + air}} \times \frac{100\%}{1}$
= $\frac{50\,cm^3}{350\,cm^3} \times 100\%$
= 14.28%

19. D (Closest option). From the given parameter
Mass of soil = (29-10)g = 19g
Mass of soil + water = (29 -18)g = 11g

% of water loss = $\dfrac{\text{Mass of water}}{\text{Mass of soil} + \text{water}} \times 100$

$= \dfrac{11}{19} \times 100$

$= 57.89\%$

20. A = All soil have humus in common in varying degree.

21. D = The spaces between clayey soil particles are very tiny because of its fine nature.

22. D = Sands have a specific heat capacity.

23. E = Fewer nutrients because of its poor aeration.

24. D

25. A = Erosion can be reduced by ridging across slope this slows down the flow of water and less soil is removed.

26. D = Mixed grazing leads to overgrazing which exposes the soil to erosion.

27. B = The process by which lime is added to clay soil is called flocculation. Lime is any material containing high amount of calcium or calcium and magnesium. Flocculation reduces soil acidity, encourages good granular soil structure for better root development.

28. C

29. C = Soil microorganisms help in the recycling of nutrients like carbon and nitrogen.

30. D = Liming makes the soil more porous by flocculation of the soil particles thereby improving capillary action.

31. A = Microbes conserve and renew soil fertility by the cycling of nutrients. They also decay dead organic matter.

32. B = Excessive application of fertilizer alters soil pH.

33. C

34. A (Closest option) = soil erosion can be controlled by:
i. Vegetation protection;
ii. Contouring of sloping ground;
iii. Terracing of slopes;
iv. Mulching;
v. Strip cropping.

35. B = Bacteria and fungi are decomposers. They are largely responsible for the decay of dead organic matter and its subsequent recycling.

36. A = Crop rotation can maintain soil fertility, it is a practice whereby farmland is planted with different crops in successive seasons.

37. B

38. D = To a soil scientist humus is the organic, non-cellular, long lasting component of the soil. It may also have some trace elements.

39. C = Soil water can be classified into 3: Gravitational water – occupies the larger pores of the soil. It is of no use to plants because is easily pushed down by gravity. Capillary water – is held in capillary pore of the soil. Plant roots are able to absorb it; hence it is the available water for plant. Hygroscopic water – The water held tightly at the surface of soil colloid particle, it can be utilized by some microorganism not by plants.

40. A = Soil contains SWOLAM = small rock particles, water, organic matter (humus), living organisms, air (oxygen and CO_2), and mineral salt.

41. A = Cover crops such as grasses check leaching.

42. C

43. C = Soil enhance plant growth, which are the primary producers that maintain life.

44. B = Leguminous cover crops enrich the soil, the residues are also rich in organic nitrogen from decomposing nodules.

45. B = Percentage of water in soil =
$\dfrac{\text{Mass of water}}{\text{Mass of soil}} \dfrac{(x-y)}{(x)} \dfrac{\times 100}{1}$

46. C

47. D = Soil pH tends to decrease (increase acidity) with the repeated use of nitrogen fertilizers.

48. B = Amount of soil water is given as: Mass of crucible and soil, X_2g – Mass of crucible and soil after heating, X_3g.

CHAPTER TWENTY ONE

1. A = Cholera can only be contacted by eating or drinking contaminated food or water.

2. E = Syphilis and gonorrhea are both caused by bacteria.

3. D = Cholera is caused by bacteria. Malaria is transmitted through the bite of female *Anopheles* mosquito. Syphilis is caused by bacteria. Sleeping sickness is caused by a protozoan (*Trypanosome*). Small pox is a viral disease, it causes skin blister and is transmitted through close contact with an infected person.

4. D = The life cycle of Bilharzia involves two hosts, man and a water snail. Man is the primary host and snail is the secondary host.

5. D = Control of Onchocerciasis is mainly by the distruction of the blackfly (vector) and by treating infected person with drugs.

6. A. = Swollen shoot disease of cocoa is a viral disease.

7. D = Blackfly, vector of Onchocerciasis breeds in fast flowing water.

8. D = Malaria – female *Anopheles* mosquito; cholera – housefly; river blindness (Onchcerciasis) – blackfly.

9. B = Bilharzias can be controlled by destroying the water snails that belong to the class of mollusca and by ensuring that water bodies are not contaminated by urine or faeces of infected persons.

10. D = Guinea worm is caused by drinking infected water. It can be controlled by provision of portable water.

11. A = Elephantiasis is transmitted through the bite female *Culex* mosquito while Taeniasis gets into man through pig or cow meat.

12. D = Vaccination or inoculation is the introduction of the mild form of the disease – causing organisms into

the body. This stimulate the production of antibodies which help to build immunity against the disease in the body.

13. D = The construction of dam may lead to an increase in the prevalence of the diseases that are associated with water such as malaria (mosquitoes breed in water), bilharzias (has water snail as the intermediate host) and onchocerciasis (blackflies breed in fast flowing streams).

14. A = Typhoid fever is caused by *Salmonella typhi*.

15. B = Bird flu is caused by virus.

16. A = The infective cercariae of (Schistosome can penetrate the bladder (*Schistosoma haematobium*) causing bloody urine.

17. B = River blindness (Onchoceciasis) is transmitted blackfly, *Simuliun damnosum*.

18. A = Poliomyelitis is caused by a virus.

19. B = Immunization is the process of rendering an animal resistant to infection artificially, by conferring either passive community or by vaccination.

20. D = *Schistosoma haematobium* penetrate the bladder. The larval stage is called Miracidiurn which swims with the aid of cilia.

21. D = Infective plasmodia of the malaria is the sporozoite.

22. D = An obligate parasite is completely dependent on its host. *Taenia* (tape worm) cannot survive without host man.

23. A = Tuberculosis is a bacteria infection transmitted by droplet infection in coughs, sneezes and conversation.

24. D = Crowded habitation favours the spread of tuberculosis.

25. C = Only meningitis (inflammation of the meninges) is not associated with water among the options.

26. D = Female *Aedes* mosquito transmits a virus causing yellow fever

27. D = *Plasmodium falciparum* causes malaria, which is transmitted by female *Anopheles* mosquitoes.

28. C = United Nations Children Fund (UNICEF) is an international organization set up to improve health and welfare of children and nursing mothers.

29. D = Refuse, when not burnt in incinerators, gives out smoke at a level that is detrimental to the health of living organisms.

30. D = Applying chemicals to water bodies would cause water pollution.

31. B = Loud disco music constitute noise pollution.

32. C = Carbon monoxide poisons tissues by combining with haemoglobin to form carboxy haemoglobin, which reduces the affinity of haemoglobin for oxygen, a condition that may lead to suffocation.

33. B = Some petrol often contain tetraethyl lead, thereby putting people who suck it at risk of increasing their blood concentration of lead.

34. C = Sulphur (IV) oxide is an acidic poisonous gas; it can pollute air, water and soil because it dissolves in moist air to form acid, when inhaled, it irritates and damages the sensitive living part of the eyes, air passages and lungs. It dissolves in water to form acid rain which dissolves the aluminum salts in soil causing them to build up to toxic levels in underground water supplies; this affects sources of drinking water.

35. C = Oil is an imported pollutant of the marine environment in Nigeria because of the perpetual oil spillage.

36. D = Chlorofluorocarbon (CFC) react with the Ozone (O_3) in the upper atmosphere converting it to oxygen, thereby depleting the ozone layer.

37. C = Agro-chemicals are used in the farms e.g. fertilizers when in excess, they are washed by rain into fresh water bodies thereby causing water pollution.

38. A

39. D = Radioactive materials can be carcinogenic (substances that cause cancer).

40. A = Carbon monoxide is issued from car exhaust, which has strong affinity for haemoglobin.

41. B = Greenhouse effect is a gradual increase in temperature of the earth due to excessive amount of heat absorbing gases such as CO_2. Trees absorb CO_2 for photosynthesis. Deforestation is cutting of trees.

42. A = Excess CO_2 in the atmosphere could lead to global warning or greenhouse effect.

43. A = The degree of water pollution can be tested using the biochemical oxygen demand (BOD). This test measures the amount of oxygen bacteria needed to break down the organic matter in a water sample in a fixed period of time.

44. A = Activities that releases excess CO_2 into the atmosphere can result in green house effect (global warming) said as burning of fossil fuel e.g. coal.

45. A

46. D = Non-biodegradable pollutants cannot be broken down into simpler harmless substances e.g. plastics, glass, tin, and metal scraps. They are better controlled by recycling.

47. D

48. C = FEPA is empowered to prevent indiscriminate destruction of the environment that may cause pollution.

49. B = Water, soil, wildlife, forest are all renewable resources. Wildlife are very important in conservation because they are under perpetual threat by man's activities.

50. A = Mineral resources are non-renewable hence the need to conserve them.

51. B = Poaching is the illegal catching or shooting of animals.

52. B = Bush burning leads to the destruction of diverse species of organisms.

53. B

54. C

55. A = Malaria is caused by – *Plasmodium*, which is a protozoan.
56. C
57. A = Recycling is used for non-biodegradable wastes such as glass, metal scraps, plastics. Organic matter is biodegradable.
58. C = River dams, acid rain, sewage are not sources of air pollutants.
59. A
60. C = Nuclear energy is powered by radioactive materials such as uranium obtained from the earth crust – nonrenewable.
61. B = Oxides of sulphur (SO_2) and nitrogen (NO_2)
62. D
63. D = Schistisoma, commonly known as blood – flukes and biharzia. *Schistosoma haematobium* is commonly referred to as bladder fluke because it enters the bladder.
64. A   65.   C   66.   C
67. B   68.   B   69.   D

CHAPTER TWENTY TWO

1. C = Variation accounts for the essential differences seen among organisms of the same species.
2. A = Individuals with unfavourable traits naturally go into extinction leaving individuals with favourable traits to dominate.
3. C = Physiological variations are variations in the functioning and activities of living organisms. These include: blood group, tongues rolling and PTC tasting. They are also discontinuous variation.
4. B = Discontinuous variation does not offer an intermediate form e.g. tongues rolling, you are either a tongue roller or not.
5. B = Left-handedness is a discontinous variation. You can either be left handed or not.
6. B = Continous variations show gradual transition between two extreme form as a result, there are intermediate e.g. size, height, weight and colour.
7. A = The different fingerprint pattern are arches, loops, whorls and compound or double whorl.
8. A = Disease resistance is a physiological variation.
9. A = See question 3.
10. B
11. E = Any character shown by an organism is due to the genes inherited from the parents, the environmental factors such as nutrition, education, climate etc
12. C = An organism is a product of its inherited genes and environment.
13. D = Identical twins inherit the same gene hence the same genotype but can show distinct variations when raised in different environments.
14. C = Sexual reproduction give rise to variation while asexual reproduction do not.

15. B = Fingerprint is unique to individuals though individuals may have same pattern.
16. A = Blood group O is a universal donor.
17. D = Agglutination occurs when contrasting antigen (e.g antigen A) combine with antibodies (e.g antibody B)
18. A = Rhesus factor was first identified in the Rhesus monkeys.
19. B = Gametes:

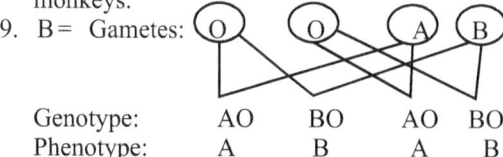

| Genotype: | AO | BO | AO | BO |
| Phenotype: | A | B | A | B |

20. B = In the ABO blood system, the recognizable groups are: A, B, AB and O (4 groups)
21. A = Parents with blood group O cannot produce a child with any blood antigen (A or B)

Parents:

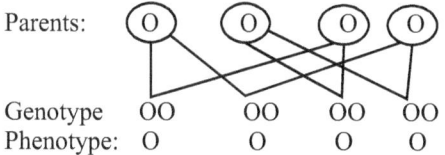

| Genotype | OO | OO | OO | OO |
| Phenotype: | O | O | O | O |

22. A = DNA analysis is the most accurate to determine paternity.
23. B   24.   D   25.   C
26. C   27.   C
28. A = Blood group O is a universal donor because it lacks blood antigen (A and B). Blood group AB is a universal recipient because it lack blood antibodies (Anti- A and anti- B).
29. A   30. B

CHAPTER TWENTY THREE

1. E = None of the children will be blind because blindness is not an inherited trait.
2. D = A monohybrid inheritance is a single characteristic that is controlled by the gene.
3. C = Let the gene for brown and blue eye "B" and "b" respectively. Genotype for homozygous brown eyed girl and blue eyed boy is BB and bb respectively to get these two genotypes, the parents must be heterozygous for brown eye i.e. Bb.

Parents: Bb   vs   Bb

Gametes:

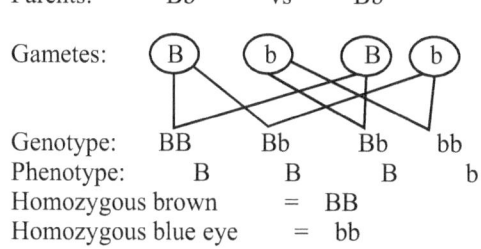

| Genotype: | BB | Bb | Bb | bb |
| Phenotype: | B | B | B | b |

Homozygous brown   =   BB
Homozygous blue eye   =   bb

4. B = Albino is a recessive trait. Therefore let the gene for normal male and albino female be AA and aa respectively.

Parents: AA   vs   aa

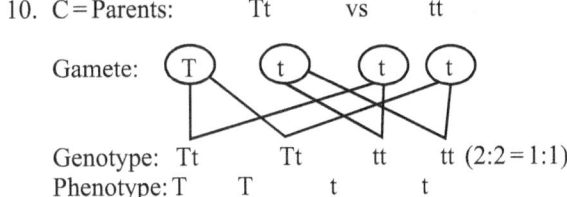

Gametes: (A) (A) (a) (a)

Genotype: Aa   Aa   Aa Aa

Phenotype: A   A   A   A

All the offspring's would be genetically carriers of albinism but phenotypically normal.

5. D = Heredity is the transmission of inherited character from parents offspring via genes. Good yield is a function of the gene and its transferable from generation to generation.

6. B = Gregor Mendel was known to be the father of genetics

7. A = The key words in Mendel's first and second law are segregation (separation) and assortment (recombination) respectively.

Mendel's first law (also called the law of segregation) states that during the formation of reproductive cells (gametes), pairs of hereditary factors (genes) for a specific trait separate so that offspring receive one factor from each parent.

Mendel's second law (also called the law of independent assortment) states that each member of a pair of homologous chromosomes separates independently of the members of other pairs so the results are random.

Mendel's third law (also called the law of dominance) states that one of the factors for a pair of inherited traits will be dominant and the other recessive, unless both factors are recessive.

8. A

9. B = For most genetic studies, fruit fly, *Drosophilia melanogaster*, was found suitable because it is small and could be kept in the laboratory in large number, it has a short lifecycle of two weeks ; it has just four homologous pairs of chromosomes; the male and the female can be easily distinguished and it has many distinguishable discontinuous characteristics.

10. C = Parents: Tt   vs   tt

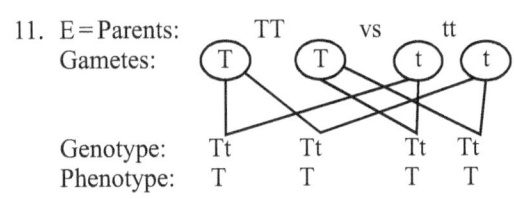

Gamete: (T) (t) (t) (t)

Genotype: Tt   Tt   tt   tt (2:2 = 1:1)

Phenotype: T   T   t   t

11. E = Parents: TT   vs   tt

Gametes: (T) (T) (t) (t)

Genotype: Tt   Tt   Tt Tt

Phenotype: T   T   T   T

12. B = An F2 generation is only possible if the parents of the F1 generation are both homozygous for a contrasting character, i.e F1 are all heterozygous. By self pollination,

Parents: Tt   vs   Tt

Gametes: (T) (t) (T) (t)

Genotype: TT   Tt   Tt   tt (1:2:1)

Phenotype: T   T   T   t (3:1)

13. D = Genotype for normal haemoglobin and sickle – cell haemoglobin are AA and SS respectively.

Parents: AA   vs   SS

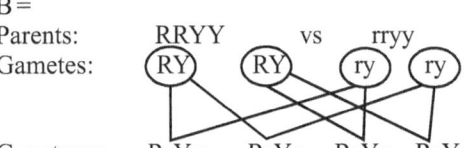

Gametes: (A) (S) (S) (S)

Genotype: AS   AS   AS   AS

14. B =

Parents: RRYY   vs   rryy

Gametes: (RY) (RY) (ry) (ry)

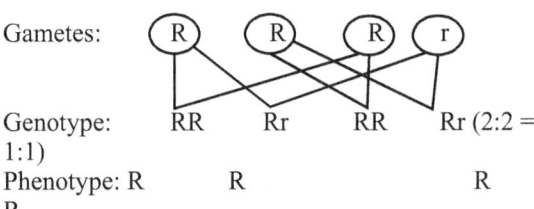

Genotype: RrYy   RrYy RrYy RrYy

Round and yellow seed, as RY is dominant over ry.

15. C = Parents: RR   vs   Rr

Gametes: (R) (R) (R) (r)

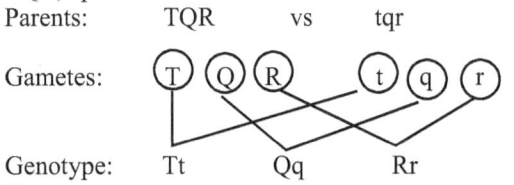

Genotype: RR   Rr   RR   Rr (2:2 = 1:1)

Phenotype: R   R   R   R

16. B = The gene content of the woman's eggs would be TQR, tqr

Parents: TQR   vs   tqr

Gametes: (T) (Q) (R) (t) (q) (r)

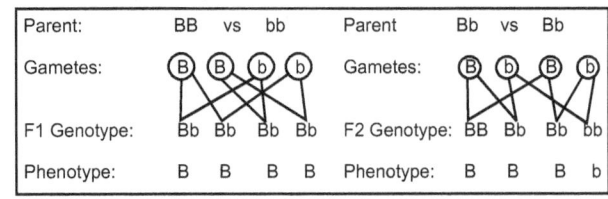

Genotype: Tt   Qq   Rr

17. B =

F1 = 1 genotype (Bb).

F2 = 3 genotypes (BB, Bb, bb)

Parent:	BB vs bb				Parent	Bb vs Bb		
Gametes:	(B)	(B)	(b)	(b)	Gametes:	(B) (b)	(B)	(b)
F1 Genotype:	Bb	Bb	Bb	Bb	F2 Genotype:	BB Bb	Bb	bb
Phenotype:	B	B	B	B	Phenotype:	B B	B	b

18. B = Blending inheritance is an example of incomplete dominance where neither allele is dominant over each other. So the heterozygous genotype has its own phenotype.

19. B
20. B = According to Mendelian inheritance, the alternative form of a gene for a character that expresses the character in a different way is called allele. E.g. blood group in man (a discontinuous variation) is controlled by a gene I, with the alleles I a I b.
21. B = Alleles are two alternative genes at the locus in homologous chromosomes.
22. D = Incomplete dominance is a situation in which both member of an allelic pair are expressed in an in an heterozygous situation. In AB blood group neither antigens is dominant over the other.
23. D = I – TT, II – Tt, III – TT, IV – tt.
24. B = 1 : 2 : 1
25. D
26. E = Genes are discrete unit of inheritance that determines hereditary traits.
27. A
28. C = Recessive genes are always suppressed in the presence of dominant genes as a result are only expressed in the homozygous form.
29. C
30. D
31. B
32. E
33. C
34. C = 2n is a diploid number, whereas in gametes, there are haploid numbers (n) of chromosome; therefore, in an organism where 2n = 16, n = 16/2 = 8.
35. C = Crossing over occurs when chromatids at meoisis exchange portions with their homologous partners. It occurs in the late prophase stage of meiosis.
36. D = DNA- Deoxyribonucleic acid is made up of repeating units called nucleotides. Each nucleotide consists of a five- carbon sugar, a phosphate group and organic nitrogen containing base; Adenine, Guanine, Cytosine and Thymine.
37. C = Somatic cells contain diploid chromosome denoted as 2N.
38. C = In a DNA strand, guanine will only pair with cytosine (i.e. G to C) and adenine with thymine (A to T). A DNA strand with a base sequence TCA will have a complementary strand of AGT.
39. D = XX – Female; XY – Male.
40. C = After meiosis the number of chromosomes changes from diploid (2n) to haploid (n).
41. D = In RNA, uracil is found instead of thymine. In DNA, A and G are purines; T and C are pyrimidines.
42. B = Couple with sickling trait will be AS
Parents: AS   vs   AS
Gametes:
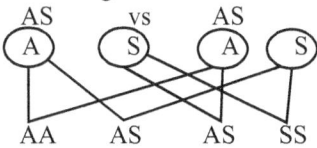
Genotype: AA   AS   AS   SS
Normal ¼ Carrier 2/4 = ½ Sickler ¼

43. A = Heterozygous parents e.g. Tt for height
Parents: Tt   vs   Tt
Gametes:
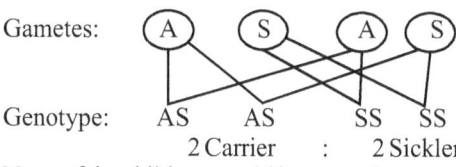
Genotype: TT   Tt   Tt   tt (1:2:1)
Phenotype: T   T   T   t (3:1);
¾ Tall (dominant) and 1/4 Short (recessive).
44. D = A child can either be a girl or a boy; hence, the probability of the fifth child being a girl is one out of two.
45. A = F1 All Tt.   F2 = TT, Tt, Tt, tt
Number of dwarf = ¼ x 120 =   30
46. D = Parents: AS   vs   SS
Gametes:
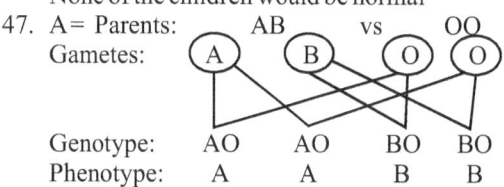
Genotype: AS   AS   SS   SS
      2 Carrier   :   2 Sickler
None of the children would be normal
47. A = Parents: AB   vs   OO
Gametes:
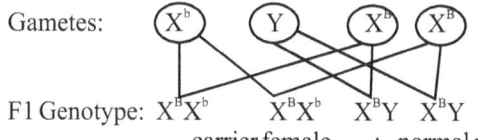
Genotype: AO   AO   BO   BO
Phenotype: A   A   B   B
O is a recessive gene, hence, BO will be B and AO with be A. Since they are six, three will fall into group B and the other three into group A
48. A = Parents:
Colour blind father (X^{b}Y) : Normal mother (X^{B}X^B)
Gametes:
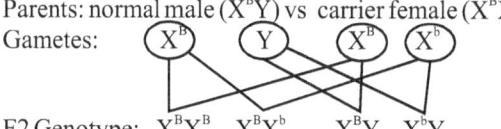
F1 Genotype: X^{B}X^b X^{B}X^b X^{B}Y   X^{B}Y
      carrier female   :   normal male
Parents: normal male (X^{B}Y) vs carrier female (X^{B}X^b)
Gametes:
F2 Genotype: X^{B}X^B X^{B}X^b X^{B}Y   X^{b}Y
Normal female, carrier female, normal male and colour blind male respectively.
∴ Ratio of sufferers, carriers and normal are 25%, 25%, and 50% respectively.
49. C = The man and his wife are both AS
Parents: AS   vs   AS
Gametes:
Genotype: AA   AS   AS   SS
The likely parentage of their offspring that will be either carriers or sickler is ¾ x 100 = 75 %

50. B

52. B = Out breeding (hybridization or crossbreeding) leads to introduction of new traits. It increase the hybrid vigour.

53. B = Inbreeding ensures the preservation of traits within a given population.

54. B

55. A = The DNA is unique to every individuals.

56. C = Blood grouping in human beings is derived from the combination three different alleles A, B and O. Blood group is a character being controlled by multiple allele.

57. A = People who have Rhesus antigen on the surface of the red blood cell are said to be Rhesus positive RR or Rr and those that lack the antigen are said to be Rhesus negative (rr). Agglutination occurs when a Rhesus negative person receives Rhesus positive blood due to the release of anti- Rhesus antibodies by the immune system of the Rhesus negative person that reacts with the antigen on the red blood cells; agglutination also occurs when a Rhesus negative mother has a Rhesus positive child; hence, genetic counselling is necessary if a Rhesus negative woman marries a Rhesus positive man since there is a probability of the couple having a Rhesus positive child.

58. D = Parents are heterozygous A and B respectively i.e.

Parents: AO vs BO
Gametes: (A) (O) (B) (O)

Genotype: AB AO BO OO

59. C = Offspring of asexual reproduction are called clones. They almost are exactly like the original organism.

60. C = Sickle cell anaemia is caused by a mutation in a gene for haemoglobin that results in an altered haemoglobin molecule.

61. D = Two tall parents can only give birth to a short child when the two parents possess recessive gene for shortness. The parent must be heterozygous tall (Tt).

62. A

63. D =

Parents: Bb vs bb
Gametes: (B) (b) (b) (b)

Genotype: Bb Bb Bb bb
2 Brown – eyed : 2 Blue – eyed = 50% : 50%

64. B 65. A 66. A
67. A 68. C 69. B
70. A 71. C 72. A

73. C = An ovum has a haploid (n) chromosome but a fertilized ovum has diploid (2n) chromosome (23 pairs or 46).

74. A = Mongolism is an obsolete name for Down's syndrome which is XXX (Trisomy 21).

75. A

76. C = Phenotype is the observable expression of the characters of an organism resulting from the interaction between its gene and the environment it is the totality of the expressed traits of an organism.

77. A = A hybrid is the cross between organisms having contrasting traits. Hybrids are produced by the process called hybridization.

CHAPTER TWENTY FOUR

1. A = XX – female XY – male. The ovum always carry the X chromosome while the sperm cell could carry X or Y chromosome. Therefore to get a female, the ovum must be fertilized by the sperm cell carrying an X chromosome.

2. E

3. A = A cross between a colour blind mother and a colour blind father will give rise to all colour blind offsprings.

4. A = River blindness is caused by a round worm, *Onchocerca volvulus*, through the bite of the blackfly, *Simulium damnosum*.

5. C = Haemophilia is a sex-linked trait. It is carried on the X- chromosome. Genotype of haemophilic man can be represented as X^hY

Genotype of normal woman is X^HX^H.

Parents: Haemophilic father (X^hY) : Normal mother (X^HX^H)

Gametes: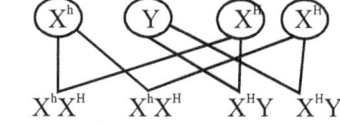

Genotype: X^hX^H X^hX^H X^HY X^HY
carrier females : normal males

6. C = The genes for sex-linked characters are recessive. When it occurs in the recessive form in a woman, its effect is suppressed by the dominant X chromosome. In man, it is readily expressed in the phenotype because they have one X chromosome, they cannot carry it in the recessive form.

7. B = Sons inherit the sex chromosomes of the mother and the sex linked trait is on the mothers X chromosome.

8. C = Sex-linked trait are carried on the X – chromosome.

9. A = Sex-linked traits are ONLY carried on the X – chromosome. Y- chromosome is empty of the traits. Also the male child cannot receive X-chromosome from his father.

10. D = The sons inherited the gene for colour blindness from their mother. As shown below:
Let gene for colour blindness be X^b
Genotype of colour blind mother is X^bX^b
Genotype of normal father is X^BY

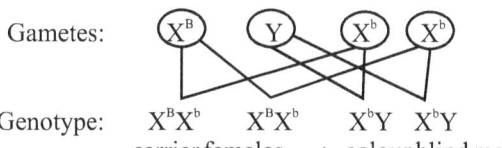

Gametes: X^B Y X^b X^b

Genotype: $X^B X^b$ $X^B X^b$ $X^b Y$ $X^b Y$

        carrier females   :   colour blind males

11. D = Colour blindness is the inability to recognize some colours usually red and green.

12. A A child with $X^N Y$ (male) will have normal vision while a child with $X^n Y$ (male) will be colour blind. Reason, sex-linked characters are expressed in the recessive form.

13. A = $X^B X^b$ – carrier female, $X^B X^B$ -- normal female, $X^b X^b$ -- bald female.

14. A = Sex is determined by the contributions of father's sex (X or Y) chromosome.

15. A = Turner's syndrome (monosony XO) is a sterile female.

16. A = Ability to grow long hair in females is sex-linked because it is carried in the X chromosome. The Y chromosome is deficient in genes. Possession of facial hair is determined by the Y-chromosome (male sexual character).

17. C = Sex is determined at fertilization (conception).

18. A = X chromosome (see question 6).

19. D = Let gene for colour blindness be X^c
Genotype of normal female is $X^c X^c$
Genotype of colour-blind male is $X^c Y$
Gametes: X^c Y X^c X^c

Genotype: $X^c X^c$ $X^c X^c$ $X^c Y$ $X^c Y$

        2 carrier females   :   2 normal males

20. A = See 14.

21. A = albinism is caused as a result of genetic mutation.

22. D = Drocunculiasis caused by *Drucunculus medinensis* is a water-borne disease and haemophilia is sex-linked.

23. B = Sex-linked characters occurs more often in men than women because it can only be expressed in the recessive form.

24. D = See 11.   25. C = Sex – linked.

26. A = See 15.

CHAPTER TWENTY FIVE

1. D

2. C = Autotrophs, especially plants, need light to manufacture food while heterotrophs depend on already manufactured food.

3. D = Intra- specific competitive will set in because they are of the same species.

4. D = Inter- specific competition occur among organisms of different species.

5. C = A worker termite and a soldier termite are of the same species; hence intra-specific competition.

6. C = Competition sets in when available resources become limited like in drought, fires out-break and other natural disasters.

7. A = Endoparasities have lost their organs of movement due to disuse. The choice of option D is wrong because not all endoparasites have piecing mouthparts e.g. *Ascaris*.

8. B = Competition will occur in a rice field when the field become weedy because the resources will become inadequate.

9. C = Early successional stages are often characterized by low productivity. Older communities are characterized by greater productivity as a result of increasing stability.

10. C = Arboreal animals are animals living on trees and tree trunks. e.g. flying birds, which lack teeth but they possess beak.

11. A = Sunbird possesses a long, slender and slightly curved beak for feeding on nectar.

12. C = Grasshoppers and cockroaches have biting and chewing mouth parts while houseflies possess sponging mouth parts. Only mosquito, tsetse flies and aphids have piecing and sucking moth parts from the lists provided.

13. D   14.   B

15. B = The keyword here is, large.

16. D = Mandibles are heavy toothed, jaw-like structures used to cut and crush food materials.

17. B = Some hydrophytes have air floats in their leaves, stem and petiole for buoyancy e.g. water hyacinth.

18. A = Camouflage helps an organism to blend with its surroundings so that it is not easily detected. Well developed limbs enchance fast movement and effective vision for good sight.

19. A = Gills – used for respiration is only efficient in water. Lateral line helps to detect movement in water. Streamlined body with no neck minimises friction during passage in water.

20. C = Forest species have buttress roots for stability.

21. A

22. B = Reptiles have dry, scaly, water proof skin which retards the loss of water.

23. B = Chameleon changes its colour to defend/ protect itself from predators.

24. C = Countershading in fish with dark dorsal surface (from top) and silver vertral surface (from bottom) protects them from predators.

25. C = Red coloured head of matured male lizard which it displays in front of a female is to attract them.

26. C   27.   C

28. A = Plants growing on dry or arid land are called xerophytes. Some of their adaptive methods are: thick succulent stems for storing water, succulent leaves for water storage, reduction of leaves to spines or scales.

29. A

30. D = Plants posses thick waxy cuticle and hairs on leaves to reduce water loss by transpiration. Animals

posses feather or modified scales (e.g) birds to prevent the evaporation of water as sweat.

31. D= Whistling pine (*Casuarina*) bears tiny scale leaves on needle-like branches to reduce water loss by transpiration thereby conserving enough water for use in the long dry period.

32. D = Feathers in birds are of different types some prevent evaporation of water as sweat some provide thermal insulatiom e.g. down feather, some are use for flight e.g contour feather.

33. A = The protoplasm of xerophytes are highly vicous and have high water content in their cells, as result, they can live with little water without being damaged.

34. D = The presence of sunken stomata or lower number of stomata and folding of leaves is to reduce water loss by transpiration.

35. A = Scales (in reptiles), feathers (in birds) and skin (in mammals) help in body temperature regulation.

36. A = Hydrophytes live directly in water; the roots are not long and the xylem tissue, which is for the transportation of water, is fearly developed.

37. C = Lung fish are capable of an amazing form of aestivation. Lungfish are primitive fish that still have lungs, allowing them to breathe air. When lung fish's lake dries up, the fish burrow into the mud, then secretes mucus until its entire body is covered. The mucus dries into a sack that holds moisture in. Even when the mud dries completely, the lung fish stays moist and breaths through a mucus tube.

38. D

39. D = The bodies of insects are covered by thick cuticle which prevents water loss by evaporation. Also in a way to conserve water they excrete their nitrogenous waste as uric acid and not as urea.

40. D = Gaping is a characteristic activity of several species of large reptiles in which the animal lies ashore with its mouth held open for prolonged periods. It is suggested as a thermoregulatory response.

41. C = Cryptic or concealing colouration helps to keep the prey and predator unnoticed. It is used in camouflage.

42. D

43. D = Disruptive colouration are patterns on the bodies of animals that break up the familiar shapes of their bodies, enabling them to go undetected e.g. spots and stripes of the leopard, giraffe, tiger and zebra.

44. B = Countershading.

45. A   46.   C

47. B = Chematophores are pigment for colour change. Also called camouflage cells present in chameleon.

48. D = Pheromones are chemicals produced as messengers that affect the behaviour of other individuals of insets or other animals.

49. C

50. C = Drones are the only males in the honey bee colony.

They mate with the queen.

51. A = The worker of the bees and termites are both sterile females. The workers termites are also blind.

52. B = The practice of removing the queen is a means of destroying the termite colony.

53. A = The protozoan helps to digest cellulose converting it into sugar which is absorbed by the worker termite. The protozoan in turn feed on the abundant nutrients of the worker termite as well as obtaining shelter in the termite's gut.

54. D

55. B = Waggle dance = food more than 100m away. Round dance = food within 100m.

56. D

57. D = Social behaviour is more complex and organized among mammals.

58. A = Aestivation = dry and hot. Hibernation = wet and cold.

59. B

60. B = III and IV are not under the control of the organism. I and II can be adopted to avoid extreme temperatures.

61. B   62.   B   63.   B

64. C

65. A = The Ladybird beetles are remarkable for shamming death as soon as they are seized.

66. C   67. A   68. A   69. D

70. A   71. A   72. C   73. B

74. C   75. B

CHAPTER TWENTY SIX

1. D = Evolution is brought about by the diferencese and similarities among organisms.

2. C = The theory of natural selection was developed by Charles Darwin and Alfred Russel Wallace in the year 1858 in a paper titled: A Theory of Evolution By Natural Selection.

3. B = Darwin proposed the theory of natural selection. Animals that are well adapted to the environment are the most successful.

4. C = Darwin's theory of natural selection suggests the survival of the fittest by better competitors

5. D = Lamarck proposed the theory of inheritance of acquired characters.

6. A = The flaw in Lamarck theory is that acquired traits are not heritable as shown by Wallace.

7. A = Long neck in giraffe was used by Lamarck to explain his theory of use and disuse. He said ancestors of giraffes had short necks; it was by stretching their necks to reach for the leaves on trees that the present-day giraffes became long-necked.

8. D = See 6.

9. B = Hugo de Vries propounded the mutation theory in 1901.

10. B = Mutation is a random change in the DNA (present in the chromosone) that alters genetic information and

so introduces new characteristics e.g. albinism, sickle cell-anaemia.

11. D = Natural selection allows organisms who are better competitors to survive by developing adaptative features.

12. A = According to Darwin, natural selection is the driving force behind evolution.

13. D = Enviromental pressure e.g. insufficient resources, climatic changes result in natural selection.

14. B = The idea of evolutionary trend is from simple to complex, unicellular to multicelular form. Therefore, older fossils will have simplest form of life.

15. B = The pentadactyl limb is modified into wings for flying in birds, flippers for swimming in whales, legs for running and walking in dogs and arms for grasping and holding in human beings and other bipeds.

16. B = The pentadactyl limb structure is absent in pisces (fishes).

17. B = Ecology is the study of organisims in relation to their environment, it has little evolutionary evidence.

18. D = Evolutionary trend in vertebrates is from Pisces à Amphibians à Reptiles à Aves (Birds) à Mammals. Therefore mammals evolved from birds.

19. D = Pisces (1 auricle and 1 ventricle), amphibians (2 auricles and 1 ventricle), reptiles (2 auricles and a partially divided ventricle), aves and mammals (2 auricles and 2 venticles).

20. D = Evidence from embryology showed that all embryos pass through a gillslit stage, establishing a common ancestry.

21. C = Radioisotope (carbon) dating is used in fossils to estimate ages; arranging the fossils according to their age shows evolutionary trend.

22. A = The relationship between living organisms and their extinct relatives can be obtained by studying the remains of organisms (extinct relatives) preserved mainly in earth's sedimentary rocks. This is called paleontology.

23. A = The appendix in man which functions as caecum in herbivores supports evolution. So also is the tail bone in man.

24. C

25. C = Comparative anatomy gives the structural affinities of organisms.

26. A = Homologous structures are e.g. pentadactyl limb pattern points to common evolutionary origin.

27. D = See 26.

28. D = Angiosperms are the most recent plants. They bear flowers and are divided into monocot and dicot. However, dicots are more advanced than monocots; and are capable of secondary thickening which monocots cannot.

29. A = Fossil records (for extinct organisms) helps to compare with presently living organism. Comparative anatomy compares similarities in forms (for modern organisms).

30. B

31. D = Simple forms before complex ones.

32. D = Natural selection suggests that organisms survive through adaptation and reproduce offsprings that are better adapted.

33. C = Evolutionary trend suggests that life begins from simple forms (unicellular) to complex forms (Multicellular).

34. A   35. C   36. D
37. C   38. A   39. A
40. B

41. A = Inheritance of acquired traits.

42. D = A common example of divergent evolution is the vertebrate limb. Whale flippers, frog forelimbs and human arms, see note on evolution.

43. B = Gene flow occurs among populations of the same species.

44. D = The disappearance of functionless organs was propounded by Jean Larmarck.

45. B = The evolutionary trend in plants is: Schizophyta, Thallophyta, Bryophyta, Pteridophyta and Spermatophyta. **Memory tip: Samuel The Biologist Planted Sugarcane.**

46. C = The tail bone is a vestigial organ in human, as it has no function.

47. B = Natural selection is a consequence of adverse condition (e.g. food scarcity) due to overpopulation.

GLOSSARY

A

Abscission Separation of an organ from the plant, as in leaf fall, fruit drop and loss of unfertilized flowers.

Accommodation Changing the shape of the lens of the eye to focus a sharp image of the objects on the retina.

Active transport The movement of substances across a biological membrane from a region of their low concentration to one of their high concentration; this movement requires energy.

Adaptation Any change in an organism that makes it better suited to survive a particular environment.

Adventitious root Roots which develop from the stems or leaves of plants.

Aerobic respiration The release of energy from food with the aid of oxygen.

Anaerobic respiration The release of energy from food without the use of oxygen.

Annual A plant that completes its life cycle from germination to flowering within one season.

Anterior The front or head end.

Antibody A protein produced by an animal to counteract the effect of the presence of a foreign protein (antigen) inside the body.

Antiseptic Substance used locally on body tissue that prevents its decay by arresting the growth of microorganism.

Anus The posterior opening of the digestive tract.

Asexual reproduction Reproduction that does not involve fusion of sex cells. It may be by spore formation, vegetative propagation, fission or budding.

Autecology The study of individual organism in relation to their environment.

Autotrophic A method of feeding in which an organism builds up its food from simple inorganic molecules.

B

Balanced diet A course of feeding contaioning all the essential food nutrients in correct proportions and in adequate amounts for a healthy life.

Binary fission Method of reproduction in unicellular organism by splitting into two.

Bilateral symmetry The property of having two similar sides with definite upper and lower surface and definite anterior and posterior ends.

C

Calorie (C) The amount of heat required to raise the temperature of 1,000 grammes of water by one degree Centigrade.

Capillarity A force causing the rise of a liquid along the surface of a tube or vessel as in the soil

Capillary Minute blood vessel which carries blood to individual cells.

Carnivore An animal whose diet consists of flesh e.g cats, leopards, dogs.

Cephalization The concentration of sense organs, nervous tissue (brain) and food catching organs at the anterior end of the body forming a head.

Chitin An insoluble polysaccharide material forming the main component of fungal cell wall and exoskeleton of arthropods.

Chromosome Thread-like structure in the nucleus that carries genes; it is made of DNA and protein.

Cloaca A common opening for the digestive, excretory and reproductive organs e.g. in roundworm, toad.

Commensalism An association between two organisms in which one of them, the commensal, derives benefit from the other, the host, which neither benefits nor suffers from the association.

Community Group of plants and animals living in a habitat.

Conduction A process of losing heat through physical contact with another object of body

Conjugation Sexual reproduction in which the contents of two cells unite as in certain algae and fungi; or exchange of nuclear materials resulting in rejuvenation of the cells as in paramecium and some bacteria.

Convection A process of losing heat through the movement of air or water molecules across the skin.

D

Deamination A process in the liver that remove (NH_2) radical from amino acid and converts the liberated ammonia into urea.

Decomposer An organism that breaks down the organic matter accumulated in the bodies of other organisms.

Diffusion The movement of molecules (liquid or gas) from a place of high concentration to one of low concentration.

Digestion The process of breaking down large insoluble food molecules into small soluble molecules that can be absorbed.

Disinfectant A chemical that frees materials fron infection by killing the microbes but not their spores. It is normally used on inanimate objects (c.f. antiseptic).

Dorsal Back or upper side of the body.

Droplet infection Infection that can be spread between persons through airborne particles containing viable microbes such as in coughing and sneezing.

E

Ecosystem The sum total of physical features and living organisms which interact in an environment to produce a stable system.

Emulsification The breaking down of fats into minute droplets.

Endoskeleton Internal skeleton of bone or cartilage typical of vertebrates.

Enzyme A protein catalyst produced by living cells which in small amount promotes chemical reactions e.g. during digestion.

Epicotyl The part of the plumule just above the attachment of seed leaves, (cotyledons).

Epigeal Type of seed germination in which the cotyledons come above soil level.

Epiphyte A plant that grows attached to the surface of another plant to gain a more beneficial position. It is naturally independent of the host. E.g orchids on tree branches, mosses on tree barks.

Epithelium A layer of cells which covers cavities, tubes and all free surface of the body.

Erosion The loss of topsoil by the action of wind or water.

Etiolation Condition of green plants when grow in darkness, characterized by thin, elongated stems and small chlorotic (yellow) leaves.

Excretion The removal of waste products of metabolism from the body.

Exoskeleton Hard outer covering which provide support for the body of certain animals especially arthropods.

F

Fertilization The fusion of a male and female nuclei to form a zygote.

Fermentation The process of anaerobic respiration in yeasts and some bacteria. In alcoholic fermentation ethanol and carbon dioxide are produced.

G

Gamete A male or female reproductive cell.

Ganglion An aggregate of nerve cell bodies located outside the central nervous system (CNS).

Gene A part of a chromosome that controls a particular hereditary feature of an individual.

Genotype The genetic characteristics of an organism.

Geotropism The growth of a parts of plant in response to gravity.

Gland A structure (cell, tissue or organ) concerned with the secretion of a specific chemical substance e.g sebaceous gland in the skin, cells in plants secreting gums, tannins, nectar. Animal glands may secrete their products through ducts (exocrine glands) or directly into the blood as in endocrine glands.

H

Habitat The place where an organism lives; the ecological characteristics in this area are relatively uniform.

Herbivore Animals whose diet consists of plant materials.

Hermaphrodite An organism with both male and female reproductive parts

Heterotroph An organism that can not build up its own food; hence it has other means of deriving its nourishment such as parasitic, saprophytic or holozoic methods.

Homeothermic Having a body temperature that remains fairly constant regardless of changes in the temperature of the environment.

Homologous organs Body organs in different organisms that have a similar origin and structure.

Homologous chromosomes A matching pair of chromosome inherited from male and female parents.

Hormone A chemical substance produced by the ductless gland in the body of organism.

Hybrid An offspring from a cross between parents differing in one or more traits.

Hybridization The crossing of two different varieties to produce a new variety with one or more desirable traits.

Hydrolysis The splitting apart of a molecule with the addition of water.

Hydrotropism The growth movement of a plant in response to water.

Hypocotyls The part of the plumule below the attachment of the cotyledons node. In epigeal germination it grows rapidly and carries the cotyledon above the ground.

Hypogeal The type of seed germination in which the cotyledons remain below soil level and inside the testa.

I

Immunity The power to resist an infective disease through natural or artificial means.

Inoculation Injecting a person or animal with ready made antibodies or weakened virus or germs to stimulate production of antibodies.

K

Kilojoule (KJ) S.I unit of energy (1 calorie = 4184 Kilojoules)

L

Larva An immature form of some animals that undergoes radical change in form to attain the adult form.

Leaching Loss of soluble mineral nutrients from the topsoil to lower levels where plant roots cannot reach as a result of the downward movement of rain or ground water.

Loam A fertile type of soil containing both sand and clay particles in roughly equal proportion.

Lumen A space with the blood vessel through which blood flow.

Meiosis Cell division that results in halving of the chromosome number occurring in the formation of gametes.

Metabolism The physiological and chemical processes in the body.

Metabolite A substance that is produced during metabolism or one that takes part in metabolism.

Metamorphosis Changes in the life cycle of an organism through one or more forms as it matures into the adult form, each form being of different appearance. In *complete metamorphosis* the immature (larval) form is markedly different from the adult form hence there is a radical or complete transformation during the maturation. In *incomplete metamorphosis* the larval form is quite similar to the adult form hence the change is subtle.

Mitosis Cell division that results in two cells identical in

277

chromosome number with the parent cell.

Moulting (*Ecdysis*) Periodic shedding and replacement of outer covering such as cuticle in arthropods, feathers in birds or skin in reptile. The shedding in arthropods allows growth to occur.

Mutualism An association between two organisms in which both benefits.

Myxoedema A disease condition resulting from low metabolic rate as a result of deficiency or lack of thyroxine.

N

Nastic movement A non-directional movement of a plant in response to a stimulus.

Nymph Young stage of an insect. It resemble the adult except that it is sexually immature and wingless.

Nerve impulse Electrical and chemical changes which pass along a nerve cell fibre (axon)

O

Omnivore An animal whose diet consists of plant or animal materials e.g. man, pig, cockcroach.

Organ A collection of different tissues grouped into an identifiable unit and performing one or more function e.g. leaf, heart, liver.

Organism A living thing.

Osmoregulation The process by which a constant balance of osmotic pressure is maintained in the body of an animal by the control of salts, ions and water content.

Osmosis The movement of water by diffusion across a semi-permeable membrance.

Oviparous Animals which lay fertilized eggs; the eggs contain poorly developed embryos with yolk to nourish it.

Ovoviviparity is when fertilized eggs are produced, developed and nourished inside the body of the mother without placenta and hatch before or soon after the eggs are laid. Examples are certain fishes and reptiles

P

Parasite An animal or plant which lives in or on another living organism called the host.

Pathogen A disease-causing organism.

Perennial A plant that continues to grow from year to year.

Peristalsis Waves of muscular contraction of the wall of the alimentary canal that helps to move the food along.

Permeable membrane A membrane that allows substances to pass through it.

Phenotype The visible physical characteristics of an organism.

Photosynthesis The process by which green plants make carbohydrates from carbon dioxide and water using light energy absorbed by the green pigment, chlorophyll.

Phototropism The growth of a plant or plant part in response to light.

Placenta 1. The part of a plant ovary to which ovules are attached 2. The tissue to which sporangia or spores are attached. 3. The structure formed from embryonic and maternal blood circulation that enables the foetus to depend on maternal systems for obtaining oxygen and nutrients as well as for removal of metabolic waste products.

Plankton The microscopic forms of life in lakes and seas; animals are called zooplankton, plants are phytoplankton.

Plasmolysis Contraction of cell protoplasm from the cell wall when placed in a highly concentrated solution due to osmotic withdrawal of water from the cell sap.

Poikilothermic Having a body temperature that varies with that of the environment. Most animals, except birds and mammals are poikilothermic.

Poison A chemical substance which when absorbed by a living body will kill or gravely harm it.

Predator An animal which hunts, captures and kills other animals for food.

Proboscis A tubular mouth in some insects (e.g butterfly, housefly) used to suck up food.

R

Radial symmetry The arrangement of the parts of an animal body around a central axis so that any two planes are exactly alike; it is characteristic of coelenterates and echinoderms. Plant or plant organs showing this type of symmetry are termed actinomorphic.

Radiation A form of heat loss through infrared rays.

Roughage Indigestible cellulose material which adds to the size of the food and thereby promotes peristalsis.

S

Saprophyte A plant which feeds by digesting dead organic material outside the body and absorbing the soluble products.

Secretion Production of chemical substances in the body a cell or a gland.

Seed The structure that develops from a fertilized ovule of a flowering plant; it contains an embryo protected by one or more coats.

Segmentation The division of the body of an animal into series of similar units, e.g. in tapeworm, earthworm.

Semi-permeable membrane A membrane which allows only water but not dissolved substances to pass through it.

Sex-linked A characteristic determined by a gene on a sex chromosome.

Sexual reproduction Reproduction that involves the fusion of the nuclei of male and female gametes.

Species A group of individuals that are alike in all characteristics and are therefore capable of production a fertile offspring.

Sperm The male reproductive cell or gamete.

Spiracle A breathing hole found in the surface of an insect's body.

Sporangium A plant part that produces spores e.g. in fungi, mosses and ferns.

Spore An asexual reproductive cell of plants produced in large numbers which may develop directly into a new organism.

Sterilization Process of rendering materials free from any microbial life.

Substrate A material on which an organism grows and secretes enzymes.

Symbiosis A natural association between two different living organisms including parasitism, mutualism and commensalisms.

Synapses A gap between two nerve cells across which the nerve impulse passes.

System A number of connected organs and tissues in the body working together for a particular major function.

T

Taxis A locomotory movement of an organism or cell in response to a directional stimulus, such as chemicals of light.

Thallus Plant body without root stem or leaves.

Tissue A group of similar cells, bound together by intercellular materials, performing a particular function.

Toxin A poison produced by certain microorganism.

Translocation The movement of dissolved food materials in the vascular tissue of plants.

Transpiration The loss of water by evaporation from the aerial parts of a plant primarily through the stomata.

Tropism A growth movement in part of a plant whose direction is determined by the direction of the stimulus.

Turgid Description of a plant cells due to presence of water.

V

Vaccination The injecting into the body of weakened pathogen to stimulate the formation of antibodies and confer immunity. The weakened pathogen is called a vaccine.

Variation The essential differences seen among organism of the same species.

Vascular tissue The tissue concerned with transporting substances within the body of an organism such as xylem and phloem in plants and blood and lymph in animals.

Vector An agent, such as an insect, capable of transferring a pathogen from one organism to another.

Vegetative propagation Reproduction by any parts of the plant except the seed.

Ventral The belly or lower surface of an organism.

Viviparous Animals which bring out their young alive and well developed; the embryo develops within and derives nourishment from the mother through a placenta. May also be used to describe plants in which the embryo develops on the parents e.g. red mangrove.

W

Watershed A hilly region, usually over a large area, which conducts surface water to streams.

Z

Zygospore A thick- walled resistant spore developing after then fusion of two similar gametes in the reproduction of some algae and fungi.

Zygote The fertilized egg cell in plants and animals.

REFERENCES

Idodo Umeh, G. (1996): College Biology, Benin-city: Idodo Umeh Publishers Limited.

IUPUI, Department of Biology 2003 Class Note.

Odunfa, S.A. (2001): Essentials of Biology, Ibadan: Heinemann.

Oyegoke O.O. et al (2008): New General Biology for Undergraduates, (Revised Edition), Rajah Dynamic Printers.

Pass At Once Biology for UTME (2009), Wisdomline International (Nig.) Limited.

Rene, F.K. et al (2010): Biology for Dummies, (Second Edition), Dummies Publishers.

Sofola, O.A. (1987): A Revision Text in Medical Physiology, (First Edition), SBO Publishers.

Taylor D.J. et al (1997): Biological Science, (Third Edition), Cambridge University Press.

Usua, E.J (1967): Biology Revision Course, University Press Limited.

www.thestudentroom.co.uk/wiki/Category:A_Level_biology_Revision_Notes

www.ingramcontent.com/pod-product-compliance
Lightning Source LLC
Chambersburg PA
CBHW080237180526

45167CB00006B/2309